THIRD EDITION

HUMAN ANATOMY and PHYSIOLOGY
LABORATORY MANUAL

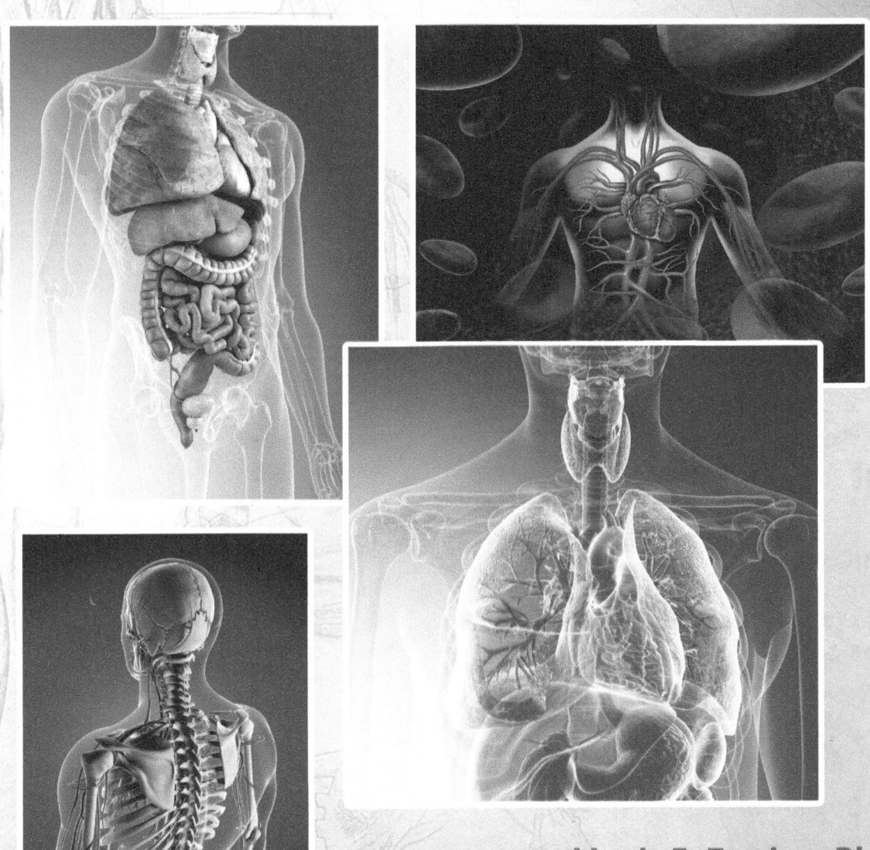

Mark F. Taylor, Ph.D.

Cover images © Shutterstock.com

Kendall Hunt
publishing company

www.kendallhunt.com
Send all inquiries to:
4050 Westmark Drive
Dubuque, IA 52004-1840

ISBN 978-1-7924-2445-8

Published in the United States of America

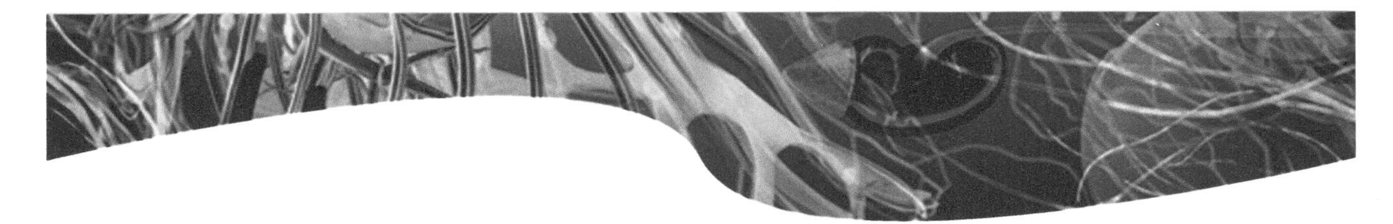

Contents

Lab 1 The Vocabulary of Anatomy

If you lived in a large city and wanted to give someone directions to your home, you might first mention an easily recognizable landmark nearby. Then you could use directional terms, such as north, south, left, and right, to guide the person from the landmark to your home. In a similar approach, anatomists can provide "directions" to any part on the body by referring to specific body regions (landmarks) and using a special list of directional terms. These directional terms and the names of specific regions of the body are part of a basic vocabulary of anatomy presented in this section.

Anatomical Position

To prevent misunderstanding when reading about and discussing anatomy, it is necessary to understand and use universally accepted terminology. You can begin building your anatomy vocabulary by learning the major surface regions of the body. We can identify these regions most effectively by looking at a body that is in **anatomical position**. In anatomical position, the person is standing erect with the head, palms, and feet facing forward, the feet slightly apart, and the arms by the sides (see **Figure 1.1**).

Body Regions

We will now describe the basic layout of the body and define the boundaries of selected regions. At the most basic level of gross anatomy, we can divide the body into two major regions: the *axial* and *appendicular* regions.

The **axial region** (AK-sē-ul), so-named because it forms the body's *axis* or central "core," includes the head, neck, and trunk regions.

The Head and Neck

The head is the **cephalic** region (sē-FOWL-ik; *cephal-*, head). It includes a dome-shaped **cranium** (KRĀ-nē-um; "skull"), which holds the brain, and a flattened **face**. The cranium includes the **frontal** ("forehead"), **occipital** (ok-SIP-i-tul; *occipit-*, back of head), and **otic** (Ō-tik; "ear") regions. The face contains the **nasal** (NĀ-zul; "nose"), **orbital** (OR-bi-tul; *orbit*, circle), **buccal** (BUCK-ul; "cheek"), **oral** ("mouth"), and **mental** (MEN-tul; "chin") regions. The neck is the **cervical** region (SER-vi-kul; "neck"), which connects the head to the trunk. The **nuchal** region (NEW-kul; "nape") is the back of the neck.

The Trunk

The **trunk** includes all regions of the body except the head, neck, and upper and lower limbs. The major regions on the *anterior* (front) side of the trunk include the **sternal** (STER-nul; "breastbone"), **thoracic** (THŌ-ras-ik; *thora-*, chest), **mammary** (MAM-a-rē; "breast"), **abdominal** (ab-DOM-i-nul; *abdomen*, belly), **umbilical** (um-BIL-i-kul; *umbilic-*, navel), **pelvic**, and **pubic** (PŪ-bik) regions. The pelvic region is named for the bowl-shaped (*pelvi-*, basin) bony *pelvis*. The pubic region is named

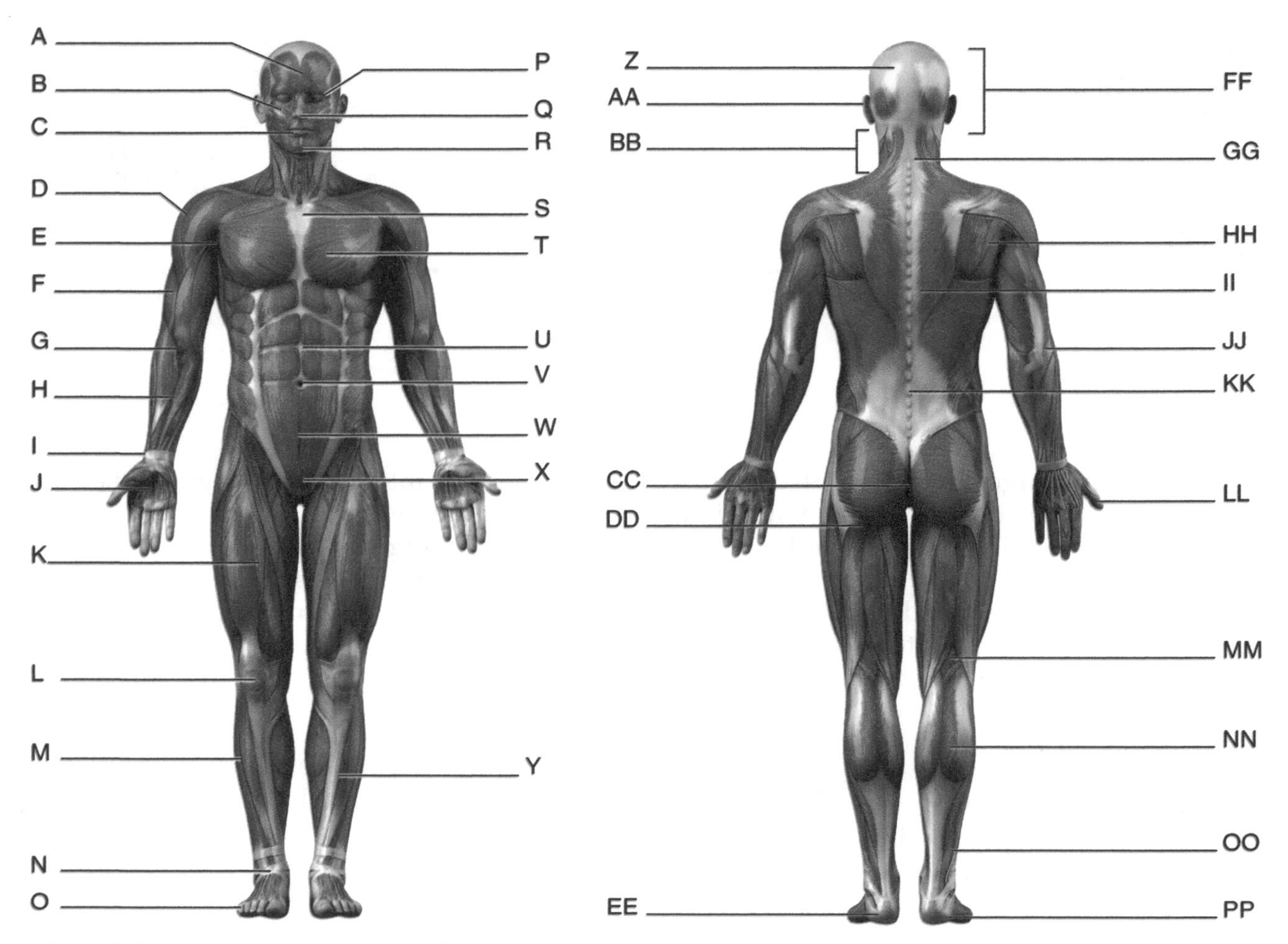

Figure 1-1 Anatomical position and body regions

for the coarse hairs, or *pubes* (PU¯ -be¯z), that appear there when a person becomes sexually mature; i.e., reaches puberty (PU¯ -ber-te¯; "grown up").

Also seen on the anterior side of the trunk are the *axillary* and *inguinal* regions, which join the axial region to the appendicular region. The **axillary** region (AK-sil-ār-ē) is the armpit or junction between the trunk and upper limbs. Similar to an axle that connects a car body to a moving wheel, the axillary region (*axil-*, axle) connects the axial region of the human body to a moving arm. The **inguinal** region (ING-gwi-nul; *inguin-*, groin) is the junction between the trunk and lower limbs.

The major regions on the *posterior* (back) side of the trunk include the **dorsum** ("back"; all parts except the lower back), **scapular** (*scapula*, shoulder blade), **coxal** (KOK-sul; "hip"), **vertebral** (VER-te-brul), **lumbar** ("loin"), **sacral** (SĀ-kral), and **gluteal** (GLŪ-tē-ul, "buttock") regions. The vertebral region is the spinal column area. However, *vertebra* means "to turn," referring to the spinal column's ability to turn or twist the back. The sacral region is between the hips. *Sacral* means "sacred," referring to the fact that some ancient cultures thought a part of the skeleton in this region would rise from the dead.

The Appendicular Region

The **appendicular region** (ap-en-DIK-ū-lar), so-named because its parts *append* (attach) to different parts of the axial region, includes the *upper limbs* and *lower limbs*.

The Upper Limbs

The **upper limbs** include the shoulders, arms, forearms, wrists, and hands. The shoulders attach to the trunk and have a point called the **acromial** region (a-KRŌ-mē-ul; *acromion*, point of the shoulder). The arm, or **brachial** region (BRĀ-kē-ul; *brachi-*, arm), extends from the shoulder and axillary regions to the elbow. The elbow has an **antecubital** region (AN-te-KŪ-bi-tul; "front of elbow") and **olecranal** region (ō-LEK-ra-nul; "point of elbow"). The forearm, or **antebrachial** region (AN-tē-BRĀ-kē-ul), extends from the elbow to the wrist, or **carpal** region (KAR-pul; wrist). The hand, or **manual** (MAN-ū-ul) region, includes the **palmar** ("palm") and **digital** (finger) regions. The digital region is also called the **phalangeal** region (fa-LAN-jē-ul; *phalanx*, line of soldiers), because it looks like a line of soldiers lined up for battle. The **pollex** (POL-eks) region is the thumb.

The Lower Limbs

The **lower limbs** include the buttocks, thighs, legs, and feet. The buttocks, or **gluteal** regions (GLŪ-tē-ul; "buttock"), attach to the trunk and are common sites for injecting drugs. The **natal cleft** is the vertical crease between the two buttocks and the **gluteal folds** are the horizontal creases between the buttocks and thighs. The thigh is the **femoral** (FEM-o-rul; "thigh") region that extends from the buttock and inguinal regions to the knee. The anterior side of the knee is the **patellar** region (pa-TEL-ar), named for the kneecap (*patella*). The posterior part of the knee is the **popliteal** region (pop-LIT-ē-ul; "back of knee"). The leg extends from the knee to the foot and has an anterior or **crural** region (KROO-rul; *crus*, leg), a posterior or **sural** region (SOO-rul; "calf"), and a **fibular (peroneal)** region (FIB-ū-lar or per-ō-NĒ-ul) on the *lateral* (outer) side. The fibular region contains the *fibula*, a long bone. The foot, or **pedal** (PED-ul) region, includes the **tarsal** (TAR-sul; "ankle"), **calcaneal** (kal-KĀ-nē-ul; "heel"), **plantar** (PLAN-tar; "sole"), **digital (phalangeal)**—toes, and the **hallux** (HAL-uks; "great toe") regions.

After learning the names and locations of the surface regions, your expanded vocabulary will provide a basis for locating numerous structures *inside* the body. For instance, the femoral region, or thigh, contains the femur (thighbone), femoral artery, femoral vein, and femoral nerve. Where would you expect to find the occipital artery, axillary nerve, and popliteal vein? If you said the back of the head, armpit, and back of the knee, respectively, then you are correct.

Directional Terms

Now that you are familiar with the body's surface landmarks, it's time to learn the anatomy-related directional terms that correlate the location of one body part to another. Using these directional terms can allow you to avoid prolonged explanations and misunderstanding when describing a particular location on or inside the body. Without this terminology, someone could interpret the statement, "The buccal region is *next to* the eye region," as meaning the forehead, nose, side of the head, or the cheek. However, saying, "The buccal region is *inferior* to the eye," eliminates the forehead, nose, and side of head as possibilities.

To ensure consistency when using anatomy-related directional terms, we always assume the body we are describing is in anatomical position. In this way, a given term always relates the location of one body part relative to another, even when the body changes its position. For example, *superior* means "above," so we can always say your cephalic region is superior to your trunk, even if you "stand on your head." Also, keep in mind that in anatomy, *right* and *left* refers to the right and left sides of the body you are viewing. **Table 1.1** lists the directional terms that you need to learn before the next lab.

Planes and Sections

Now that you are familiar with the body's surface anatomy, we will prepare to look at some of the body's internal features. To do this, we must consider the angles at which anatomists cut the body they want to study. This is important because many images in this text show flat cut surfaces within three-dimensional body parts. In anatomy, a *section* is a cut that follows a straight path along an imaginary flat surface called a *plane*. Cutting the body along sectional planes

Table 1.1 Directional Terms

Term	Description	Example
Anterior	Toward the front	The toes are anterior to the heel.
Posterior	Toward the back or behind a structure	The heel is posterior to the toes.
Caudal	Toward the inferior end of spine	The lumbar region is caudal to the thoracic region.
Contralateral	On opposite sides of the midline	The right lung and left kidney are contralateral.
Ipsilateral	On the same side of the midline	The left lung and left kidney are ipsilateral.
Deep	Away from the body's surface or part	Muscles are deep to the skin.
Superficial	Toward or on the body's surface or part	The skin is superficial to the muscles.
Distal	Away from the point of origin of a part	The elbow is distal to the shoulder.
Proximal	Closer to the point of origin of a part	The shoulder is proximal to the elbow.
Dorsal	Back or toward the back	The olecranon is dorsal to the antecubital region.
Ventral	Toward the front	The antecubital region is ventral to the olecranon.
Inferior	Below or toward the feet	The elbow is inferior to the shoulder.
Superior	Above or toward the head	The shoulder is superior to the elbow.
Intermediate	Between two structures	The heart is intermediate to the lungs.
Lateral	Away from a midline or toward the side	The thumb is lateral to the little finger.
Medial	Toward a midline or the inner side	The little finger is medial to the thumb.

allows for studies of deep anatomical structures. Most anatomical studies utilize sections made along three planes that intersect the body at right angles. These include the *sagittal*, *frontal*, and *transverse* planes. A section made along a given plane has the same name as that plane; for example, to make a sagittal section, one cuts along a sagittal plane. **Figure 1-2** shows the three major planes of the body.

A **sagittal plane** (SAJ-i-tul; "arrow") runs vertically and divides the body or structure into right and left sides. We can be more specific when considering a sagittal plane's location relative to the body's *midline* (an imaginary vertical line running along the exact center of the body). A **midsagittal,** or **median, plane** lies on the midline and divides the body or structure into two *equal* halves. A **parasagittal plane** runs alongside (*para-*, beside) the midline and divides the body or structure into two *unequal* sides. Most images showing sagittal sections of the body or one of its parts are actually midsagittal sections.

A **frontal**, or **coronal**, **plane** (kō-RŌ-nul) runs vertically and divides the body into anterior and posterior parts. The word *corona* means "crown," so imagine you are placing a crown on your head and look at the orientation of your

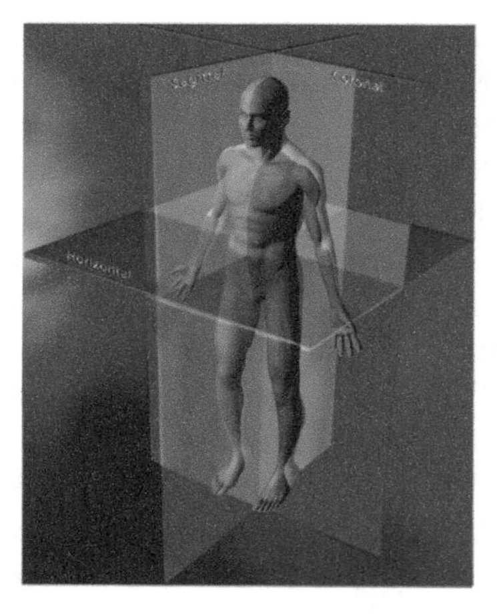

Figure 1-2 Body planes.

thumbs. Since they point toward your head from either side, they can remind you of a coronal plane's orientation. There are many frontal planes, but most images of frontal sections show the body or an organ divided into nearly equal anterior and posterior parts.

A **transverse**, or **horizontal**, **plane** runs perpendicular (at right angles) to the sagittal and frontal planes and, therefore, runs horizontally *across* the body's long axis. For this reason, we sometimes call a transverse section a **cross section**. Any section made between the transverse plane and either a sagittal or a frontal plane is an **oblique section** (ō-BLĒK; "slanting").

Body Cavities

The axial region of the body contains cavities that hold and protect vital organs. Making either a midsagittal or a frontal section through the axial region of the body reveals two major body cavities, the *dorsal* and *ventral* cavities. These cavities have no connections with the outside of the body but serve three major functions: They (1) provide a location to hold vital organs; (2) protect vital organs from the external environment; and (3) allow room for certain internal organs, such as the heart and lungs, to expand without compressing adjacent organs. The major body cavities are shown in **Figure 1-3**.

Dorsal Body Cavity

The **dorsal body cavity** is the more posterior of the body's two major cavities and it consists of two parts: a *cranial cavity* and a *vertebral cavity*. The **cranial cavity**, which is located inside the cranium of the cephalic region, is the more superior cavity and holds the brain. The **vertebral (spinal) cavity** is in the vertebral region and holds the spinal cord. The cranial and vertebral cavities are continuous with one another; that is, there is no barrier between them. Thick bones and tough connective tissue surround the dorsal cavity and protect the brain and spinal cord against external blows. In addition, the dorsal cavity contains a watery fluid that cushions the vital organs against the jarring effects of walking and running. This fluid also helps maintain a stable temperature within the dorsal cavity.

Ventral Body Cavity

The **ventral body cavity** is much larger than the dorsal cavity and is anterior to it. The ventral cavity has two major subdivisions: a *thoracic cavity* and the *abdominopelvic* cavity, separated from one another by a dome-shaped muscle called the *diaphragm* (DĪ-uh-fram, "partition"). In general, anatomists refer to all organs within the ventral body cavity collectively as **visceral organs**, or the **viscera** (VIS-er-ul or VIS-er-a; *viscus*, "internal organ").

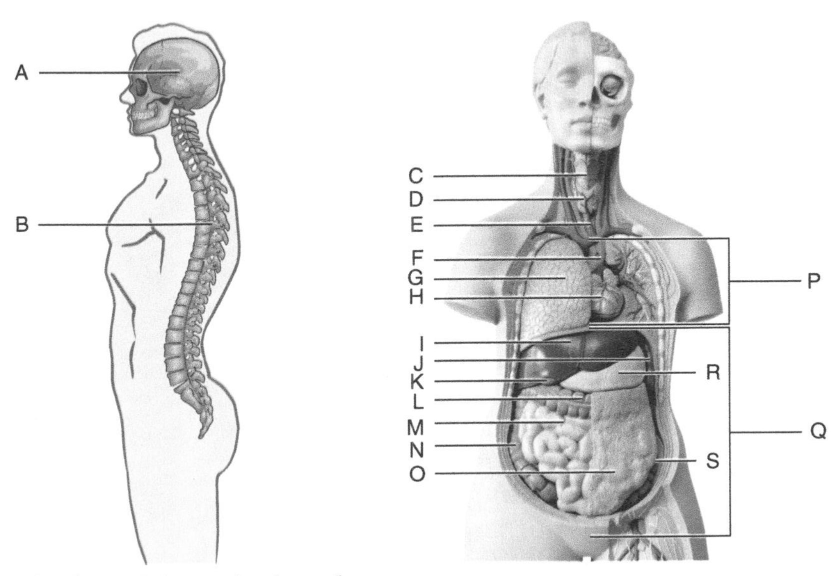

Figure 1-3 Major body cavities and selected organs.

The **thoracic cavity** is superior to the diaphragm and contains four smaller cavities. Two of these, the right and left **pleural cavities** (PLU¯R-ul), each hold a lung and are so-named because they occupy the lateral portions (*pleura*, "side") of the thoracic cavity. Located intermediate to the right and left pleural cavities is the **mediastinum** (ME¯-de¯-as-TI¯-num; "middle"). It contains the trachea (windpipe), esophagus, thymus gland, major blood vessels, and the **pericardial cavity** (per-i-KAR-de¯-ul), which encloses the heart (*peri*, "around"; *cardia-*, "heart").

The **abdominopelvic cavity** (ab-dom-i-no¯-PEL-vik) is inferior to the diaphragm and has two major parts: the *abdominal cavity* and the *pelvic cavity*. These two cavities are actually not truly separate but are continuous with one another. The **abdominal cavity** extends from the diaphragm to the pelvic region and contains the stomach, liver, pancreas, and other organs, including most of the intestines. The **pelvic cavity** is inferior and slightly posterior to the abdominal cavity and lies within the bony pelvis (mentioned earlier). Visceral organs within the pelvic cavity include the urinary bladder, internal reproductive organs, and the rectum (part of the large intestine).

To reinforce your knowledge of different body cavities, practice describing the location of various internal organs, beginning with the most general cavity and ending with the most specific cavity or vice versa. For example, you could say, "The brain is within the cranial cavity, which is within the dorsal cavity," or "The dorsal cavity contains the vertebral cavity, which contains the spinal cord." In addition, you could say, "The heart is within the pericardial cavity, which is in the mediastinum, which is in the thoracic cavity, which is in the ventral cavity."

Serous Membranes

A thin membrane, called a **serous membrane** (SĒR-us), protects the walls of the ventral cavity and the surfaces of visceral organs as these organs carry out their normal functions. Some visceral organs, such as the heart and lungs, change shape constantly while other organs, including the stomach, intestines, and urinary bladder, change shape on a regular basis. These movements generate friction between adjacent organs and between the organs and walls of the ventral cavity. The serous membrane produces a slippery **serous fluid** that reduces this friction. The word *serous* relates to "whey" or the watery part of curdled milk; whereas, the "curd" is the solid part of curdled milk used to make cheese.

There are several serous membranes in the ventral cavity, and each has two layers. The *parietal* (pa-RĪ-e-tul) layer attaches to the wall of the cavity (*paries*, "wall"), and the *visceral* layer surrounds individual visceral organs. Serous fluid is between the parietal and visceral layers of each serous membrane. You can make a model of a serous membrane by inserting your fist into a balloon filled with water. The outer part of the balloon represents the parietal layer and the inner part represents the visceral layer. Your fist represents a visceral organ and the fluid in the balloon represents serous fluid (see **Figure 1-4**).

Two serous membranes, the *pleura* and *pericardium*, lie within the thoracic cavity and have even more specific names, depending on their precise location. The serous membrane associated with the lungs is the **pleura** (PLU¯R-uh), so named because it occupies the lateral (*pleura*, "side") portions of the thoracic cavity. The two layers of the pleura include the **parietal pleura**, located on the walls of the thoracic cavity, and the **visceral pleura**, on the surface of the lungs. The pleura cavities lie between the parietal and viscera pleura. The serous membrane associated with the heart is the *pericardium* (per-i-KAR-de¯-um;

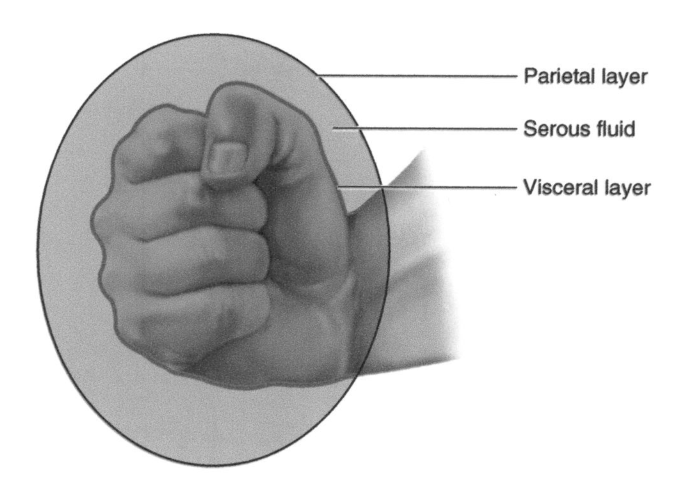

Figure 1-4 Model of a serous membrane.

peri-, "around"; *cardi*, "heart"). The **parietal pericardium** lines the inner wall of the *pericardial sac*, a bag-like structure that encloses the heart. The **visceral pericardium** covers the surface of the heart. The pericardial cavity is between the parietal and visceral pericardium.

The most extensive serous membrane, the *peritoneum* (per-i-tō-NĒ-um; "stretched around"), lies within the abdominal cavity. The **parietal peritoneum** lines the walls of the abdominal cavity and the **visceral peritoneum** surrounds most of the visceral organs within the cavity. The parietal and visceral peritonea are compressed together into several double-layered membranes, including the *mesentery* and *omenta*. **Mesentery** (MEZ-en-tair-ē; "between intestine") is the double membrane located between parts of the small intestine. The **greater omentum** (ō-MEN-tum; "bowel membrane") connects the stomach to part of the large intestine, while the **lesser omentum** connects the stomach and parts of the small intestine to the liver and diaphragm (see **Figure 1-5**).

A few organs, such as the kidneys, adrenal glands, pancreas, and parts of the small and large intestine lie posterior to the parietal peritoneum. For this reason, we say these organs are *retroperitoneal* (*retro*, "behind"). The inferior portion of the parietal peritoneum extends into much of the pelvic cavity to cover portions of certain reproductive organs and the urinary bladder. The space between the parietal and visceral peritoneum is the *peritoneal cavity*. In much of the abdominopelvic cavity, the peritoneal cavity is only a "potential" space because the peritoneal membranes lie next to one another.

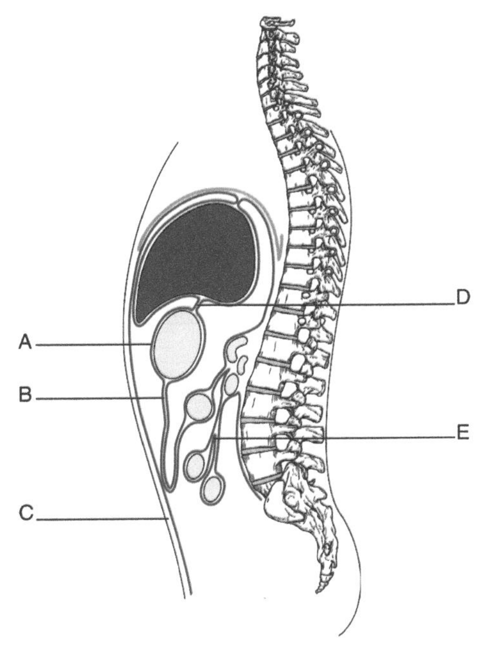

Figure 1-5 Serous membranes in the abdominal cavity.

Mucous Membranes

Mucous (myew-kus) **membranes** line passageways in the head and torso (excludes the limbs) that have connections to the outside of the body. These include the inner lining of the nasal cavity and tubes of the digestive, respiratory, urinary, and reproductive systems. Mucous membranes contain cells that secrete a viscous liquid called **mucus** (note the spelling). Mucus can trap foreign particles in the respiratory system and make it easier for movement of materials through tubes in the other three systems.

Abdominopelvic Regions

Just as a grid on a city map can help you locate a particular street, an imaginary grid on the abdominopelvic cavity can help you locate a particular visceral organ. The simplest grid, and the one used mostly by healthcare professionals, has one vertical line and one horizontal line intersecting at the *umbilicus* (navel). This simple grid forms four **abdominopelvic quadrants** (*quad-*, "four"). The names of the quadrants refer to their location relative to the subject's right and left sides and include the **right upper quadrant** (**RUQ**), **left upper quadrant** (**LUQ**), **right lower quadrant** (**RLQ**), and the **left lower quadrant** (**LLQ**). See the abdominopelvic quadrants in **Figure 1-6a**.

Anatomists prefer a more elaborate grid to compartmentalize the abdominopelvic cavity. This grid resembles a tic-tac-toe grid with the two parallel vertical lines oriented just medial to each nipple. One of the horizontal lines is oriented at the bottom of the ribcage and the other is along the top of the hip bones. The intersecting lines divide the region below the diaphragm into nine **abdominopelvic regions**.

- The **right** and **left hypochondriac regions** (h ī-pō-KON-drē-ak) are the most superior, lateral regions. The word *hypochondriac* denotes

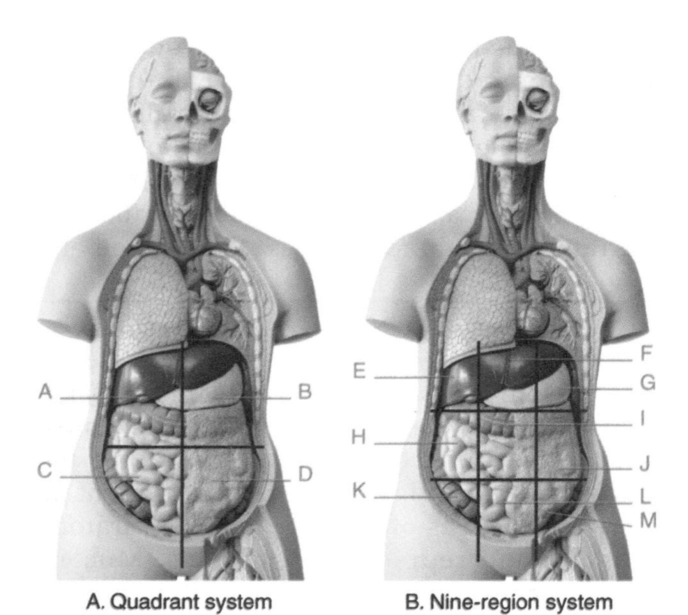

Figure 1-6 Abdominopelvic regions.

a part of the abdominopelvic cavity deep to cartilage (a type of connective tissue), which is on the anterior, inferior portion of the rib cage (*hypo*, "below"; *chondr-*, "cartilage").

- The **epigastric region** (ep-i-GAS-trik) is intermediate to the right and left hypochondriac regions and is so-named because it contains most of the stomach (*epi-*, "above"; *gastr-*, "stomach").
- The **right** and **left lumbar regions** are immediately inferior to the right and left hypochondriac regions.
- The **umbilical** region is in the center of the grid (in the vicinity of the *umbilicus* or navel), immediately inferior to the epigastric region and medial to the right and left lumbar regions.
- The **right** and **left inguinal (iliac) regions** are the most inferior lateral regions and are immediately inferior to the right and left lumbar regions, respectively.
- The **hypogastric (pubic) region** is intermediate to the right and left inguinal regions. See the nine abdominopelvic regions in **Figure 1-6b**.

Table 1.2 Abdominopelvic Regions and Their Internal Organs

Abdominopelvic Quadrants	
Upper Right	Gallbladder, right kidney, liver, stomach, colon
Upper Left	Spleen, left kidney, stomach, small intestine and colon, pancreas
Lower Right	Small intestine, colon, reproductive organs, urinary bladder
Lower Left	Small intestine, colon, reproductive organs, urinary bladder
Nine Abdominopelvic Regions	
Right Hypochondriac	Gallbladder, right kidney, liver, small intestine, colon
Epigastric	Stomach, liver, pancreas, colon
Left Hypochondriac	Spleen, stomach, pancreas, small intestine, left kidney, colon
Right Lumbar	Small intestine, colon, right kidney
Umbilical	Small intestine, right and left kidneys
Left Lumbar	Small intestine, colon, left kidney
Right Inguinal (Iliac)	Small intestine, colon (including appendix)
Hypogastric (Pubic)	Small intestine, colon, urinary bladder, reproductive organs
Left Inguinal (Iliac)	Small intestine, colon

Name ______________________ Course Number: _________ Lab Section _______

Lab 1: Vocabulary of Anatomy Worksheet

A. Body Cavities

The (1) ______________ cavity holds the central nervous system while the (2) ______________ cavity holds the lungs, heart, esophagus, trachea, thymus gland, stomach, liver, intestines, kidneys, spleen, adrenal glands, urinary bladder, and other organs. Body cavity (1) is subdivided into the (3) ______________ cavity that holds the brain, and the (4) ______________ cavity that holds the spinal cord. The largest body cavity contains (5) ______________ membranes, which do not have connections with the outside of the body. The watery (6) ______________ fluid secreted from these membranes reduces friction as internal organs move. The largest body cavity is subdivided into two major cavities by the (7) ______________ a dome-shaped, muscular partition. The most superior of these two cavities is the (8) ______________ cavity, which is surrounded by the chest wall. This cavity is divided into three smaller cavities, including the right and left (9) ______________ cavities, which hold the lungs, and a middle cavity called the (10) ______________ which houses the trachea, esophagus, thymus gland, and major blood and lymphatic vessels. Two membranes are associated with the lungs: the (11) ______________ ______________ lines the outer walls of each cavity around a lung, and the (12) ______________ ______________ adheres to the surface of each lung. The middle cavity contains a smaller (13) ______________ cavity that holds the heart. This cavity's outer wall is lined with a membrane called the (14) ______________ ______________ while the heart's outer surface is covered with a membrane called the (15) ______________ ______________ The largest body cavity inferior to (7) is the (16) ______________ cavity, which is subdivided into two major cavities. The (17) ______________ cavity is the largest and it holds the stomach, liver, most spleen, kidneys, and most of the intestines. The smaller (18) ______________ cavity is more inferiorly located and holds the urinary bladder, reproductive organs, and parts of the intestines. The membrane that covers the outer walls of these two cavities is the (19) ______________ ______________, while the membrane that adheres to the surface of most organs within (17) is called (20) ______________ ______________. A few organs, including the kidneys, adrenal glands, pancreas, and parts of the intestines lie posterior to (19) are their location is said to be (21) ______________.

B. Planes and Sections

22) A ______________ plane divides a structure into anterior and posterior parts.
23) A ______________ plane is another name for plane 22.
24) A ______________ section cuts across the long axis of a structure.
25) A ______________ is another name for section 24.
26) A ______________ is another name for section 24.
27) A ______________ plane divides a structure into right and left parts, which may be equal or not.
28) A ______________ plane divides a structure into right and left equal parts.

C. Organ Location: List the most specific cavity in which you would find the organ.

29) Heart: ______________________
30) Lung: ______________________
31) Tongue: ______________________
32) Brain: ______________________
33) Urinary bladder: ______________________
34) Gallbladder: ______________________
35) Stomach: ______________________
36) Kidney: ______________________
37) Spleen: ______________________
38) Adrenal gland: ______________________
39) Thymus gland: ______________________
40) Esophagus: ______________________
41) Transverse colon: ______________________
42) Pancreas: ______________________
43) Appendix: ______________________
44) Teeth: ______________________

D. Membranes in the Ventral Cavity

45) General name for type of membrane on the surface of a lung: ______
46) General name for type of membrane lining the inside of the stomach: ______
47) General name for type of membrane covering the wall of the body's largest cavity: ______
48) Most specific name for membrane that covers the wall of the cavity around a lung: ______
49) Most specific name for membrane that covers the surface of the intestines: ______
50) Most specific name for membrane that covers the anterior wall of a kidney: ______

E. Directional Terms (Fill in the blank or circle the correct choice)

51) The ankle is superior/inferior to the knee.
52) The ankle is also distal/proximal to the knee.
53) The navel is medial/lateral to the hips.
54) The heel is anterior/posterior to the toes.
55) The hand is inferior/superior to the elbow.
56) The hand is also proximal/distal to the elbow.
57) The elbow is superior/inferior to the hand.
58) The elbow is also distal/proximal to the hand.
59) The muscles are deep/superficial to the skin.
60) The little finger is medial/lateral to the thumb.
61) The spine is dorsal/ventral to the breastbone.
62) The skin is deep/superficial to the breastbone.

F. Organ Location: Which abdominopelvic quadrant(s) contain(s) the organ listed?

63) Urinary bladder: ______
64) Most of liver: ______
65) Spleen: ______
66) Appendix: ______
67) Heart: ______
68) Gallbladder: ______

G. Organ Location: Which of the NINE abdominopelvic regions contains the organ listed?

69) Urinary bladder: ______
70) Most of liver: ______
71) Spleen: ______
72) Appendix: ______
73) Heart: ______
74) Gallbladder: ______

H. Anatomical Regions: What is the anatomical name for the following regions?

75) Point of elbow: ______
76) Sole of foot: ______
77) Neck: ______
78) Back of neck: ______
79) Groove intermediate to buttocks: ______
80) Wrist: ______
81) Point of shoulder: ______
82) Back of head: ______
83) Cheek: ______
84) Chin: ______
85) Shin: ______
86) Heel: ______
87) Arm: ______
88) Forearm: ______
89) Fingers: ______
90) Toes: ______
91) Lower back: ______
92) Back of knee: ______
93) Thigh: ______
94) Back of leg: ______
95) Side of leg: ______
96) Ear: ______
97) Head: ______
98) Thumb: ______
99) Big toe: ______
100) Ankle: ______

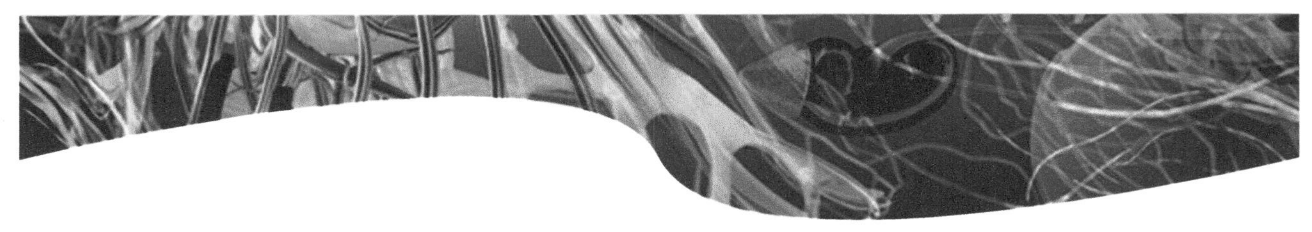

Lab 2 Microscope, Cells, and the Cell Cycle

Human anatomy covers many structural aspects of the human body, including features we can see with the naked eye and those we cannot readily see. The structures that are so small that we need special magnifying instruments to see them are said to be **microscopic** (mī-krō-SKOP-ik; *micro*, "small"; *scop*, "to view"). In this lab, you will be using a **microscope** (MĪ-krō-skōp) to view some of these extremely small components of the body.

Becoming Familiar with the Microscope

While all microscopes magnify tiny objects, some microscopes are more sophisticated than others. A *simple* microscope has only one lens, or glass disc, through which light can pass. A hand-held magnifying glass is an example of a simple microscope. In contrast, a *compound* microscope has a *set* of lenses; i.e., it has more than one lens. The microscopes used in this lab are compound, *light* microscopes, meaning they have their own light source. The light's intensity can be adjusted so that appropriate contrast can be made among different structures being viewed.

Before using your microscope, it is important to learn some basic rules for its transport and care. If you must carry the microscope, the lab instructor will demonstrate the proper procedure for doing so. Always use one hand to support the microscope's base and the other to hold firmly onto the microscope's arm. Before proceeding to the next section, take some time to review the parts of a compound microscope in **Figure 2-1** and **Table 2.1**.

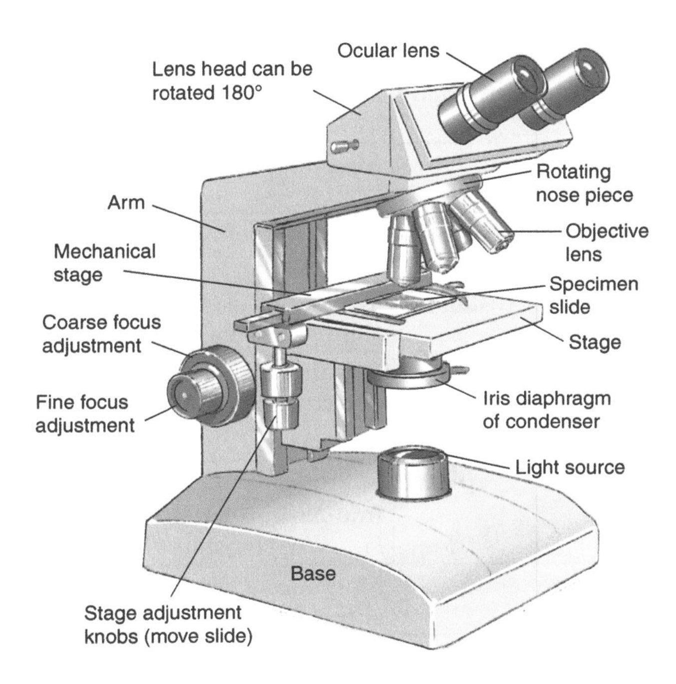

Figure 2-1 Compound microscope.

Using the Microscope

When you are ready to use the microscope for the first time, the instructor will describe the procedure. Be sure to follow these directions precisely, as improper technique can cause costly damage. If you forget a step, ask your instructor for help before proceeding.

Table 2.1 Parts of a Compound Microscope

Part	Description
Arm	"Backbone" to which the microscope's head, nosepiece, and base attach; it is also one of the parts held when transporting the microscope
Condenser	Located beneath the stage aperture, it contains the iris diaphragm and focuses light through the slide. Some condensers are adjustable with a small knob.
Coarse adjustment knob	Largest focusing knob used for making dramatic adjustments in focus with minimal turning; it should only be used with the scanning and low-power objectives and never with the high power or oil immersion objectives
Fine adjustment knob	Smaller focusing knob used for making minimal adjustments; it can be used with any objective, but it should be the only knob used when focusing on high power and oil immersion objectives
Head	Superior portion of the microscope to which the ocular and arm attach
High power objective	Magnifies ~40X (excluding ocular); it is often marked with a yellow band
Iris diaphragm	Located within the condenser, it regulates the amount of light passing up through the stage aperture and microscope slide; adjusted with a small lever.
Light control knob	Adjusts the amount of light emitted from the light source; it should be turned to the lowest level prior to turning on the light switch
Low power objective	Magnifies ~10X (excluding ocular); it is often marked with a blue band
Mechanical stage	Platform that holds a slide in place while allowing it to be moved in various directions
Mechanical stage knobs	Two knobs used to move the mechanical stage; one knob moves the stage left and right while the other one moves it forward and backward
Nosepiece	Rotating disc that holds the objective lenses; it can be rotated to bring different objectives into position over the microscope slide
Objective	Tubular device containing a set of lenses and mounted on the nosepiece
Ocular or eyepiece	Magnifies ~10X; it is the set of lenses closest to the viewer's eyes; *monocular* (mō-NOK-ū-lar; *mono*, "one") microscopes have one ocular; *binoculars* (bī-NOK-ū-lar; *bi*, "two") have two
Oil immersion objective	Magnifies ~95X (excluding ocular); its end is submerged in a drop of oil on the cover slip; it has the smallest field of vision and shallowest depth of focus
Scanning objective	Magnifies 4X (excluding ocular); it has greatest field of vision and deepest depth of focus
Stage	Platform that supports the microscope slide; it has an aperture (opening) through which light passes up through the microscope slide
Stage aperture	Opening in the stage that allows light to pass up through the microscope slide
Stage clips	Metal structures on the stage used to keep a microscope slide in place

a. Plug the microscope's cord into the electrical outlet at your table. To increase the life of the bulb, make sure the light control knob is positioned for the lowest level of light before turning on the light.

b. Click either the scanning or the low power objective into place above the stage. Using the coarse adjustment knob, raise the nosepiece (or lower the stage) so there is a generous distance between the stage and objective.

c. Place a microscope slide on the mechanical stage and clamp it into place. Turn the mechanical stage knobs until the object to be viewed on the slide moves directly above the stage aperture and below the objective.

d. Using the course adjustment knob, bring the objective as close to the slide as possible. (**CAUTION**: When you begin to feel resistance in the knob, stop turning it to prevent damage to the gears.) It may seem like the objective is going to touch the slide, but if you are using either the scanning or low-power objective then the lens should not touch the slide.

e. While looking through the ocular(s) and using the course adjustment knob, bring the object into focus, then the fine adjustment knob can be used to bring the object into more clear focus. The area on the slide that you can see through the ocular is called the **field of vision** (FOV), and it decreases as the magnification increases. Therefore, the scanning objective has a greater FOV than the low-power or higher-power objectives. This is important to remember for the following reason: If you are viewing a specimen in clear focus on one objective, but the specimen is not in the center of the FOV, then you may not see it when you click to the next higher-power objective.

   Not only do the various objectives differ in their FOV, they also differ in their **depth of focus** (DOF), which is the amount of depth on the slide that can be seen at one time. Lower-power objectives have a greater DOF than higher-power objectives.

f. After you have the object in clear focus on one objective, turn the nosepiece to bring the NEXT highest power objective into place above the slide. (**CAUTION**: Do not skip over an objective and do not move the next objective farther away from the slide prior to changing to the next objective. Although it may look like the next objective might touch the slide, it will not, provided you have the object in clear focus on the current objective.)

Parfocal Aspect of Microscopes

Compound light microscopes are said to be **parfocal** (par-FŌ-kul; literally, "near focus"). This means that an object in clear focus on one objective will be in clear (or almost clear) focus on the next highest-power objective. Only minimal focusing with the fine adjustment knob should be required with each change of the objective. If you cannot find the object of interest within a few turns of the fine adjustment knob, go back to the previous objective and get it in clear focus. (**CAUTION**: Do not turn the fine adjustment knob more than a few revolutions in either direction.)

Calculating Magnfication

Determine the **total magnification** of the object you are viewing by multiplying the magnification of the ocular lens by the magnification of the objective being used. Most oculars have a magnification of about 10X; i.e., by themselves, they can magnify an object ten times larger than actual size. If you are using the low-power objective that also has a magnification of 10X, then the total magnification of the object when viewed through the ocular and low power objective is 100 (or 10 X 10). If you are using a high-power objective having a magnification of 40X, the total magnification is 400 (or 10 X 40), and so on.

Drawing of the "e" as seen on LOW power

Drawing of the "e" as seen on HIGH power

Viewing a Prepared Slide

The lab instructor will provide for you a prepared microscope slide containing a small *e* printed on a piece of paper and covered with a thin piece of glass or plastic called a **coverslip**. Follow the steps below to understand a unique aspect of viewing objects through a set of glass lenses:

a. Place the prepared slide on the stage with the coverslip up and oriented so that the *e* is readable from your position when viewed with the naked eye. Clip the slide in place using the stage clips.

b. Click the scanning or low power objective into place above the letter on the slide. Looking through the ocular, bring the letter into clear focus, first using the course adjustment knob then using the fine adjustment knob. In the box below, draw what you see exactly how you see it on scanning or low power.

Notice the difference in orientation of the *e* when viewed with the naked eye versus being viewed through the microscope. This difference is due to a phenomenon called **inversion**, in which the image appears upside-down and backwards. Inversion also affects other aspects of using a microscope. While viewing the *e*, gently move the stage adjustment knobs back and forth. You will see that when the stage moves in one direction, the image appears to move in the opposite direction. For example, moving the stage to the right causes the image to move to the left, and moving the stage away from you causes the image to move toward you. This may make using the stage adjustment knobs confusing at first, but with experience it becomes easier to make the desired adjustments.

c. Make sure the *e* is clearly focused and in the center of the FOV on scanning and/or low power objective, then rotate and click the high power objective into place. You will not be able to see the entire letter "e" under high power, but that is because the FOV is significantly smaller. In the box below, draw what you see.

Preparing a Wet-Mount Slide

The slide with the letter *e* did not contain any liquid, but some slides contain wet specimens and are called **wet-mount slides**. In order to protect the objective lenses from the liquid a cover slip must always be in place on top of the specimen. Your instructor will show you how to properly place a cover slip on a wet-mount slide in such a way to prevent trapping air bubbles (see **Figure 2-2**).

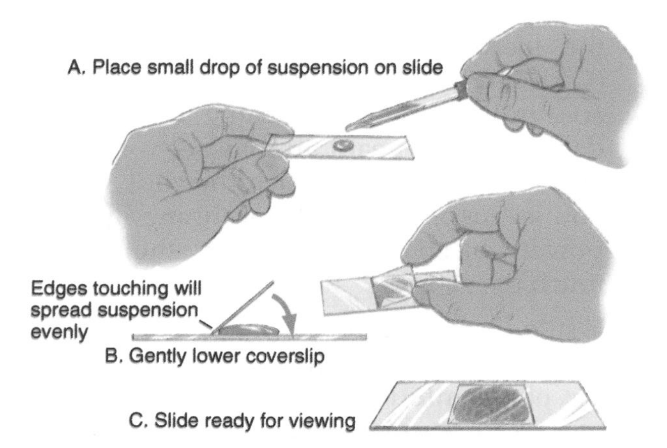

Figure **2-2** How to prepare a wet-mount slide.

Prepare a wet-mount slide using some of your own cells as follows:

a. Place one drop of methylene blue stain on a clean slide. Be careful not to get this stain on your skin or clothing, as it will turn anything it touches blue.

b. Using a clean toothpick, gently scrape the inside of your cheek and mix the saliva and its contents with the stain on the slide. (**Caution**: Do not put the stained toothpick back into your mouth!)

c. Place a cover slip over the mixture, then place the slide on the microscope stage.

d. Observe the cells with the scanning objective. Look for tiny blue specks.

e. Find one cell or a clump of cells and using the stage adjustment knobs move them to the center of the field of vision (FOV). Bring the cells into clear focus on the scanning objective. Draw several of these cells in the box below.

f. Next rotate the nosepiece to bring the low-power objective into place above the slide. Make sure the objective clicks into place.

Find a single cell (it should have a blue spot in its center) and center it in the FOV. Bring it into clear focus using the course adjustment knob first, then the fine adjustment knob. Make a sketch of some of the cells as seen on low power.

g. Rotate the nosepiece to bring the high-power objective into place above the slide. At this point, you must no longer use the course adjustment knob. (**CAUTION**: Using the course adjustment knob while on high power may cause a sickening "crunch" sound as the cover slip, slide, and/or the objective lens break!). When in doubt about how to use the microscope, ASK FOR HELP. Make a sketch of some of the cells as seen on high power. These cells are called *epithelial cells* (ep-ih-THĒ-lē-ul), which are part of the *mucous membrane* lining the cheek inside the oral cavity.

Drawing of cheek cells seen on LOW power

Drawing of cheek cells seen on HIGH power

Overview of a Generalized Cell

Using a compound microscope to view cheek cells can reveal the incredibly small nature of cells and, at the same time, can reveal some of the cell's basic features. For example, you should have seen a distinct line forming the edge of the cell; this is the *plasma membrane* and it holds in the cell's internal material. The most notable internal feature that you should have seen was the dark, pea-like *nucleus*. There are over 200 different kinds of cells in your body, and these come in a wide variety of shapes. However, most cells have several basic features in common. **Figure 2-3** shows a generalized human cell, and its components are listed in **Table 2.2**. No particular cell in the body would necessarily have all these components, but this figure allows one to appreciate the remarkably complex nature of a cell's internal anatomy. You should be able to recognize a generalized cell and its components as depicted in diagrams, photographs, and on models. Also, be able to describe the function of each part listed.

The Cell Cycle

The **cell cycle** includes all events in the life of a cell from the time the cell forms to the time when it divides and becomes two new cells. The cell that divides is the **parent cell**, and the two new cells are **daughter cells**. The cell cycle includes three major stages: *interphase*, *mitosis*, and *cytokinesis*.

Interphase

Interphase (IN-ter-fāz) refers to the stage of the cell cycle between cell divisions (*inter*, "between"), and it is subdivided into three specific phases: G_1, S, and G_2.

G_1 phase Known as the first *gap* phase, G_1 is the time when the cell is carrying out basic metabolic activity and not preparing to divide. In most cells, G_1 phase is the longest stage of the cell cycle. However, some cells, such as neurons (specialized cells in the nervous system), do not divide. Instead of saying these cells are permanently in G_1 phase, which would imply that other phases are forthcoming, they are said to be in **G_0 phase.**

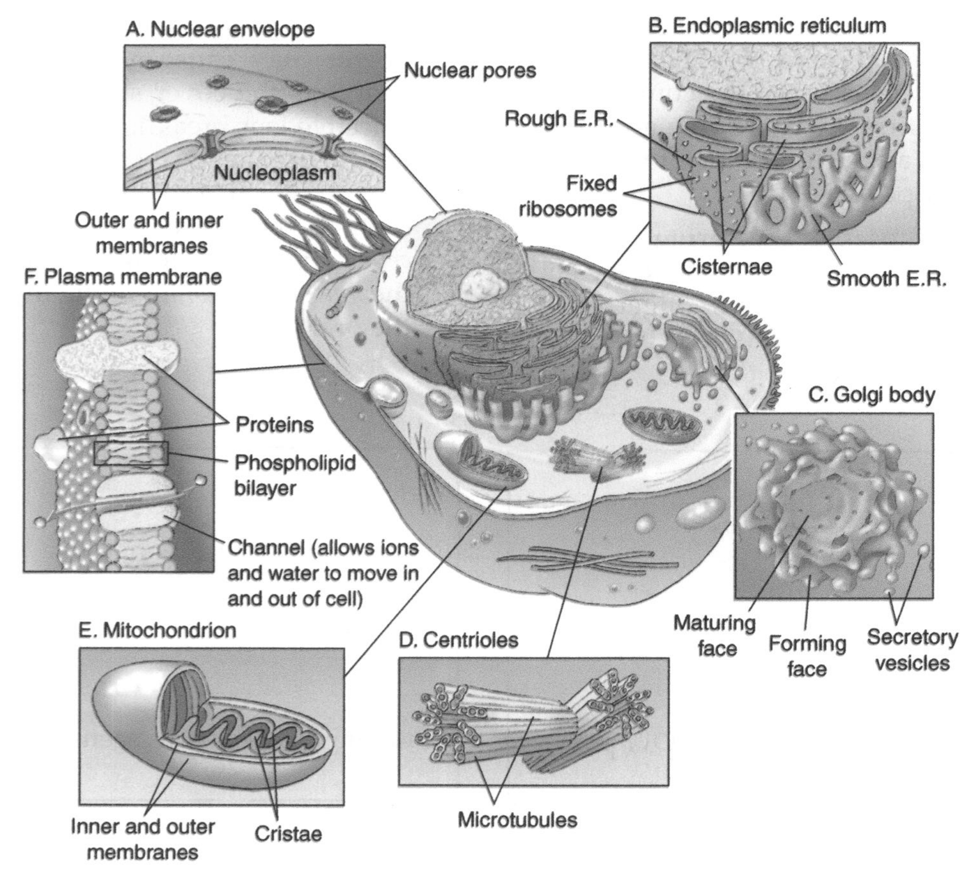

Figure 2-3 A generalized animal cell.

Table 2.2 Components of a Generalized Cell

Structure	Characteristics
Basal body	Microtubule-organizing center at the base of a cilium or flagellum; forms microtubules inside these structures; identical to centrioles with 27 microtubules arranged in 9 triplets; gives rise to a 9+2 arrangement of microtubules (9 pairs—called doublets—surrounding one pair)
Centriole	Organelle within a centrosome; consists of two structures arranged at right angles to each other; each structure contains 27 microtubules arranged in 9 triplets (9 sets of 3); plays a role in cell division by giving rise to **spindle fibers** that attach to kinetochores on chromatids
Centrosome	Region that contains centrioles and is located near the nucleus
Cilia	Short, hair-like projections with a 9+2 arrangement of microtubules (9 pairs of microtubules surrounding a single pair); arise from basal bodies; propel mucus on mucous membranes
Cytoplasm	All material between the cell membrane and nucleus; includes cytosol and cytoplasmic organelles
Cytoskeleton	Network of **microfilaments** (actin), **microtubules** (tubulin), and **intermediate filaments** (usually keratin); forms the internal framework of the cell; keeps organelles in place; allows cell to modify its shape; functions in cell division; serves as "highway" for moving vesicles via motor molecules
Cytosol	Intracellular fluid outside of the nucleus; contains mostly water with a colloid consistency

Table 2.2	Components of a Generalized Cell *(Cont'd)*
Flagellum	Long projection of the cell membrane containing nine pairs of microtubules surrounding one pair; interaction of dynein with microtubules allow flagellum to move sperm cells
Golgi body	Membranous organelle that processes and packages items for the cell; receives ER's transport vesicles at its *cis* face; synthesizes glycoproteins including *hyaluronic acid* and *chondroitin sulfate*; gives rise to secretory vesicles and lysosomes at its *trans* face
Inclusions	Collections of substances in the cytoplasm not contained within an organelle; includes **glycogen** in liver and muscle cells, **triglyceride** droplets in adipocytes, **melanin** in melanocytes, **keratin** in keratinocytes of the skin, and **hemoglobin** in red blood cells
Lysosomes	Specialized vesicles arising from Golgi bodies; contain *hydrolase* enzyme that digests contents of endocytic vesicles
Microvilli	Tiny extensions of the plasma membrane that greatly increase the cell's surface area for absorbing nutrients.
Mitochondria	Double-membrane organelles that synthesizes most of a cell's ATP; has an outer compartment (or intermembrane space) between the two membranes; the inner compartment (or *matrix*) is separated from the outer compartment by membranous folds called *cristae*
Nuclear membrane	Double membrane that encloses the nucleoplasm of the nucleus; continuous with ER's membrane
Nucleoli	Dark-staining regions in the nucleoplasm where ribosomal RNA molecules and proteins come together to form large or small ribosomal subunits
Nucleoplasm	Liquid material inside the nucleus
Nucleus	Largest organelle in the cell; contains nucleoplasm and nucleoli
Organelles	Structures within the cell that are specialized for particular functions; includes the nucleus and cytoplasmic organelles
Peroxisomes	Small spherical organelles that may bud off the ER or may arise from other peroxisomes; contain *oxidase* enzymes that bind free radicals with hydrogen atoms to form *hydrogen peroxide*; also contains *catalase* that reduces hydrogen peroxide to form water and oxygen
Ribosomes	Non-membranous organelles consisting of protein and rRNA; sites where all protein synthesis occurs; consist of large and small subunits
Endoplasmic reticulum	Double-membrane channel system that serves as intracellular "highway" for transporting materials; continuous with the nuclear membrane; **rough ER** has ribosomes and is a site for protein synthesis and modification; **smooth ER** lacks ribosomes and is the site of lipid synthesis and hydrolysis of alcohol and drugs; smooth ER in muscle cells stores Ca^{2+} ions
Vesicles	Membranous organelles formed at the following: **transport vesicles** from the ER, **secretory vesicles** from the Golgi complex; and **endocytic vesicles** from the cell membrane

S phase The *S* stands for synthesis and refers to the time when the cell makes exact copies of its linear DNA molecules in a process called *replication* or *duplication*. Recall that linear DNA molecules are found in the cell's nucleus, while circular DNA molecules are found inside mitochondria. Both types of DNA are double-stranded, consisting of two single DNA polymers held together in a helical structure by hydrogen bonds between complimentary base pairs. Staining with special dye allows DNA to be more easily seen through a microscope, and this is why the word **chromosome** (KRŌ-mō-sōm; *chrom*, "color"; *some*, "body") is often used synonymously with DNA.

After replication, each linear chromosome exists as a *replicated* chromosome consisting of two double-stranded DNA molecules. The two complete DNA molecules within a replicated chromosome are called **sister chromatids** (KRŌ-muh-tidz), and they are temporarily held together by a **centromere** (SIN-trō-mēr; *centro*, "center"; *mere*, "part"). As the name implies, some replicated chromosomes have their centromere located near the center of the chromatids, but other replicated chromosomes may have a centromere nearer one end. A **karyotype** (KĀR-ē-ō-tīp; *karyo*, "kernel") is a picture of an organism's full complement of chromosomes. **Figure 2-4** shows a close-up view of a generic replicated chromosome and a human male's normal karyotype.

G2 phase Known as the second gap phase, this represents the time when the cell is making final preparations for division. During this stage the replicated chromosomes are condensing, becoming shorter and thicker due to excessive coiling of the chromatids. Linear DNA molecules are able to condense by wrapping repeatedly around special globular proteins called *histones*. Additionally, centrioles replicate and move to opposite ends of the cell. Spindle fibers, made of microtubules, radiate out from the centrioles and attach to specific sites, called *kinetochores* on the chromatids. Each chromatid displays one kinetochore in the region of the centromere.

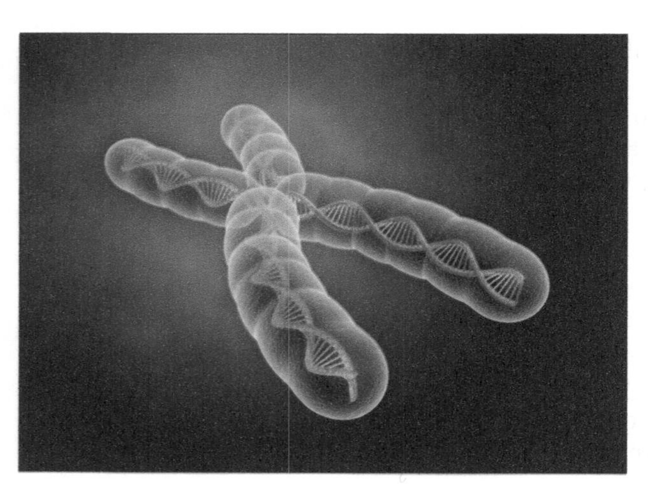

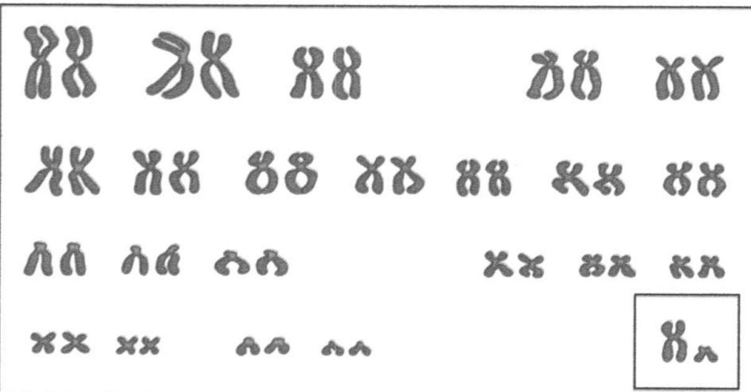

Figure 2-4 Replicated chromosomes in a human male's karyotype

Mitosis

Mitosis (mī-TŌ-sis) is the process leading up to a parent cell's replicated, nuclear DNA molecules becoming equally distributed at opposite ends of the cell. Succinctly put, mitosis is the process of *nuclear* division. The equal division of nuclear material ensures that the daughter cells formed when the parent cell divides will have exactly the same genetic material. Mitosis is so-named because during this process the DNA molecules appear as tiny thread-like structures (*mito*, "thread"; *osis*, "condition") when viewed through a microscope. Mitosis includes four major stages: *prophase, metaphase, anaphase,* and *telophase*.

Prophase **Prophase** (PRŌ-fāz; *pro*, "before") immediately follows G2 of interphase. Each chromatid of a replicated chromosome is fully condensed and has a spindle fiber attached to its kinetochore. The replicated chromosomes are randomly distributed within the cell.

Metaphase **Metaphase** (MET-a-fāz; *meta*, "between") is the stage in which replicated chromosomes align themselves along the parent cell's equator. From a side view, the chromosomes appear to form a line through the center of the cell, but actually they align around the cell's periphery as shown below:

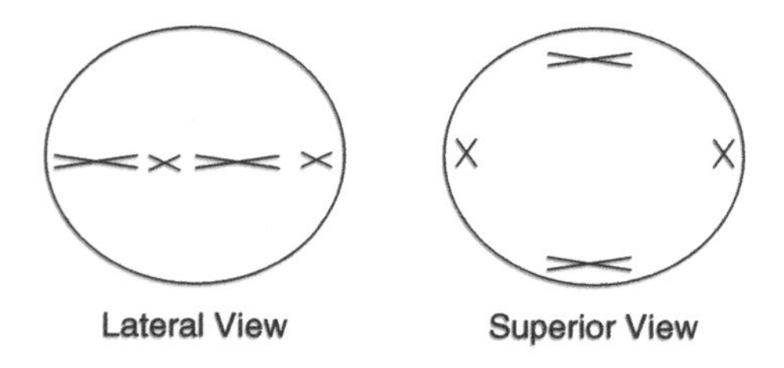

Anaphase **Anaphase** (AN-a-fāz; *ana*, "apart") is the time when sister chromatids separate from one another and move toward opposite poles of the cell. The separated chromatids are now called **sister chromosomes**.

Telophase **Telophase** (TĒ-lō-fāz; *telo*, end) is the last stage of mitosis when the condensed chromosomes have reached opposite poles of the parent cell and begin to uncoil. During this stage a new nuclear membrane forms around the elongating strands of chromatin. **Figure 2-5** shows the stages of mitosis.

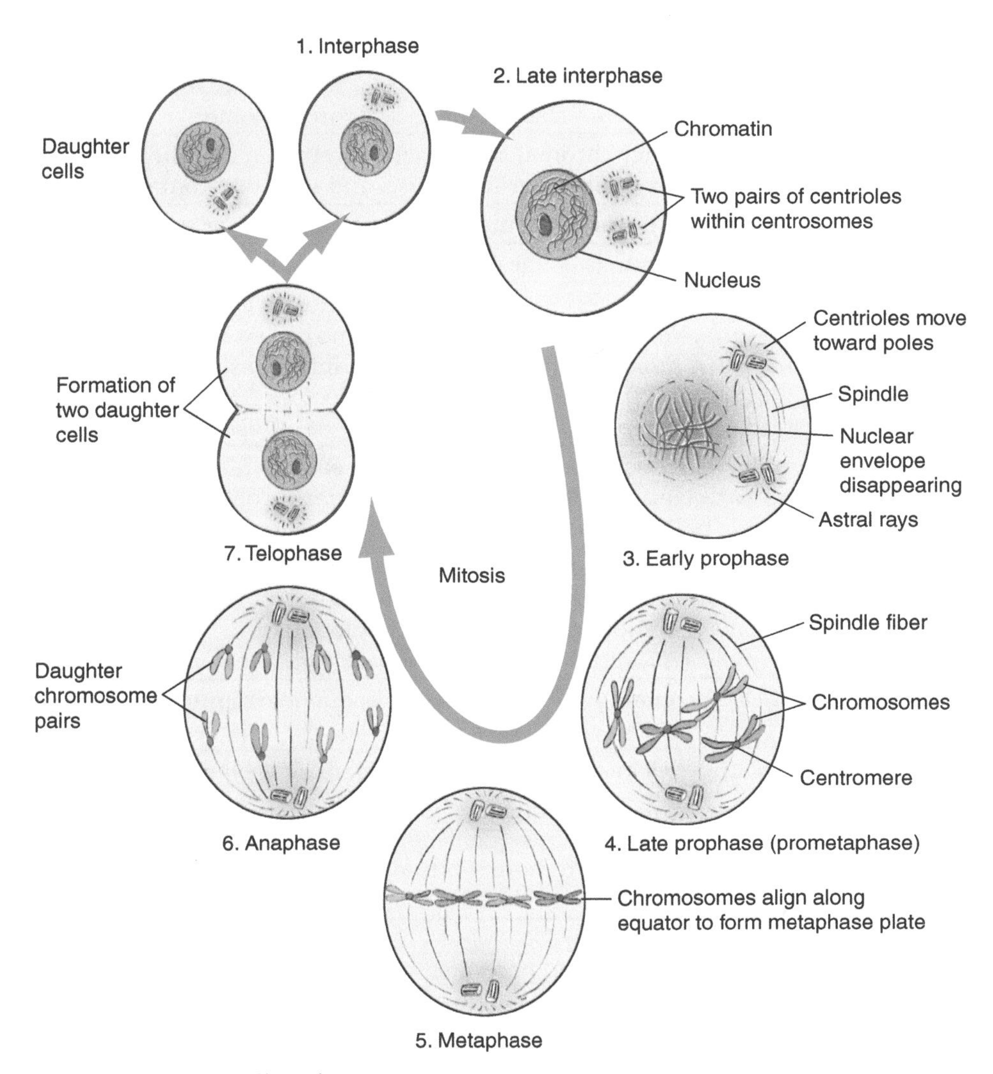

Figure 2-5 The cell cycle.

Cytokinesis

Cytokinesis (sī-tō-ki-NĒ-sis) is the final stage of the cell cycle when the parent cell divides to form two daughter cells. Cytokinesis begins when a *constriction ring,* made of microfilaments, tightens around the cell's periphery causing the formation of a **cleavage furrow**, or groove. The contraction ring continues to squeeze inward, like over-tightening a belt around one's waist, until the parent cell is split into two daughter cells. Cytokinesis often begins during telophase of mitosis, but these two stages are distinct processes.

In addition to some of the cell cycle terms mentioned above, other terms that you need to learn are included in **Table 2.3**.

Table 2.3 Terms Related to the Cell Cycle	
Autosome	Any linear chromosome not classified as a sex chromosome
Centromere	Region where two sister chromatids are held together in a replicated linear chromosome; centromeres are visible during mitosis and each one contains two kinetochores
Chromatid	One of two identical linear DNA molecules held together at a centromere
Chromatin	Linear DNA plus protein (histones) that coil to make a visible chromosome
Chromosome	Term used for linear DNA in different stages of the cell cycle and for circular DNA in the mitochondria. For linear DNA, a chromosome can be in two forms: (1) A single, double helix DNA molecule during G_1 phase of interphase and during anaphase and telophase of mitosis. (2) Two identical DNA molecules (*sister chromatids*) during G2 of interphase and during prophase and metaphase of mitosis
DNA molecule	*Linear*: double helix nucleic acid that comprises each non-replicated chromosome in the nucleus and each chromatid in a replicated nuclear chromosome *Circular*: double helix nucleic acid found within a mitochondrion
Diploid number	Condition in which a nucleus has two sources of each type of linear chromosome; in humans, these two sources are the mother (maternal chromosomes) and father (paternal chromosomes); diploid nuclei have a homologous pair for each type of chromosome; the diploid (2N) number for humans = 46 (or 23 chromosomes from each parent)
Gene	Segment of DNA containing information for sequencing amino acids in a specific protein
Genome	Complete set of genes in an organism; for humans, it is all genes found on all 46 chromosomes
Haploid number	Number of *different* types of linear chromosomes in a cell's nucleus; different types can be identified by length, centromere location, and/or banding patterns; the haploid (1N) number for humans is 23
Histones	Proteins around which linear DNA molecules wrap prior to mitosis; allows linear chromosomes to condense and become more visible during mitosis; not present with mitochondrial DNA
Homologous chromosomes	Linear chromosomes that contain information responsible for the same genetic traits; humans have 23 pairs of homologous chromosomes in each diploid nucleus
Karyotype	Photograph of an organism's linear chromosomes
Kinetochore	Region on a chromatid where a spindle fiber attaches; it consists of DNA and protein and is located in the centromere region
Maternal chromosome	Any linear chromosome donated to the offspring by the mother
Nucleosome	Cluster of 8 globular, histone proteins around which a segment of linear DNA coils
Paternal chromosome	Any linear chromosome donated to the offspring by the father
Replicated chromosome	*Replicated nuclear chromosome*: A linear chromosome derived from one parent (i.e., it is either maternal or paternal) consisting of two double-helix DNA molecules (chromatids) joined by a centromere *Replicated mitochondrial chromosome*: A circular chromosome within a mitochondrion and consisting of two circular, double helix DNA molecules prior to binary fission
Sex chromosome	Linear chromosome responsible for determining the sex of the individual; designated "X" and "Y" in humans; a female has two X chromosomes and a male has an X and a Y
Somatic cells	Any cell not considered a sex cell; sometimes called a general "body" cell
Spindle fibers	Microtubules (tubulin polymers) that aid in the separation of sister chromatids during mitosis; originate from centrioles within centrosomes

Name ______________________ Course Number: ________ Lab Section ______

Lab 2: Chromosome Models Worksheet

Chromosome Models

In order to test your knowledge of the cell cycle and the terms in Table 2.3, you will be asked to make drawings related to (1) a set of laminated paper chromosomes and (2) examples of cells in different stages of the cell cycle from hypothetical organisms. Work with a partner to answer the following questions.

Determining the Haploid and Diploid Number

1. Remove the chromosome models from the bags and determine the number of *different kinds* by looking for differences in length and/or centromere location. An example of one kind, #1 autosome, is shown below. Each strip of paper represents a single, double-stranded DNA molecule (dsDNA), as depicted by the helix design on the back of each strip.

Front of paternal #1 — Number 1 (Autosome)

Back of paternal #1

Front of maternal #1 — Number 1 (Autosome)

Back of maternal #1

2. Based on the total number of different kinds of chromosomes, 1N = ______ and 2N = ______.

3. Simulate G_1 phase of interphase using the chromosome models for a 2N parent cell. Draw a sketch of the chromosomes below.

4. Simulate S phase of interphase using the chromosome models. You will need to use all strips of paper to depict replication. Use a paperclip to hold together the sister chromatids that make up a replicated chromosome. (You will not be drawing this phase).

5. Simulate G2 of interphase and prophase of mitosis using the chromosome models. Make a sketch below.

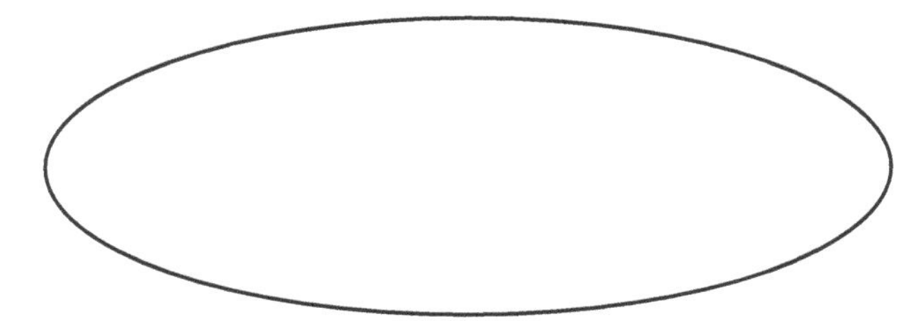

6. Simulate metaphase of mitosis using the chromosome models and make a sketch below.

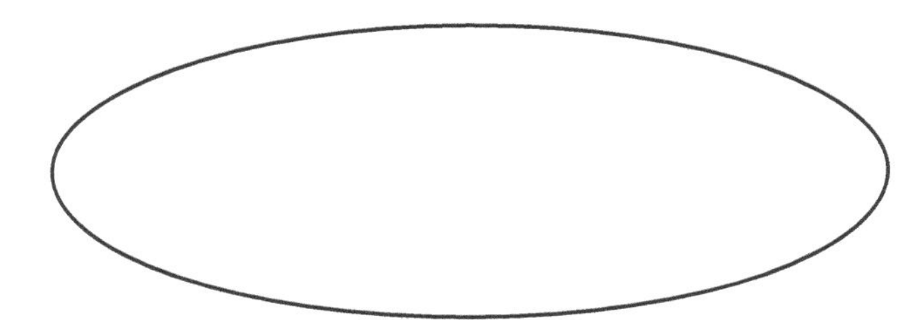

7. Simulate anaphase of mitosis using the chromosome models and sketch their positions below.

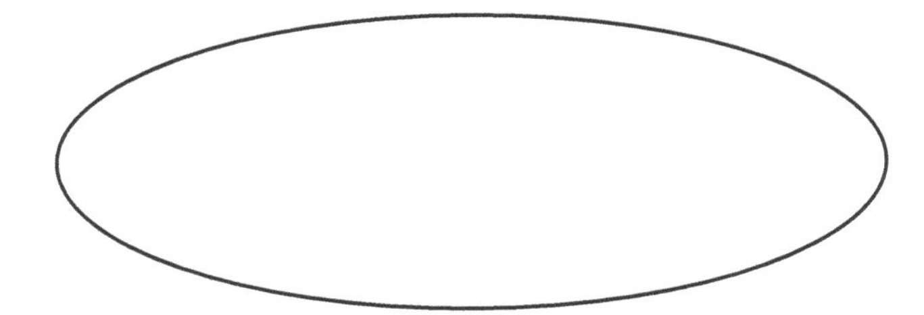

8. Immediately after they form, the two new daughter cells are in which specific stage of the cell cycle?

Additional Drawings

9. Draw a **2N** (diploid) cell during prophase of mitosis in an organism with **2N = 8.**

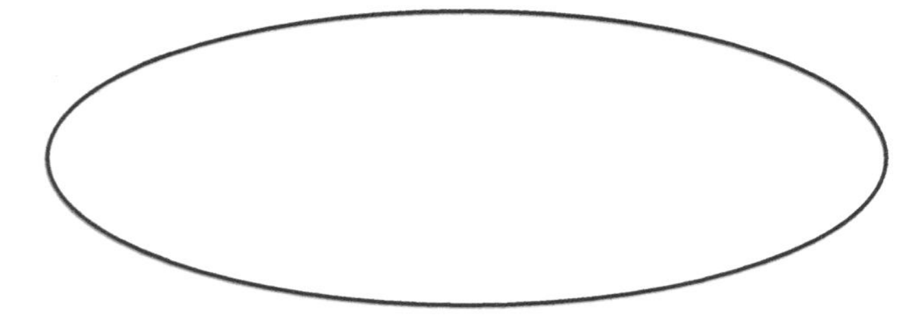

10. Draw a **1N** (haploid) cells metaphase of mitosis from an organism with **1N = 2.**

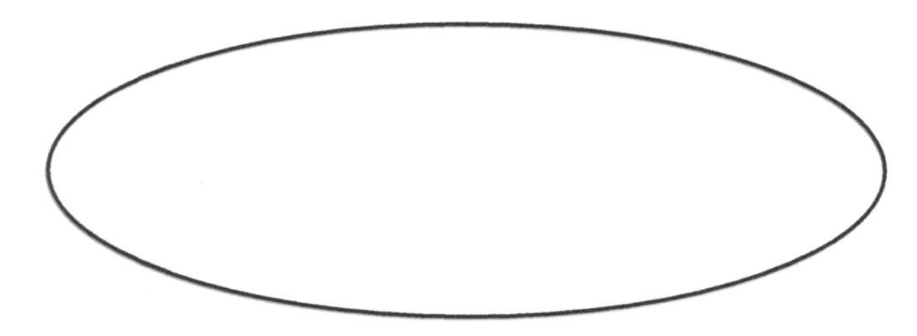

11. Draw a **2N cell** during G_2 from an organism with **1N = 4.**

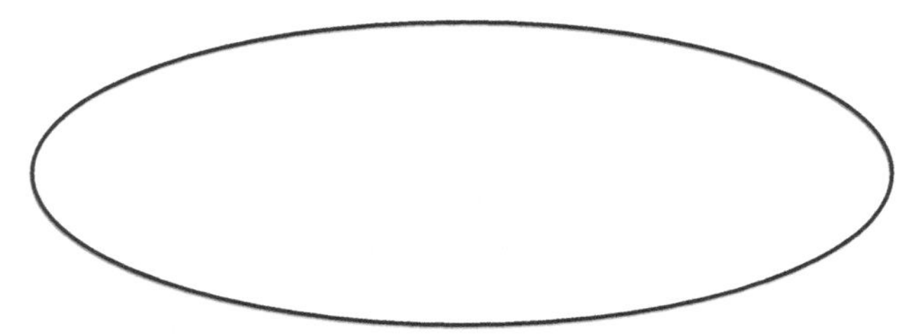

12. If an organism's 1N number = 12, how many chromatids are present in one of its 2N cells during metaphase? ______

13. How many linear, double-helix DNA molecules are present in one of your 2N cells during prophase of mitosis? ______

14. How many kinetochores are present in one of your 2N cells during metaphase of mitosis? ______

15. How many replicated autosomes are present in one of your 2N body cells during G2 phase? _______

16. How many centromeres are present in one of your 2N body cells during G1? ______ G2? _______ Prophase? _______

17. Why must cells replicate their DNA prior to entering mitosis?

LAB

3 MEMBRANE DYNAMICS

Membrane Vocabulary

In the last lab you studied the structural components of a generalized animal cell and how a cell divides its genetic material and cytoplasm through mitosis and cytokinesis, respectively. In this lab we focus on how materials can pass into and out of a cell through the cell (or plasma) membrane. More detail about a membrane's structure is presented in lecture, so here we center our attention on the diffusion of substances through the membrane; hence, the title membrane *dynamics*. **Table 3.1** summarizes key terms related to the diffusion through a cell membrane.

Table 3.1 Vocabulary Related to Membrane Dynamics

Term	Definition
Bound water	Water molecules bound to a solute particle; these are *non-diffusible* water molecules
Crenation	Shriveling of a cell due to a loss of water when surrounded by a hypertonic solution
Dialysis	Separation of small and large solutes by diffusion of the smaller particles through a semipermeable membrane
Diffusion	Movement of a substance from a region where it is in a higher concentration to a region where that substance is in a lower concentration
Free water	Water molecules not bound to a solute particle; these are *diffusible* water molecules
Hydration shells	Layers of polar water molecules surrounding solute particles in a solution; hydration shells consist of *bound, non-diffusible* water molecules
Hydrostatic pressure	Pressure exerted by a fluid on the inside of a tube or other cavity; it is the tendency of a fluid to push its way out of a confined area.
Hyperosmotic	Solution having a higher osmolarity (solute particle concentration); around a cell it is called a *hypertonic* solution and causes the cell to crenate
Hypoosmotic	Solution having a lower osmolarity (solute particle concentration); around a cell it is called a *hypotonic* solution and causes the cell to swell
Isoosmotic	Solutions having the same osmolarity (solute particle concentration); around a cell it is called an *isotonic* solution and causes neither crenation or swelling

(*Continued*)

Table 3.1 Vocabulary Related to Membrane Dynamics (Cont'd)	
Isotonic glucose solution	5% glucose (5 g glucose in 100 mL) is approximately 280 mOsm/L
Isotonic saline solution	0.9% NaCl (0.9 g NaCl in 100 mL) is approximately 280 mOsm/L; this is also known as *physiological saline*
Lysis	Rupturing of a cell due to an influx of water from a surrounding hypotonic solution
Osmosis	Diffusion of water through a semipermeable membrane
Osmolarity	Measure of the concentration of solute particles per unit^3 of solution; usually reported in milliosmoles per liter (mOsm/L)
Osmotic pressure (OP)	Pressure required to prevent osmosis; it can be thought of as a solution's tendency to gain water due to its osmolarity; solutions with a higher concentration of solute particles have a higher OP than solutions with a lower OP
Semipermeability	Ability to allow some materials to pass through while preventing others; also called *selective* or *differential* permeability (allowing different things to pass through)
Solute	Particles dissolved in a solvent
Solution	The combination of a solvent in which one or more types of solutes are dissolved
Solvent	Substance in which solute particles are dissolved and dispersed

Effect Of Temperature On Diffusion

1. Add 200 ml of cold water into a beaker labeled "cold."
2. Add 200 ml of room-temperature water into a beaker labeled "room temperature."
3. Add 200 ml of warm water into a beaker labeled "warm."
4. Let the beakers stand at your workstation for approximately 2 minutes with no agitation.
5. After 2 minutes, place one drop of food coloring into each beaker.
6. After 5 minutes, record your observations about the extent of food coloring movement that takes place in each beaker.

The coloring in (cold, room temperature, warm) water had the fastest rate of diffusion? Circle your observation and explain why this happened.

Effect Of Particle Size On Diffusion

At the same temperature and in the same type of environment, will a particle's size affect how fast it will diffuse? To find out, you will add three different-size particles to a three-way divided Petri dish containing clear gelatin. Gelatin is a colorless and tasteless water- soluble protein prepared from collagen and has a semi-liquid consistency. **CAUTION:** To avoid staining of your skin or clothing, carefully follow the directions given by your instructor when placing the dye in your Petri dish.

The smaller diffusible particles are derived from liquid yellow food dye, which is a combination of Yellow #5 and Yellow #6 and have the molecular formulae of $C_{16}H_9N_4Na_3O_9S_2$ and $C_{16}H_{10}N_2Na_2O_7S_2$, respectively. The molecular weight of Yellow #5 is _________ and the weight of Yellow #6 is _________ (refer to the Periodic Table). The larger molecules will be derived from liquid blue and liquid red food dyes. The molecular formula of Blue #1 is $C_{37}H_{34}N_2Na_2O_9S_3$ and

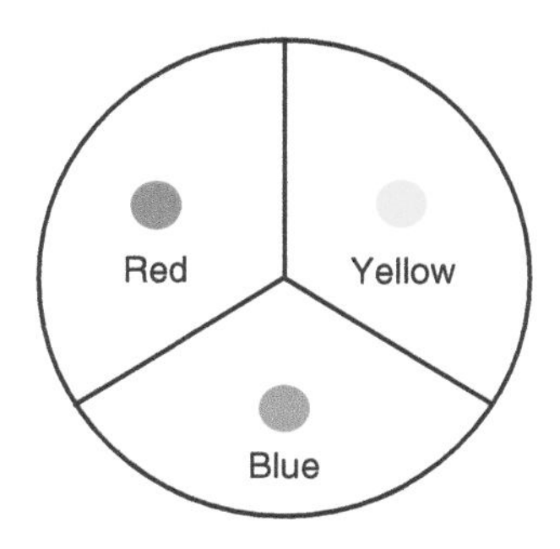

Figure 3-1 Agar plate for diffusion demonstration

its molecular weight is ________. The red dye is a combination of Red #40 and Red #3, and their molecular formulae are $C_{18}H_{14}N_2Na_2O_8S_2$ and $C_{20}H_{12}N_2Na_2O_7S_2$, respectively. The molecular weight of Red #40 is ________ and the weight of Red #3 is ________.

1. Place a drop of yellow food dye into one of the three sections of the Petri dish as close to the center of the triangle as possible.
2. Repeat using the blue and the red food dyes so that each section of the Petri dish has a different color drop (see **Figure 3-1**).
3. After 1 hour, carefully use a piece of paper towel to absorb the remaining liquid sitting on the surface of the gelatin. Sit the Petri dish on top of a small ruler to accurately measure (in mm) the diameter of the diffusion rings around the dye and record your results below.

   Red dye moved ________ mm/30 min.

   Yellow dye moved ________ mm/30 min.

   Blue dye moved ________ mm/30 min.

Plotting the data on a relational graph

Label the axes on the relational graph and then draw a relational line to illustrate the effect of particle size on diffusion rate.

Osmosis Experiment

In the following exercise you will observe osmosis, which is the diffusion of water through a semipermeable membrane. The membrane you will be using is *dialysis tubing*, which initially is a clear, flat plastic strip. Dialysis tubing contains ultra-microscopic pores through which free water molecules can pass. On the other hand, bound water molecules, which are bound to solute particles, cannot pass through these tiny pores. For osmosis to occur, the membrane must separate two solutions containing different concentrations of solute particles.

The solute particles in this demonstration will be molecules of *sucrose*, the disaccharide commonly known as table sugar. On one side of the membrane will be the sucrose solution and on the other side will be *deionized water*, which has virtually no solute particles. Working with a partner, prepare two beakers and two strips of dialysis tubing as follows (also see **Figure 3-2**).

1. Fill beaker A with 200 mL of distilled water.
2. Fill beaker B with 200 mL of sucrose solution.
3. Fold over one end of a strip of dialysis tubing and have your partner tie a piece of dental floss or string tightly in a knot around the fold so that liquid will not be able to pass through this end of the tube. Your instructor will demonstrate how to do this.
4. Using a graduated pipet, squeeze 4 mL of sucrose solution into the open end of the tube, then fold over that end and tie it with a string in the same way you tied off the other end. This will be bag A. Put it aside and proceed to the next step.
5. Following steps 3 and 4 with another piece of dialysis tubing, but instead of using a sucrose solution you will squeeze 4 ml of deionized water into the tube. This will be bag B.
6. Add tap water to the 90-ml mark of a 100-ml graduated cylinder. Determine the volumes of bags A and B by submerging them separately into the graduated cylinder and recording the volume of water displaced. In the table below, record the initial volumes for each bag.
7. Place bag A (containing the sucrose solution) into beaker A (containing deionized water)

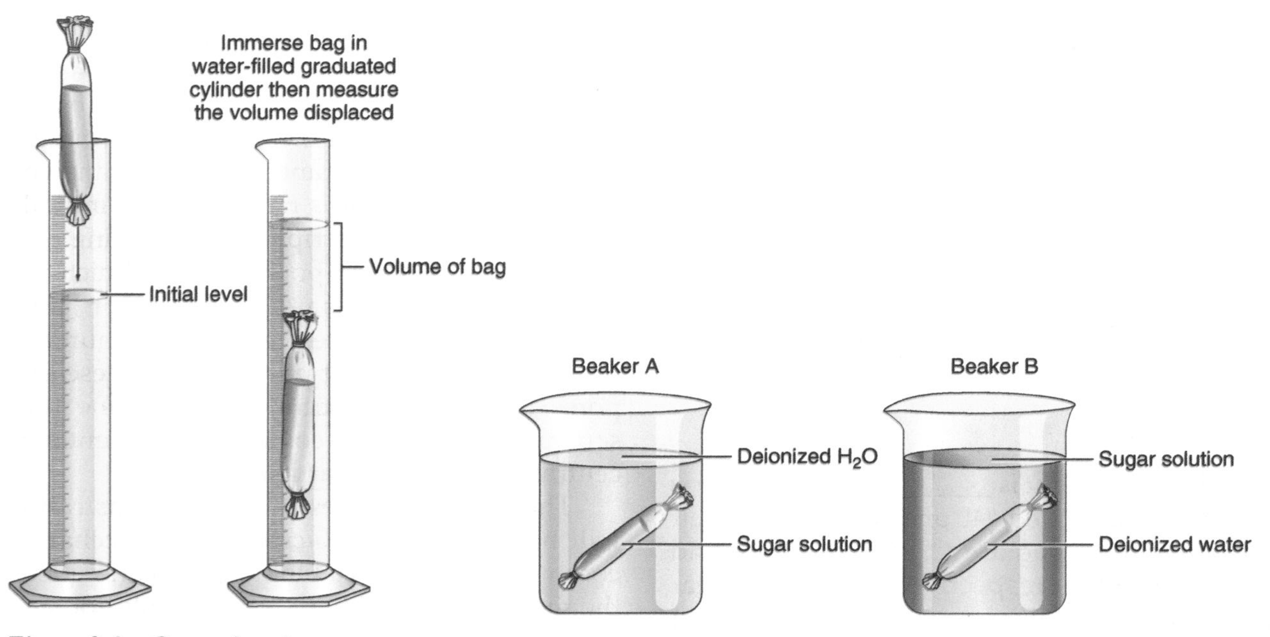

Figure 3-2 Osmosis setup

Dialysis Bag	Initial volume (mL)	Final volume (mL)	Volume Change (mL)
Containing sucrose			
Containing deionizing water			

and place bag B (containing deionized water) into beaker B (containing sucrose solution).

8. After 30 minutes, follow step 6 to determine the final volumes of each bag. Record your results below, using a plus sign (+) if the bag's volume increased or a minus sign (–) if the bag's volume decreased.

Questions About Osmosis Experiment

1. Bag ____ contained a solution that was *hyperosmotic* to the surrounding fluid, while Bag ____ contained a solution that was *hypoosmotic* to the surrounding fluid.
2. Bag ____ contained a solution with a *higher osmotic pressure* compared to the solution surrounding it, while Bag ____ contained a solution with a *lower osmotic pressure* compared to the solution surrounding it.
3. The volume of the (sugar bag, water bag) increased due to osmosis. Circle the correct answer.
4. Why did the bag in question 3 gain water?

If over time the bag in question 3 gained so much water via osmosis that it became very **turgid** (TUR-jid; "swollen"), it would not likely burst because the tubing is too strong. However, osmosis would eventually cease and free water molecules would then move into and out of the bag at equal rates. Would this be because the osmotic pressure inside the bag is now equal to the osmotic pressure around the bag? No, because sugar cannot pass through the dialysis tubing. Instead, a net movement of free water into the bag would cease when the bag's (osmotic, hydrostatic) pressure, which attempts to bring water into the bag, becomes equally opposed by the bag's (osmotic, hydrostatic) pressure, which attempts to push water out of the bag. Circle the correct answers.

Osmolarity

Osmolarity (oz-mo¯-LAR-i-te¯) is a measure of the concentration of solute particles in a solution, and knowing its value allows you to predict whether water will move into or out of a cell by osmosis. The standard units for osmolarity are milliosmoles of solute per liter of solution (or mOsm/L). Normal human tissue fluid contains a solute particle concentration of about 280–300 mOsm/L. When comparing two solutions that are separated by a semipermeable membrane, the solution with the higher concentration of solute particles has a higher osmolarity.

Calculating osmolarity requires a quick review of some basic chemistry. The number of milliosmoles of a substance equals the number of millimoles of the substance multiplied by the number of particles liberated when each molecule of the substance dissolves. Recall that 1 mole of a substance is the molecular weight of the substance expressed in grams. In turn, a millimole is the molecular weight of the substance expressed in milligrams (mg). For example, the molecular weight of glucose ($C_6H_{12}O_6$) is 180, which means 1 mole of glucose is 180 grams of glucose and 1 mmole of glucose is 180 mg of glucose. If the dissolving substance does not ionize, then 1 mmole of the substance is equal to 1 mOsm. In our example, glucose is a *nonionizing* substance; therefore, 1 mmole of glucose equals 1 mOsm of glucose.

If the dissolving substance ionizes, then 1 mmole of the substance equals the number of ions formed when the substance dissociates. NaCl (table salt) is an *ionizing* substance that dissociates to form two ions, Na^+ and Cl^-, each of which is a solute particle when in a solution. Therefore, 1 mmole of NaCl equals 2 mOsm of NaCl. Several examples showing how to calculate osmolarity are provided here. Look over these equations to make sure you understand how to set them up in the future.

Calculating Osmolarity For Solutions With One Solute

1. Calculate the osmolarity of a 200-mL solution containing 3 grams of glucose, a *nonionizing* solute?

$$3\ \cancel{\text{g glucose}} \times \frac{1000\ \cancel{\text{mg glucose}}}{1\ \cancel{\text{g glucose}}} \times \frac{1\ \cancel{\text{mmole glucose}}}{180\ \cancel{\text{mg glucose}}} \times \frac{1\ \text{mOsm glucose}}{1\ \cancel{\text{mmole glucose}}} \times \frac{1}{200\ \cancel{\text{mL}}} \times \frac{1000\ \cancel{\text{mL}}}{\text{L}} = \frac{83.3\ \text{mOsm glucose}}{\text{L}}$$

2. What is the osmolarity of a 200-mL solution containing 3 grams of NaCl, an *ionizing* solute?

$$3\ \cancel{\text{g NaCl}} \times \frac{1000\ \cancel{\text{mg NaCl}}}{1\ \cancel{\text{g NaCl}}} \times \frac{1\ \cancel{\text{mmole NaCl}}}{58\ \cancel{\text{mg NaCl}}} \times \frac{2\ \text{mOsm NaCl}}{1\ \cancel{\text{mmole NaCl}}} \times \frac{1}{200\ \cancel{\text{mL}}} \times \frac{1000\ \cancel{\text{mL}}}{\text{L}} = \frac{517.2\ \text{mOsm NaCl}}{\text{L}}$$

Notice that 3 mg of NaCl dissolved in 200 mL of deionized water contains **6.2 times** more mOsm of solute/L than 3 mg of glucose dissolved in 200 mL of deionized water.

Calculating Osmolarity For Mixed Solutions

When considering solutions that contain two or more different kinds of solute, you must add the osmolarities for each solute to determine the total osmolarity of the solution. For example, if you placed 1 mmole of glucose and 1 mmole of NaCl into a beaker and added enough distilled water to make a 1-liter solution, the total osmolarity would be calculated as follows:

$$\underline{1\ \cancel{\text{mmole glucose}}} \times \frac{1\ \text{mOsm glucose}}{1\ \cancel{\text{mmole glucose}}} + \underline{1\ \cancel{\text{mmole NaCl}}} \times \frac{2\ \text{mOsm NaCl}}{\cancel{\text{mmole NaCl}}} \times \frac{1}{\text{L}} = \frac{\mathbf{3\ mOsm}}{\mathbf{L}}$$

Clinicians consider **0.9% NaCl (physiological saline)** and **5% glucose** solutions to be isotonic to human cells. Note that these numbers are not exactly what you calculated above, but slight variations occur because living cells have the ability to pass materials selectively through their membranes. Moreover, glucose is readily taken in by living cells, making it a *permeating* solute. Therefore, a 5% glucose solution around a cell is isotonic only for a short while, but it becomes hypotonic to the cell as glucose enters the cell.

While glucose is a permeating solute, sodium and chloride ions are *nonpermeating solutes* because they do not enter the cell as easily as do permeating solutes. Furthermore, living cells have the ability to expel Na^+ and Cl^- through their cell membrane. To help you visualize how similar weights of different kinds of substances can affect osmotic pressure differently, we will describe how a hypothetical salt (XY) and a hypothetical sugar (Z) dissolve in a solution (see **Figure 3-3**).

The top of Figure 3-3 shows our hypothetical substances dissolve to yield solute particles. Assume that each molecule of salt XY can ionize to form two ions, X and Y, both of which are solute particles, but a molecule of sugar Z does not ionize. The figure shows that 1 gram of XY yields 8 solute particles; therefore, we would assume there must have been 4 molecules of XY present in 1 gram of XY. In contrast, one gram of sugar Z yields two solute particles, and since molecules of Z do not ionize, 1 gram of Z must have contained two molecules of Z.

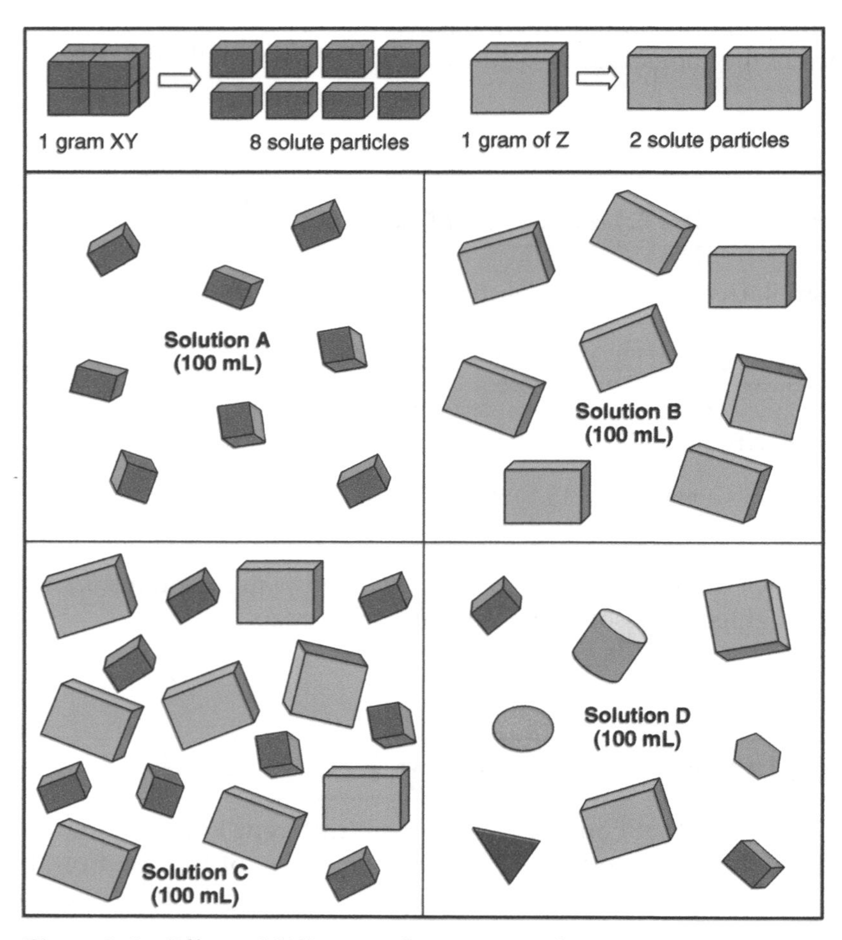

Figure 3-3 Effect of different solute on osmolarity.

Dialysis

Dialysis (dī-AL-i-sis; "to separate") is a passive process involving the separation of different-size solute particles using a semipermeable membrane. Only smaller solute particles can pass through the membrane. Dialysis is used in artificial kidneys to cleanse the blood of patients whose kidneys no longer function efficiently.

In the following exercise, you will observe the effect of particle size on the movement of particles through a semipermeable membrane; in this case, dialysis tubing. One of the solute particles is **starch**, a polysaccharide consisting of glucose molecules linked together. The other solute particle is iodine found in a special mixture called **Lugol's solution**. When the brownish iodine comes in contact with starch, the starch turns blue. Set up the dialysis experiment as follows:

1. Fill a beaker with 200 mL of deionized water and add 20 drops of Lugol's iodine solution to the water.
2. Fold over one end of a strip of dialysis tubing and have your partner tie a piece of dental floss or string tightly in a knot around the fold so that liquid will not be able to pass through this end of the tube.
3. Use a pipet to add 4 mL of starch solution to the bag, then fold over the end and seal it tightly with string.
4. Submerge the bag of starch solution in the beaker containing Lugol's solution and allow it to stand for 30 minutes.

A blue color developed in the (beaker, bag). Circle your answer.

Explain how the movement of solute particles accounted for the color change in the location (beaker or bag) that you observed.

Name ____________________ Course Number: ________ Lab Section _______

Lab 3: Worksheet

Osmosis And Osmolarity Problems

1. Calculate the number of grams of NaCl needed to prepare 100 mL of a 280 mOsm/L NaCl solution. (The molecular weight of NaCl is 58.) Show detailed steps in your calculations.

2. Calculate the number of grams of glucose needed to prepare 100 mL of a 280 mOsm/L glucose solution. (The molecular weight of glucose is 180.) Show detailed steps in your calculations.

3. If you placed 3 grams of glucose and 2 grams of NaCl into a graduated cylinder and added enough distilled water to make a 1-liter solution, what is the total osmolarity of this solution in **mOsm/L**? Show your detailed calculations below:

4. Based on your calculations, a _____ % NaCl solution and a _____ % glucose solution are isotonic to one another.

Questions About Figure 3-3

Assume that all solutions in Figure 3-3 consist only of the solutes shown and pure water. If each mL of pure water weighs 1 gram, then based on weight-to-weight (w/w) ratios, we can also assume that 1 gram of a substance dissolved in a 100 mL solution containing pure water would be a 1% solute solution. Based on this information, answer the following questions.

5. How many grams of XY are in solution A?

   ______

6. Solution A contains a ____% solution of XY.
7. How many grams of Z are in solution B?

   _____

8. Solution B contains a ____% solution of Z.
9. Solution C contains ____% XY and ____% Z.
10. Solution D contains ____% XY and ____% Z.

For the following questions, circle the correct answers.

11. Solution A is (hyperosmotic, hypoosmotic, isotonic) to solution B.
12. Solution A is (hyperosmotic, hypoosmotic, isotonic) to solution C.
13. Solution D is (hyperosmotic, hypoosmotic, isotonic) to solution A.
14. Solution A is (hyperosmotic, hypoosmotic, isotonic) to solution D.
15. Solution B is (hyperosmotic, hypoosmotic, isotonic) to solution D.
16. If solution A were 0.9% NaCl and solution B were 5% glucose, then solution (A, B, C, D) would be most like the chemical makeup of a human cell.
17. If 0.9% NaCl around a human cell is isotonic to that cell, then based on Figure 3-3 we can be sure that a human cell contains (exactly, less than, more than) 0.9% NaCl.
18. If 5% glucose around a human cell is isotonic to that cell, then based on Figure 3-3 we can be sure that a human cell contains (exactly, less than, more than) 5% glucose.
19. Assume that the four compartments in the next figure are separated by a membrane that is impermeable to glucose, Na^+, and Cl^-, but it will allow free water to pass through. Assume there is only one type of solute in each compartment and it is dissolved in deionized water. Draw arrows vertically, horizontally, and/or diagonally to show the direction that **water** will travel.

2% Glucose	0.9% NaCl
1.2% NaCl	5% Glucose

20. In the following example, 3 dialysis bags containing varying amounts of saline (salt) are placed into a 0.9% saline solution. The only solute inside the bags and the beaker is NaCl. The concentrations indicate the weight of the solute per total weight of solution (w/w). Indicate with arrows the net direction that **water** will diffuse through the membranes.

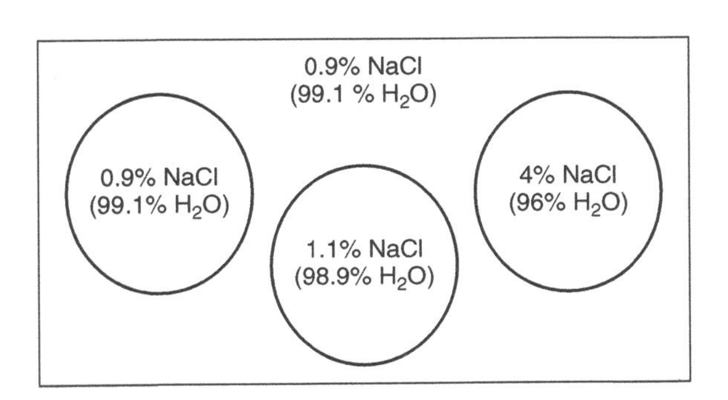

21. Consider four compartments separated by a membrane that is impermeable to glucose, Na^+, and Cl^-, but will allow water to pass through freely. There is only one type of solute present in each compartment. Draw arrows vertically, horizontally, and/or diagonally to show the direction that **water** will travel.

95% H_2O	0.9% NaCl
99.1% H_2O	5% Glucose

22. The dotted line in the following figure is a semipermeable membrane through which only water can pass. This membrane separates a glucose/NaCl solution from a red blood cell. The volume of solution is the same on both sides of the membrane. Place the following boldface labels on the appropriate side of the membrane. The side on which you place the first label is not important, but the location of all other labels will depend on the location of the first label.

HYPER (hyperosmotic), **FSP** (fewer solute particles per unit volume), **SGW** (side gaining water), **MHS** (more hydration shells), **MDW** (more diffusible water), **LOP** (lower osmotic pressure), **6% glucose** (concentration of glucose solution), **2% NaCl** (concentration of the salt solution), **LBW** (less "bound" water), **280 mOsm/L** (relative solute concentration), **CRBC** (crenating red blood cell).

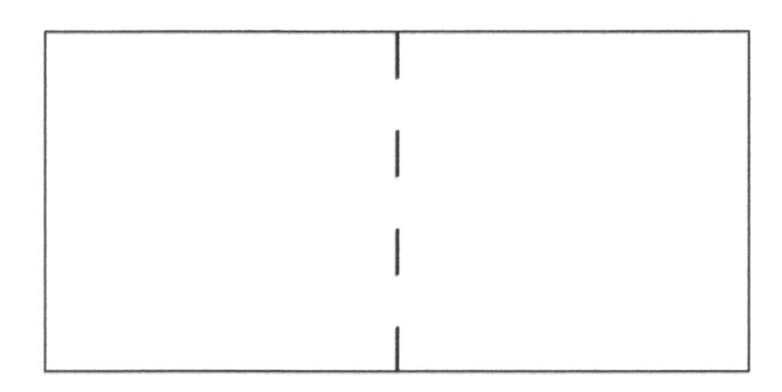

Dialysis Problem

23. Four compartments are separated by a membrane that is permeable to Na^+, Cl^-, and water, but NOT to glucose. There is only one type of solute present in each compartment. Remember that a particular solute will diffuse down its own concentration gradient independently of other solute particles. Draw arrows vertically, horizontally, and/or diagonally to show the direction that Na^+ and Cl^- will travel.

2% Glucose	0.9% NaCl
1.2% NaCl	5% Glucose

LAB 4 TISSUES

Overview Of Tissues

In the last two labs you studied parts of individual cells and how certain materials pass into and out of them. In this lab, we will look at different kinds of tissues, which are groups of cells. There are four major types of tissue in the body: **connective**, **epithelial**, **muscular**, and **nervous** tissue. While you will not be looking at every specific type of tissue in the body, you will examine a number of representative samples. You should be able to recognize these samples on microscope slides, drawings, and photographs. Be able to recognize each specific type of tissue and describe their functions and locations in the body. The microscope slides containing tissue samples are provided at your table.

Membranous Epithelium

Epithelial tissue (ep-ē-THĒ-lē-ul; "covering") includes groups of cells that make up glands (glandular epithelium) and those that make up most multicellular membranes (membranous epithelium). **Membranous epithelium** covers most of the body's free surfaces; i.e., surfaces that are not directly contacting another tissue. The body's free surfaces include the outer surface of the skin; the inner surfaces of visceral body cavities; and tubes (ducts) that transport secretions of certain types of glands.

Membranous epithelium is avascular (lacks blood vessels), has a basement membrane, has a free surface, may contain nerves, and can be classified by cell arrangement, cell shape, and location. The **basement membrane** is the glycoprotein foundation on which a membranous epithelium rests. It can be seen with the microscope as a thin line separating the epithelium from the connective tissue.

Table 4.1 summarizes basic terms related to the classification of membranous epithelial tissues. **Table 4.2** describes specific types of membranous epithelium and their locations in the body.

Table 4.1 Classification and Location of Membranous Epithelium

Classification	Specific Types	Description
Cell arrangement	**Simple**	Single layer of cells
	Stratified	Two or more layers of cells
	Pseudostratified	Single layer of cells in which all cells attach to the basement membrane, but not all cells reach the free surface; gives the illusion of being stratified
Cell shape	**Squamous**	Flat, scale-like cells
	Cuboidal	Cube-shaped cells
	Columnar	Column-shaped cells

(Continued)

Table 4.1 Classification and Location of Membranous Epithelium *(Cont'd)*

	Transitional	A stratified epithelium in which the apical cells change shape (between cuboidal and squamous) throughout the day due to pressure applied to them by overlying liquid. Found in the inner lining of the ureters and urethra (tubes that transport urine)
Location	**Endothelium**	Simple squamous epithelium that forms the inner lining of vessels
	Serous membrane	Simple squamous epithelium lining cavities that have no connection with the outside of the body; It secretes a watery *serous fluid.* Serous membranes include the pleura (around the lungs), pericardium (around the heart), peritoneum (around the abdominopelvic organs), mesentery (between loops of the intestines)
	Mucous membrane	Lines cavities that have connections with the outside of the body; Contains *goblet cells* that secrete viscous fluid called *mucus;* Forms inner linings of four organ systems (respiratory, digestive, urinary, and reproductive)
	Cutaneous membrane	Skin, including the epidermis, which is stratified squamous epithelium overlying the true skin or dermis

Table 4.2 Specific Types of Membranous Epithelium

Type	Figure	Description	Location
Simple squamous	4-1a,b	Single layer of flat cells	Endothelium, serous membranes, alveoli (sacs in lungs), and glomerular capsules (filtering structures in kidneys)
Simple cuboidal	4-1c	Single layer of cuboidal cells	Kidney tubules, pigmented layer of eye's retina, surface of ovaries and eyeball lens
Simple columnar	4-1d,e	Single layer of columnar cells; some contain cilia	Lining of fallopian tubes and uterus and spinal cord canal; ciliated versions form mucous membrane of the uterus and bile ducts
Pseudostratified	4-1f	Single layer of columnar cells but not all reach the apical surface; apical cells may be ciliated	Ciliated version forms mucous membrane within respiratory passageways; nonciliated version forms the lining of the epididymis (tube that holds sperm in the testes), part of male urethra, and some ducts
Stratified squamous	4-2a,b	Multiple layers; keratinized version has apical layer of dead squamous cells	Keratinized version forms epidermis (outer covering) of the skin; nonkeratinized version forms inner lining of mouth, esophagus, epiglottis, and vagina; covering of cornea on eyeball
Stratified cuboidal	4-2c	Multiple layers with cuboidal apical cells	Ducts of glands (sweat, mammary glands, salivary); male's urethra
Stratified columnar	4-2d	Multiple layers with columnar apical layer	Conjunctiva (covers anterior surface of eyeball) and part of urethra (tube that transports urine)
Transitional	4-2e	Stratified layers with changing apical cells	Lining of ureters, urinary bladder, and part of urethra

Connective Tissue

When first hearing the name *connective* tissue, one might think of a supporting tissue that helps hold the body together. Indeed, this is a major function of certain types of connective tissue, but "holding things together" is not what defines a connective tissue. Unlike epithelial tissues, which have tightly packed cells and contain no blood vessels, most connective tissues have widely scattered cells and an extensive matrix that contains abundant fibers and blood vessels (cartilage is an exception, being avascular). Interestingly, blood itself is a connective tissue. Therefore, connective tissues have a wide array of important functions in the body. Depending on the specific type of connective tissue, these functions include support, protection, reduction of friction, delivering of nutrients and removal of wastes from tissue fluids, and storage of nutrient molecules.

The extracellular material around most connective tissue cells is called the **matrix**, and it includes a liquid component, called **ground substance**, and abundant fibers. With few exceptions, the extracellular fibers consist of filamentous proteins intertwined like fibers in a rope. Like all proteins, these filamentous proteins are made at ribosomes inside cells. After their synthesis, they are secreted into the extracellular matrix where they intertwine to form the fibers. **Collagenous fibers** (kō-LAJ-en-us) are made of the protein *collagen*, which is the most abundant protein in the body. **Reticular fibers** (re-TIK-ū-lur, *retic*, "net") are also made of collagen protein but they are thin, highly branched strands. **Elastic fibers** consist of the protein *elastin*, which is a highly elastic material that can recoil after being stretched.

Connective tissues are classified into five major groups, some of which have several subgroups. **Loose connective tissue**, as the name implies, has a loosely organized arrangement of cells and fibers. It has an extensive matrix and a variety of fibers and cells. **Dense connective tissue** has more tightly packed fibers, but the cells are relatively widely scattered. **Cartilage**, sometimes called *gristle*, contains a firm matrix with cells trapped within tiny pockets called *lacunae*. The last two major groups have the most different matrix in terms of consistency. **Blood** (or **vascular**) **tissue** is a fluid tissue that travels throughout the body in tubes called blood vessels. **Osseous** (or **bone**) **tissue**, on the other hand, has a hard, solid matrix. Its cells, like those in cartilage, are trapped within lacunae.

Several suffixes may be applied to various connective tissue cells. The suffix *–blast* refers to a cell that is actively helping to generate more of the connective tissue of which it is a part. The suffix *–cyte* refers to mature connective tissue cells that may be helping to maintain the tissue but are not actively producing more of it. **Table 4.3** summarizes the major connective tissues in the body.

Table 4.3 Classification and Location of Connective Tissues

Type	Figure	Location
Loose Connective		
Areolar tissue	4-3a	Supports epithelial tissues; found in part of the skin, in subcutaneous tissue, blood vessel walls, and nerves; forms *adventitia* on the surface of trachea and esophagus; forms synovial membranes in movable joints
Adipose tissue	4-3b	Subcutaneous tissue, around kidneys and bone joints, surface of heart, yellow marrow in bones, abdominal cavity, breasts
Mesenchyme	4-3c	Undifferentiated tissue that gives rise to connective tissues
Reticular tissue	4-3d	Spleen, liver, lymph nodes, red marrow

(Continued)

Table 4.3 Classification and Location of Connective Tissues (Cont'd)		
Dense Connective Tissue		
Dense irregular	**4-4a**	Reticular skin, bone coverings, submucosa, dura mater, heart valves, capsules around bone joints and various organs
Dense regular	**4-4b**	Tendons, ligaments, aponeuroses (broad, flat tendons)
Elastic tissue	**4-4c**	Blood vessel walls, vocal cords, trachea, vertebral ligaments, lungs, penile ligament
Cartilaginous Tissue		
Elastic cartilage	**4-5a**	Found in the ear, epiglottis, and part of larynx
Fibrocartilage	**4-5b**	Intervertebral discs (between vertebrae), pubic symphysis (between pubic bones), and menisci (crescent-shaped wedges in certain joints)
Hyaline cartilage	**4-5c**	Costal (rib) cartilages, articular cartilages on ends of bones, growth plates of bones, trachea, bronchi, larynx (voice box), and nose
Osseous (Bone) Tissue	**4-6a**	Bones
Blood (Vascular) Tissue	**4-6b**	Blood vessels

Muscle and Nervous Tissues

The last two major groups of tissues, muscle and nervous, will be introduced here but more detail will be given in later chapters. Muscle tissue has the ability to produce tension (a pulling or squeezing) force resulting from a process called contraction. **Skeletal muscle tissue** is found in muscles that can be contracted voluntarily. Skeletal muscles get their name because they attach to parts of the skeleton and produce movement of skeletal structures. **Cardiac muscle tissue** is found only in the heart and its contractions generate the force that propels blood through blood vessels. **Smooth muscle tissue** is found in many visceral organs and the walls of many vessels.

Nervous tissue includes specialized cells called *neurons* that can generate electrical-like signals called *impulses*. These signals can influence many different kinds of cells in the body. Neurons are found in the brain, spinal cord, and nerves. **Table 4.4** provides more detail about muscle cells and neurons.

Table 4.4 Muscle and Nervous Tissue		
Muscle Tissue		
Cardiac muscle	4-7a	Branched cells that have striations; uninucleated and involuntarily controlled; located only in the heart and responsible for pumping blood
Skeletal muscle	4-7b	Long, cylindrical cells that have striations (stripes); multinucleated and voluntarily controlled; attached to bones and able to produce body movements; generates heat for regulating body temperature
Smooth muscle	4-7c	Tapered on ends and lack striations; uninucleate and involuntarily controlled; located in walls and tubes of visceral organs and vessels
Nervous Tissue	4-7d	Brain, spinal cord, and nerves

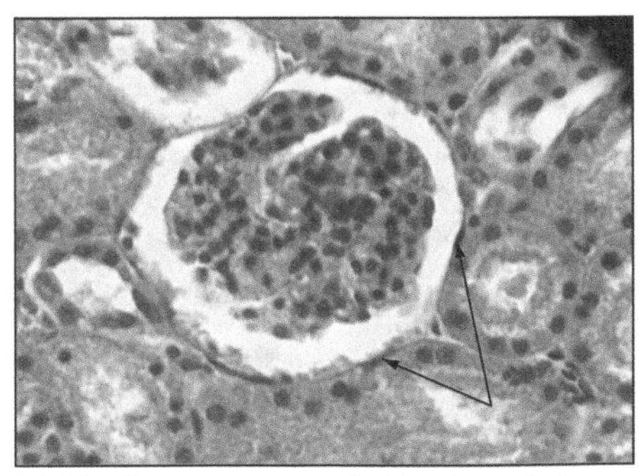

(a) Simple squamous epithelium in wall of kidney glomerulus (filtering structure) (400x)

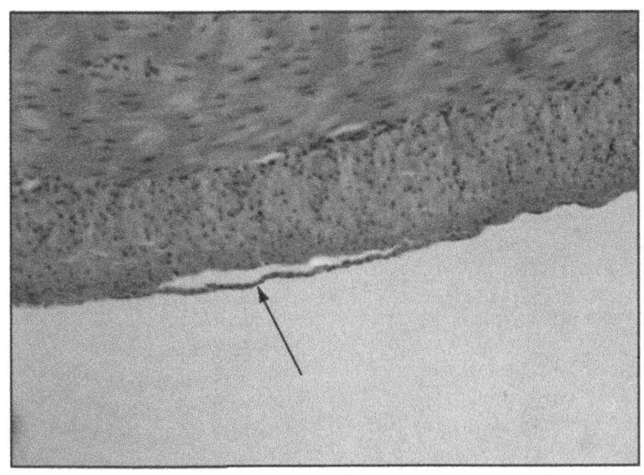

(b) Simple squamous epithelium of serous membrane (peeling off) of intestine (400x)

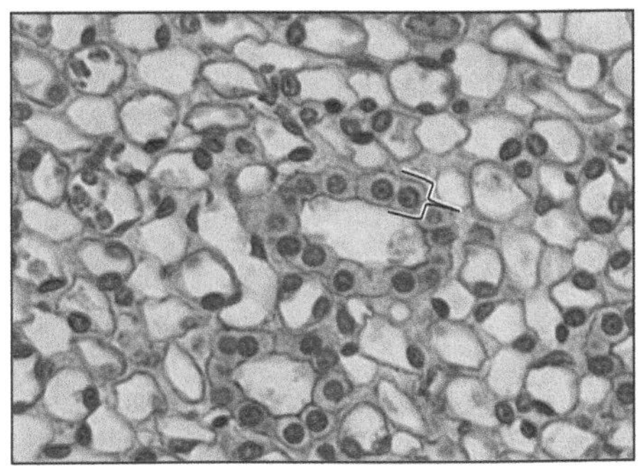

(c) Simple cuboidal epithelium: Cross-section of nephron tubule inside the kidney (400x)

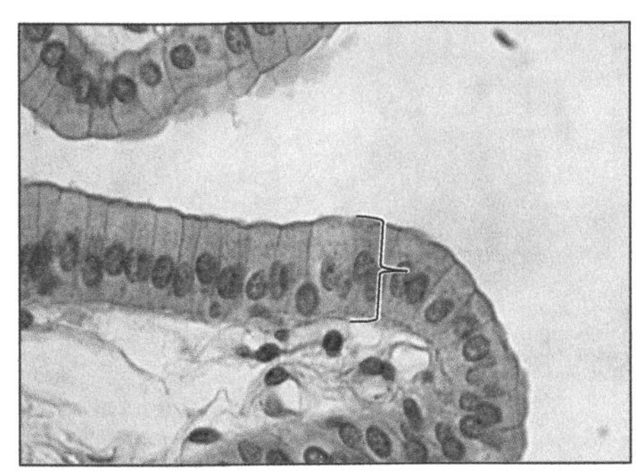

(d) Simple columnar epithelium of mucous membrane inside the gall bladder (400x)

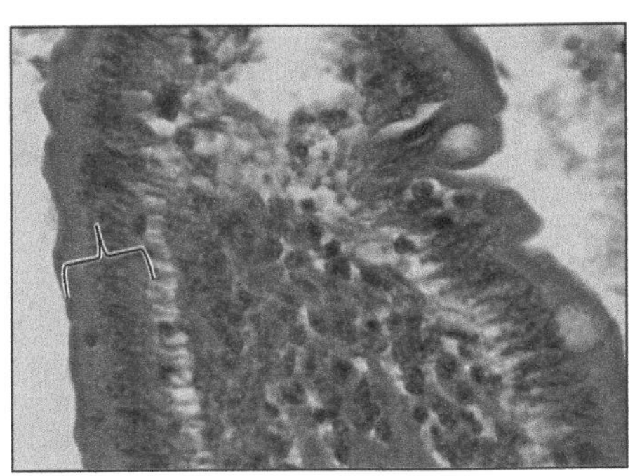

(e) Simple columnar epithelium of mucous membrane inside small intestine (400x)

(f) Pseudostratified ciliated columnar epithelium of mucous membrane in trachea (400x)

Figure 4-1 Simple epithelium.

All images source: Mark Taylor

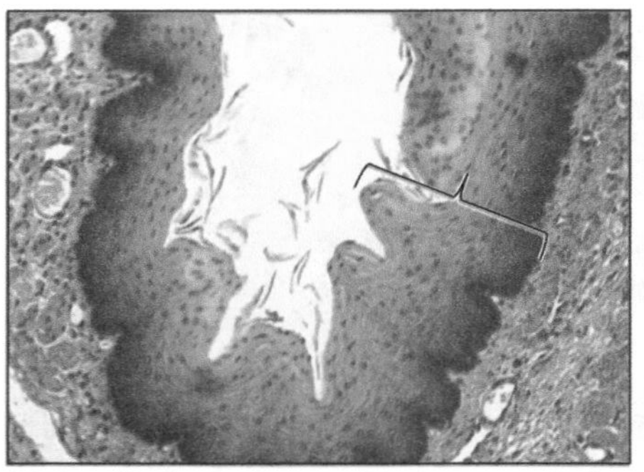

(a) Stratified squamous epithelium in wall of urethra (400x)

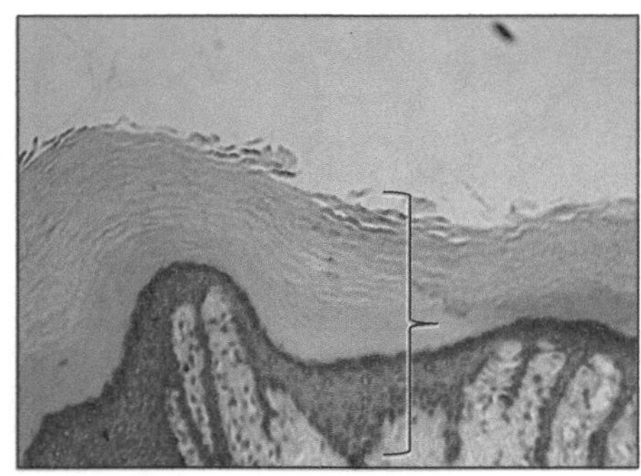

(b) Stratified squamous epithelium of thick skin epidermis (400x)

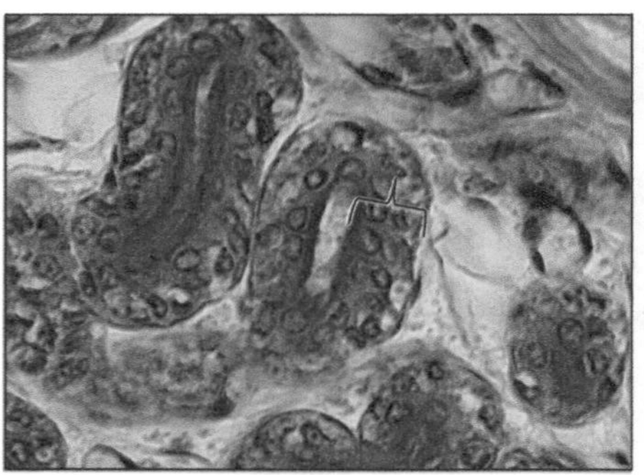

(c) Stratified cuboidal epithelium: Cross-section of sweat gland duct (400x)

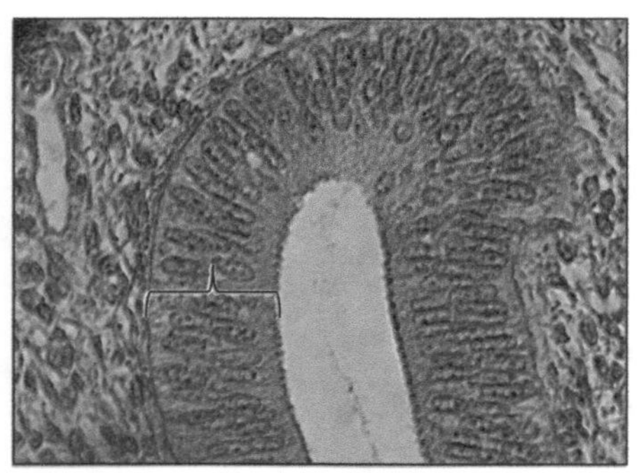

(d) Stratified columnar epithelium: Cross-section of uterine gland duct (400x)

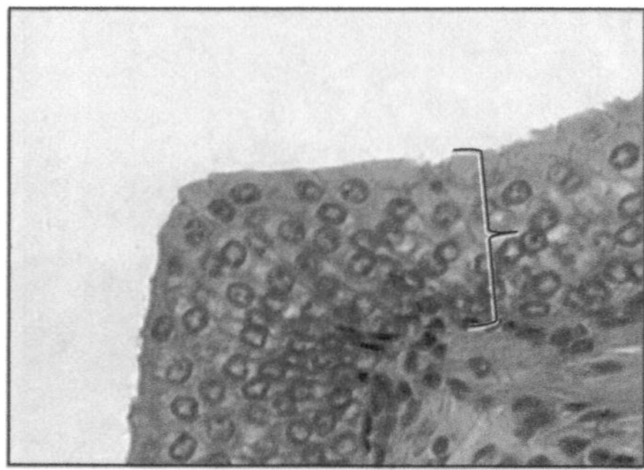

(e) Transitional epithelium in inner lining of ureter (tube from kidney) (400x)

Figure 4-2 Stratified epithelium.

All images source: Mark Taylor

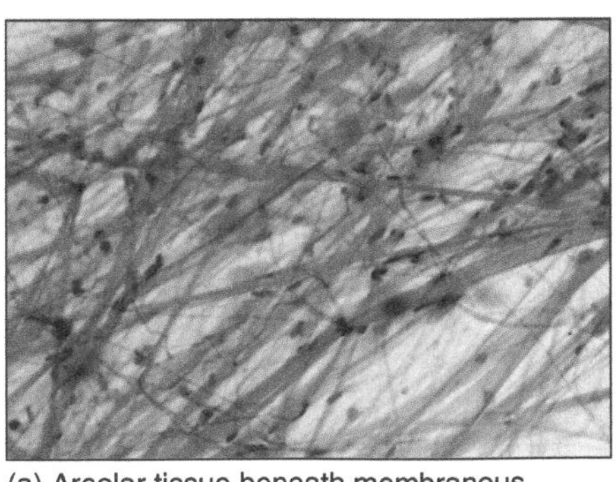

(a) Areolar tissue beneath membranous epithelium (400x)

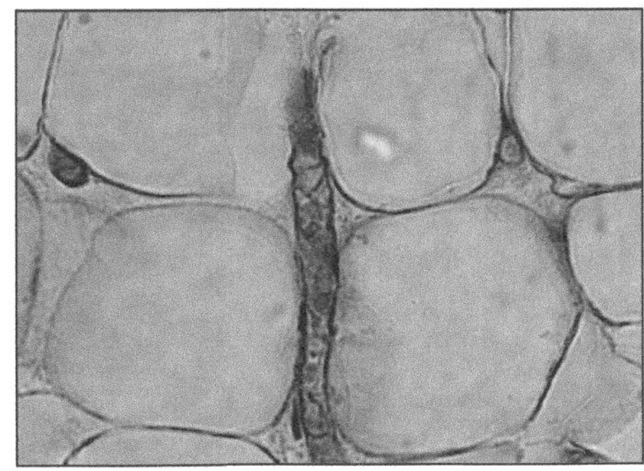

(b) Adipose tissue in subcutaneous region; red tube is a blood vessel (400x)

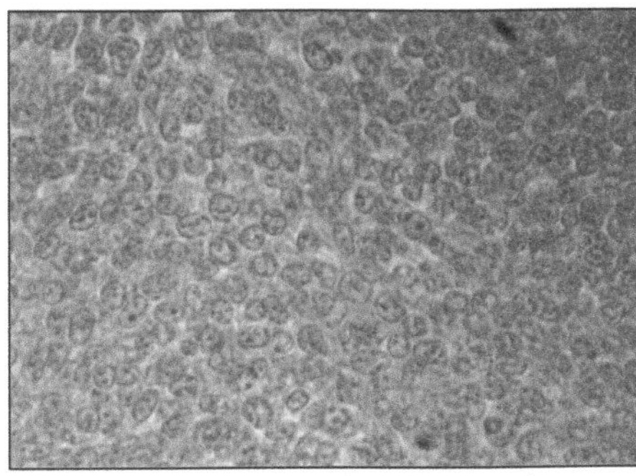

(c) Mesenchyme tissue that develops into connective tissues (400x)

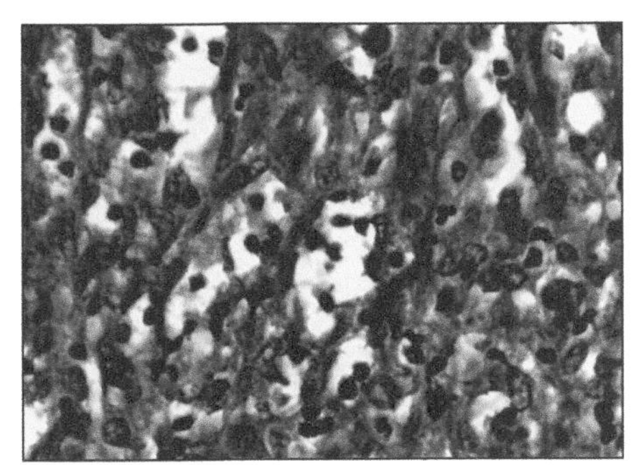

(d) Reticular tissue in the liver, spleen, and lymph nodes (400x)

Figure 4-3 Connective Tissue Proper: Loose Tissues.

All images source: Mark Taylor

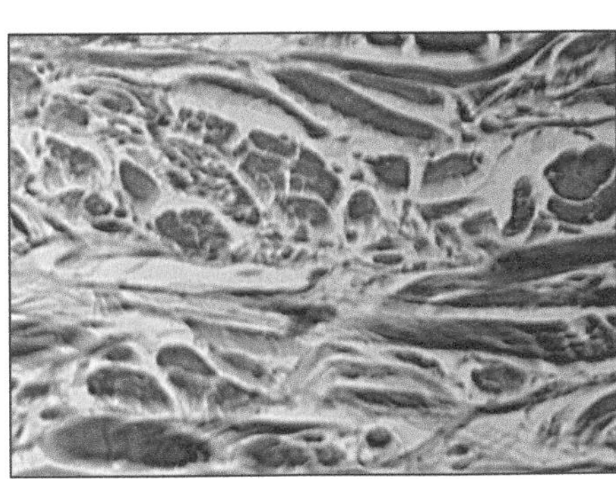

(a) Dense irregular tissue in the dermis (400x)

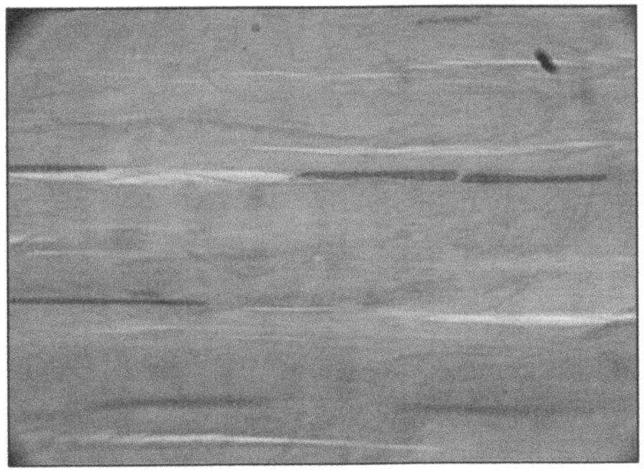

(b) Dense regular tissue in a tendon (400x)

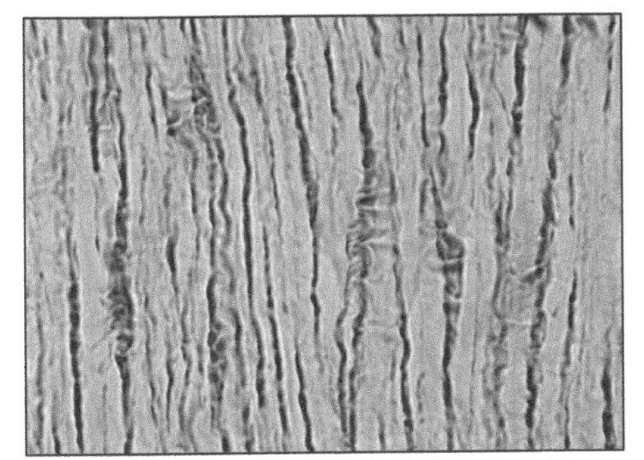

(c) Elastic tissue in the nuchal ligament of the neck (400x)

Figure 4-4 Connective Tissue Proper: Dense Tissues.

All images source: Mark Taylor

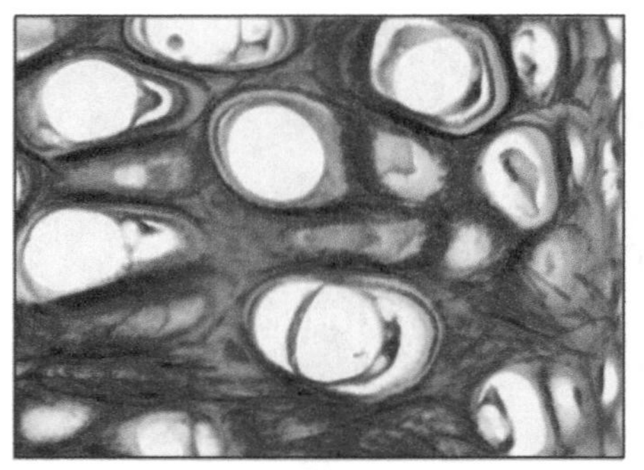

(a) Elastic cartilage from the ear (400x)

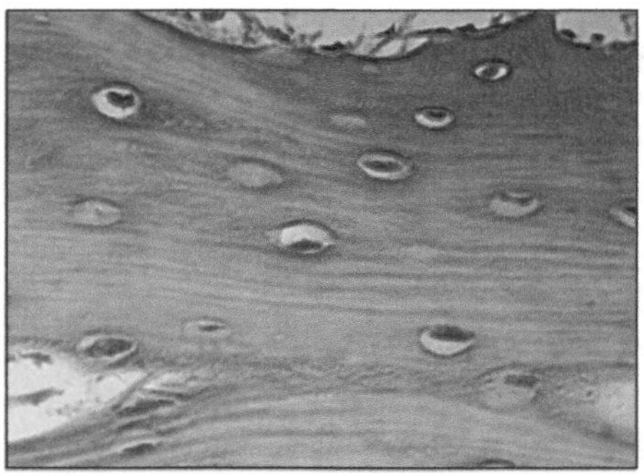
(b) Fibrocartilage from intervertebral disc (400x)

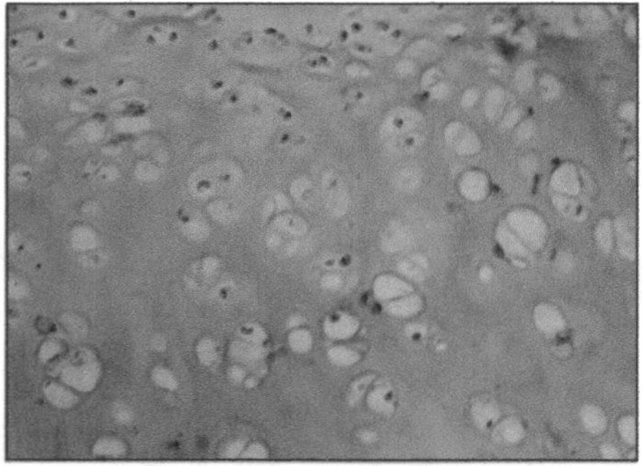
(c) Hyaline cartilage from the trachea (400x)

***Figure* 4-5** Cartilaginous tissues.

All images source: Mark Taylor

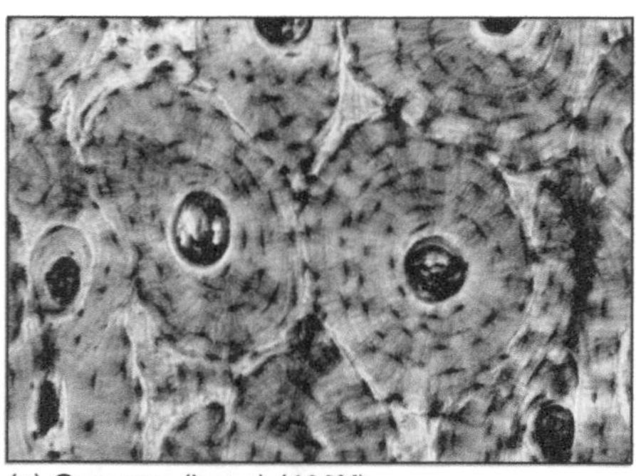
(a) Osseous (bone) (400X)

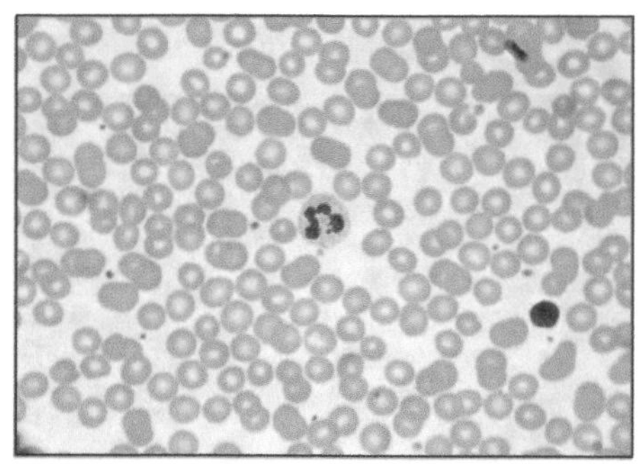
(b) Blood (400X)

***Figure* 4-6** Osseous and Blood Tissue.

Both images source: Mark Taylor

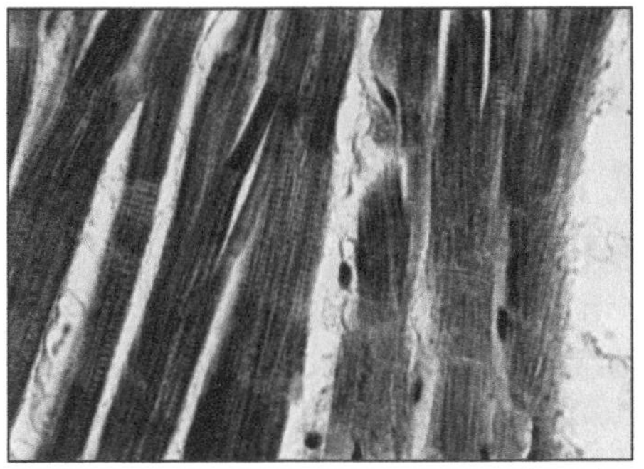

(a) Cardiac muscle tissue in the heart (400x)

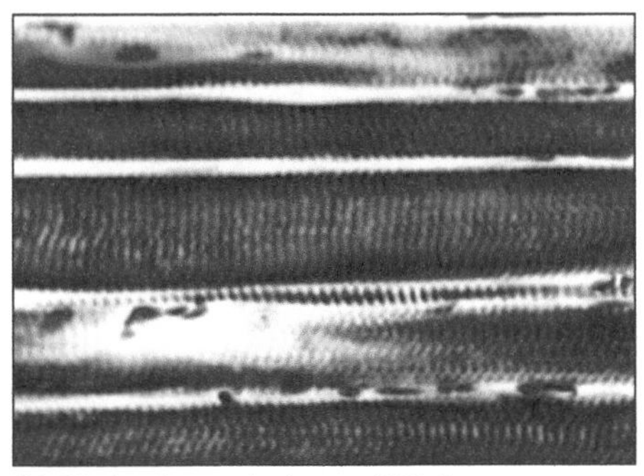

(b) Skeletal muscle tissue (400x)

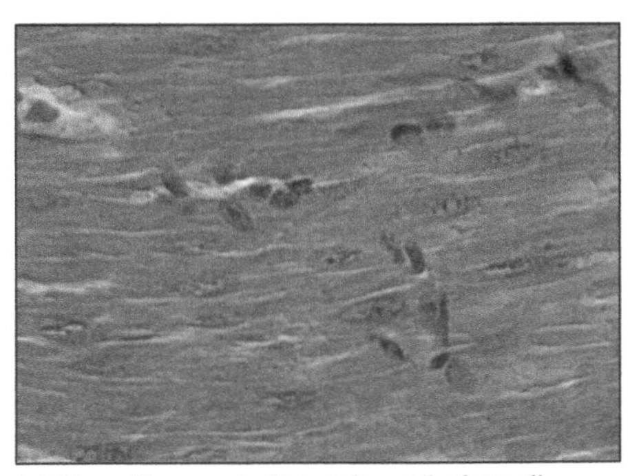

(c) Smooth muscle tissue in wall of small intestine (400x)

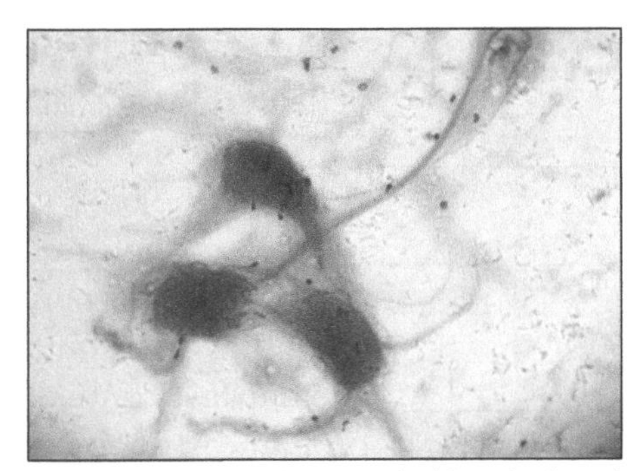

(d) Nervous tissue: Neurons inside spinal cord (400x)

Figure 4-7 Muscle and nervous tissues.

All images source: Mark Taylor

Name ____________________________ Course Number: _________ Lab Section _______

Lab 4: Tissues Worksheet

Use terms in this chapter to fill in the blanks.

1. ____________________ Embryonic tissue that becomes connective tissue
2. ____________________ A membrane that contains goblet cells
3. ____________________ The viscous secretion on the inner wall of the nasal cavity
4. ____________________ Another name for any serous membrane
5. ____________________ One of the criteria used to classify membranous epithelium
6. ____________________ Denotes a single-layered membranous epithelium
7. ____________________ Connective tissue abundant in the liver and spleen
8. ____________________ Cells in this type of epithelial tissue contain basal bodies
9. ____________________ Connective tissue that provides compression strength between vertebrae (bones of the back)
10. ____________________ Specific epithelium that forms endothelium
11. ____________________ One of the criteria used to classify membranous epithelium
12. ____________________ Epithelial cells that are ""taller"" than they are wide
13. ____________________ The hardest connective tissue that contains lacunae
14. ____________________ Type of cartilage found in the external ear
15. ____________________ Specific epithelium that forms mesothelium
16. ____________________ Only type of connective tissue that does not stay in one place
17. ____________________ Cartilage known as gristle and found on the ends of bones
18. ____________________ Epithelium that appears to be layered, but is actually a single layer
19. ____________________ Cells that produce most of the connective tissue matrix in the body
20. ____________________ The fluid component of connective tissue matrix
21. ____________________ Name for a body surface in contact with only one tissue
22. ____________________ Cell shape, cell _______, and location are used to classify membranous epithelium
23. ____________________ Connective tissue abundant in tendons and ligaments
24. ____________________ Connective tissue abundant in the skin
25. ____________________ Most abundant protein in the body
26. ____________________ Simple squamous epithelium lining the inside of blood and lymph vessels
27. ____________________ An epithelium consisting of two or more layers of cells
28. ____________________ Denotes a lack of blood vessels
29. ____________________ Apical cells that change shape throughout the day
30. ____________________ Another name for everything extracellular in a connective tissue

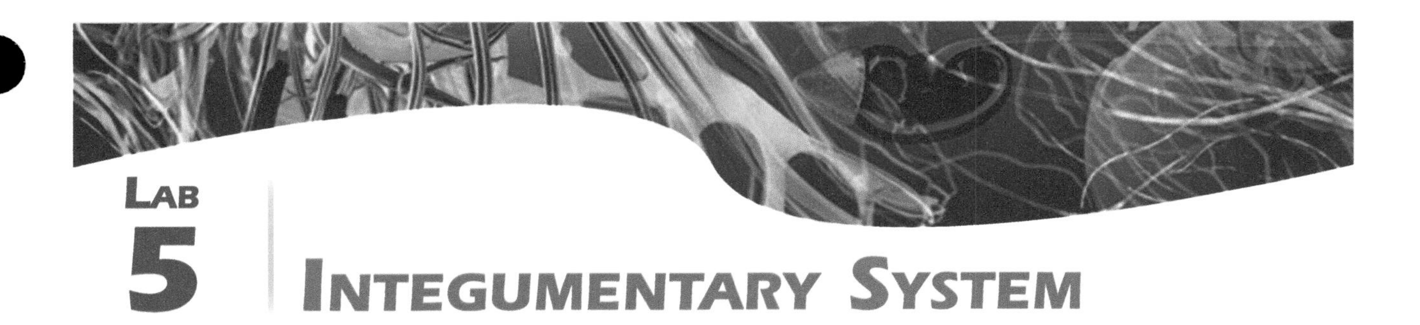

LAB 5 INTEGUMENTARY SYSTEM

Integumentary System

The integumentary system includes (1) the skin; (2) skin glands; (3) hairs, hair follicles, and arrector pili muscles; and (4) nails. Coursing through the skin are several other organs, including blood vessels in the cardiovascular system, lymphatic vessels in the lymphatic system, and sensory receptors and nerves in the nervous system. The skin includes the *dermis*, or true skin, and the *epidermis*, which is the superficial, epithelial covering of the dermis. Lying deep to the dermis is the *hypodermis* (also called *subcutaneous tissue* or *superficial fascia*). In this lab you will be studying the anatomical features of the skin, hair, and nails.

The Skin

Table 5.1 summarizes the major regions of the skin and their components. You should be able to identify all of these items on diagrams and models. In addition, the items marked with an asterisk (*) should also be identified on microscope slides and photographs of slides. **Figure 5-1** provides an overview of the skin.

Table 5.1 Regions of the Skin

EPIDERMIS*: Outer covering of the skin; consists of stratified squamous epithelium; forms downward projecting *epidermal pegs* that interdigitate with upward projecting dermal papillae of the dermis	
Basement membrane*	Glycoprotein foundation on which epidermis rests; includes a **basal lamina** secreted by the epidermis and a **reticular lamina** secreted by the dermis
Keratinocyte*	Most abundant cell type in the epidermis; produced in the stratum basale and gradually pushed toward the apical surface; produces abundant keratin protein
Langerhans cell	Specialized WBC that attacks foreign particles in the epidermis and dermis
Melanocyte	25% of cells in stratum basale; produce melanin
Merkel's cell	Part of Merkel's disc that responds to sustained touch/pressure
Stratum basale*	Basal layer of epidermis; cells capable of dividing and give rise to other strata; secretes the basal lamina part of the basement membrane
Stratum corneum*	Apical layer of epidermis; dead, keratinized squamous cells
Stratum granulosum*	Between stratum spinosum and stratum corneum (or lucidum in soles or palm)
Stratum lucidum*	Thin layer visible only in soles or palms; located between stratum corneum and stratum granulosum
Stratum spinosum*	Between granulosum and basale

(Continued)

Table 5.1 Regions of the Skin *(Cont'd)*

DERMIS*: "True skin" located between the epidermis and hypodermis	
Hair root plexus	Free nerve ending wrapped around a hair follicle that responds to hair movement
Krause end bulb	Lamellated (layered) tactile receptor found in the oral cavity's mucosa, the lips, and conjunctiva (thin membrane on the eye)
Meissner's corpuscle	Tactile receptors within dermal papillae; respond to vibration and change in texture
Merkel nerve ending	A button-like nerve ending lying beneath a Merkel cell; together they form the *Merkel's disc*; functions as tactile receptor responding to sustained touch/pressure
Pacinian corpuscle	Tactile receptor that responds to vibration
Papillary region*	Forms upward projecting *dermal papillae* that interdigitate with downward projecting epidermal pegs; consists of areolar connective tissue
Reticular region*	Major portion of dermis; consists of dense irregular connective tissue; contains skin glands, hair follicles (see Table 5-2), and various sensory receptors
Ruffini's end organ	Tactile receptor that responds to stretch deep in the skin
Sebaceous gland*	Oil glands; *free sebaceous glands* have ducts leading directly to the skin's surface; found in axillary and pubic regions; *pilosebaceous units* have sebaceous glands that empty oil directly into hair follicles
Sudoriferous gland*	Sweat glands; *eccrine sweat glands* secrete watery sweat for thermoregulation and have ducts that lead to skin's surface; *apocrine sweat glands* secrete sweat into hair follicles in axillary and pubic regions
HYPODERMIS*: Located deep to the dermis and consists mostly of adipose tissue	

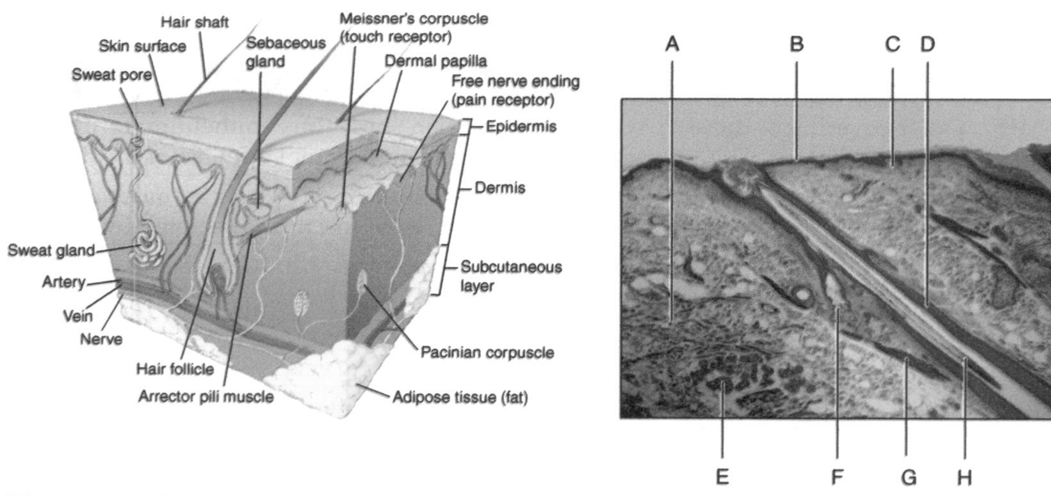

Figure 5-1 Overview of the skin.

(a) Skin diagram © Kendall Hunt Publishing Company
(b) Source: Mark Taylor

Identify the labeled structures in Figure 5-1.

A: ____________________ E: ____________________

B: ____________________ F: ____________________

C: ____________________ G: ____________________

D: ____________________ H: ____________________

Table 5.2 Hair and Hair Follicles	
Hair Follicle: Structure that gives rise to a hair and provides a passageway for its growth to the skin's surface	
Arrector pili muscle	Attached to hair follicle and pulls it, causing goose bumps or hair to rise
Dermal root sheath	Outer connective tissue layer of hair follicle
Dermal papilla	A superficial projection of the dermis in contact with an epidermal peg
Epithelial root sheath	Adjacent to the hair and consisting of cells derived from the epidermis
External sheath	Extension of stratum basale of epidermis; external to Henle's layer
Glassy membrane	Basement membrane between the epithelial and dermal root sheaths
Hair papilla	Dermal papilla at base of hair follicle; nourishes hair matrix
Internal sheath	Extension of stratum spinosum; the *internal cuticle* is in contact with the hair cuticle; may be pulled out and seen around the base of a hair; *Huxley's layer* is external to the internal cuticle; *Henle's layer* is external to Huxley's layer
Hair: Elongated collection of tightly compressed hair cells (keratinocytes) that originate near the base of a hair follicle	
Hair bulb	Swollen basal portion of a hair; overlying the hair papilla
Hair cortex	Middle, thickest portion of a hair shaft
Hair cuticle	Outer layer of overlapping dead cells of a hair shaft
Hair matrix	Modified stratum basale over the hair papilla; gives rise to hair cells
Hair medulla	Innermost portion of a hair shaft
Hair root	Part of hair shaft that extends from the hair papilla to the level of the sebaceous gland
Hair shaft	Part of hair located superficial to the sebaceous gland

Hair Follicles And Hair

Table 5.2 summarizes the major features of a hair follicle and the hair it produces. Arrector pili muscles are included in this table because they pull on hair follicles causing the hairs to "stand up." This pulling effect also elevates the skin around the hair pore to produce a "goose bump." You should be able to identify all of these items on diagrams and models. In addition, the items marked with an asterisk (*) should also be identified on microscope slides and photographs of slides, including those in **Figure 5-2**.

Anatomy Of Nails

Table 5.3 summarizes the major features of a nail and its associated structures. You should be able to identify all of these items on diagrams and models. In addition, the items marked with an asterisk (*) should also be identified on microscope slides and photographs of slides, including those in **Figure 5-3**.

Observing Your Own Integument

You will gain a better appreciation for the complexity of your own integument by observing it with a dissecting microscope. This type of scope allows you to look at larger, yet still small, objects without having to use a microscope slide and coverslip. Your instructor will show you how to use the dissecting microscopes set up at your table or at the side of the room.

1. Observe the skin on the back of your hand and note the tiny lines and grooves that form the **surface tension lines**. Make a sketch below.

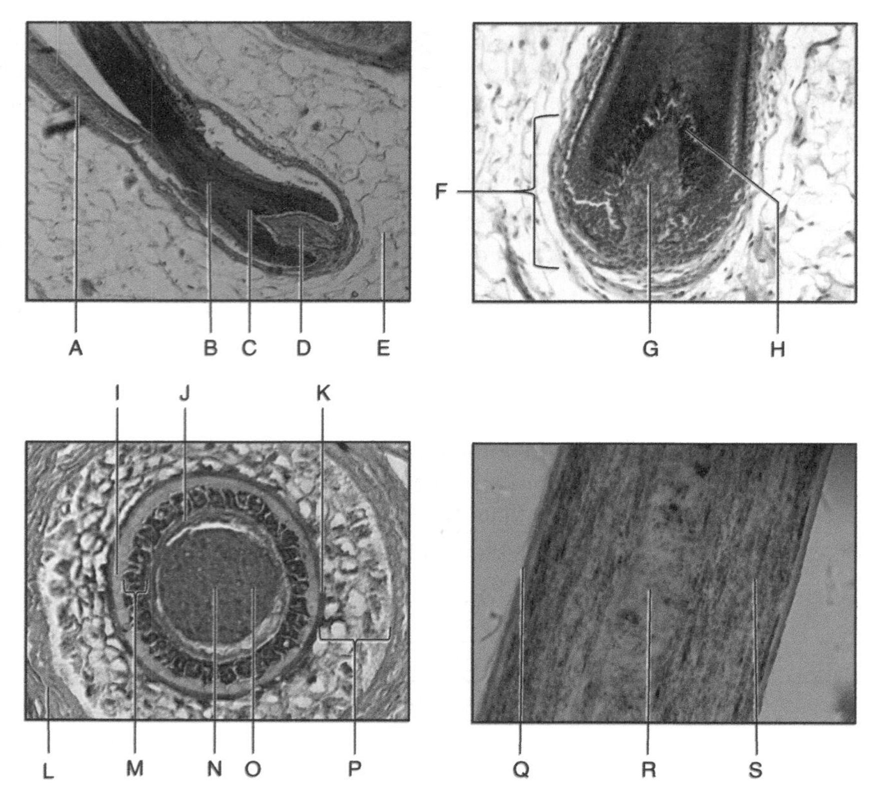

Figure 5-2 Hair follicle and hair.

All images source: Mark Taylor

Identify the labeled structures in Figure 5-2.

A. ______________________

B. ______________________

C. ______________________

D. ______________________

E. ______________________

F. ______________________

G. ______________________

H. ______________________

I. ______________________

J. ______________________

K. ______________________

L. ______________________

M. ______________________

N. ______________________

O. ______________________

P. ______________________

Q. ______________________

R. ______________________

S. ______________________

Table 5.3 Nails and Associated Structures	
Nail body	Hard, visible part of nail; derived from nail matrix; dead, highly keratinized keratinocytes; Also called the nail plate
Nail bed	Lies immediately deep to nail body
Nail root	The proximal end of a nail beneath the skin
Dorsal matrix	Generates nail cells on dorsal side of nail body's proximal end
Ventral matrix	Generates nail cells on ventral side of nail body's proximal end; partly visible as the *lunula*
Lunula	Light-colored region at the proximal end of nail; the visible part of the ventral nail matrix
Eponychium	Cuticle skin on the proximal end of nail; consists of stratum corneum
Hyponychium	The "quick" located under the free edge of a nail
Nail fold	A ridge of skin where the nail body and the skin connect
Nail groove	A "crease" between the nail fold and nail body

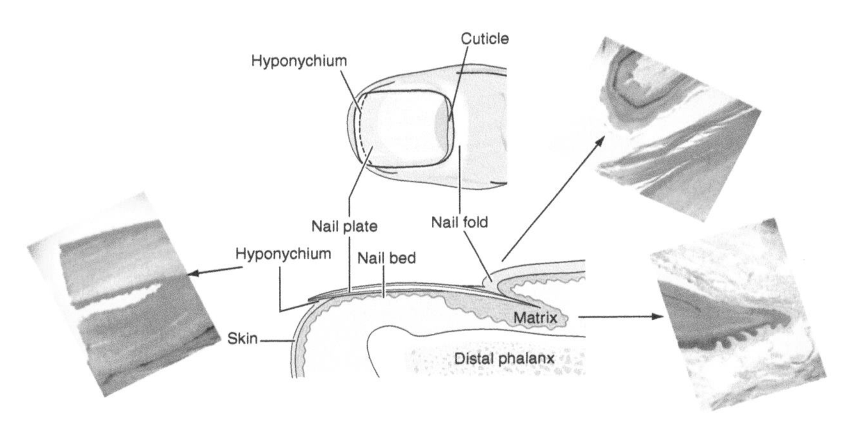

Figure 5-3 Nail and associated structures.

2. Observe the **friction ridges** (or **epidermal ridges**) that form your fingerprints. Note the basic pattern as being *whorl*, *loop*, or *arch*. Make a sketch below.

3. Look at the **flexure lines** on your palms and on the back of your wrist. Also, look closely to see the detail around the pores from which *hair* arises on the back of your hand.

4. Look at your **fingernails** to see their uneven surface. Hopefully, you will not see any dirt under your nails, but if there is some there then it is interesting to view as well.

5. If you or a lab partner have a scab or cut on the skin, observe the rough nature of it. You might also be able to see granulation tissue, but don't remove the scab in order to do so.

Prepared Slides

1. Make sure you have the following four integumentary slides at your table.

   - Skin from scalp (follicles in longitudinal section)
   - Thick skin
   - Scalp skin (follicles in cross section)
   - Mature nail

2. Look at the slides of thin and thick skin and locate the items marked by an asterisk in **Tables 5.1** and **5.2**.
3. Look at the slide of the mature nail and locate the items marked by an asterisk in **Table 5.3**.

Measuring Subcutaneous Fat

Although not officially part of the skin, the hypodermis is considered part of the integumentary system. Consisting primarily of adipose tissue, it serves as an insulator, shock absorber, and lipid storage site. In most cases, by measuring the amount of fat in the hypodermis, which is subcutaneous fat, one can make a reasonable estimation of the person's total body fat.

Measuring subcutaneous fat can be done in several ways, but only two will be used in this lab. The instructor will show you how to use (1) skin calipers to measure the thickness of subcutaneous fat in selected parts of the body and (2) an electronic meter to measure the amount of electrical resistance in the body. Since water is a good conductor of electrical current, tissues high in water, such as muscle tissue, conduct electrical currents faster than does fat, or adipose tissue, because these latter tissues contain much less water. The electronic meter emits a minute electrical current into one hand and measures the amount of time required for it to pass through the body to the other hand. Based on the person's size (determined by height and weight input), the meter determines the person's relative percent body fat.

Location	Caliper reading	Calculated body fat using caliper data	Electronic reading for % body fat	Electronic reading for BMI* (High, Normal, Low)
Back of arm				
Side of abdomen				
Anterior thigh				
Under scapula				

***Body Mass Index**: a calculation based on a person's weight and height

Name ____________________ Course Number: ________ Lab Section ______

Lab 5: Integumentary System Worksheet

Use terms in this chapter to fill in the blanks.

1. ____________________ Part of a hair that surrounds a papilla
2. ____________________ Deep tactile receptor that responds mainly to skin vibration
3. ____________________ Layer that comprises most of a hair root and shaft
4. ____________________ Light-colored region of a nail attributed to the underlying matrix
5. ____________________ Dermal projection within a hair bulb
6. ____________________ Epidermal cell that stimulates a button-like nerve ending in the dermis
7. ____________________ Ridge of skin adjacent to a nail
8. ____________________ Part of a hair deep to a sebaceous gland
9. ____________________ Region of the dermis immediately deep to the epidermis
10. ____________________ Sweat (sudoriferous) gland that secretes a watery fluid for cooling the skin
11. ____________________ Shortest part of the nail matrix
12. ____________________ Part of the nail matrix responsible for the lunula
13. ____________________ Also called the nail cuticle
14. ____________________ Soft tissue deep to the nail plate
15. ____________________ Deepest stratum of the epidermis
16. ____________________ Hard, visible part of a nail
17. ____________________ Oil-secreting skin gland
18. ____________________ Part of the integumentary system immediately deep to the dermis
19. ____________________ Most superficial stratum in the epidermis
20. ____________________ Part of the epithelial root sheath in contact with a hair root
21. ____________________ Another name for the outer, connective tissue root sheath of a follicle
22. ____________________ Lies between the nail fold and nail plate
23. ____________________ Tactile receptor that responds mostly to stretching of the skin
24. ____________________ The central core region of a hair
25. ____________________ Cells in the epidermis that engulf foreign particles
26. ____________________ Tube containing a hair root
27. ____________________ Part of a nail to which new cells are added
28. ____________________ Tightly compressed keratinocytes produced by a matrix next to a papilla
29. ____________________ Third deepest epidermal stratum in thick skin
30. ____________________ Epidermal stratum visible mainly in thick skin
31. ____________________ Part of the epithelial rooth sheath in contact with the glassy membrane
32. ____________________ Also called the nail quick
33. ____________________
34. ____________________ Sweat (sudoriferous) gland that secretes into a hair follicle
35. ____________________ Name for basement membrane separating the dermal and epithelial root sheaths of a follicle
36. ____________________ Part of a hair superficial to a sebaceous gland
37. ____________________ Largest region of the dermis
38. ____________________ Cells that comprise most of the epidermis

39. ______________________ Tactile corpuscles mainly within dermal papillae
40. ______________________ Layer of glycoproteins that separates epidermal pegs from dermal papillae
41. ______________________ Muscle that pulls on a hair follicle
42. ______________________ Visible part of a hair shaft
43. ______________________ Nerve endings allowing detection of hair follicle movement
44. ______________________ Third deepest epidermal stratum in thin skin
45. ______________________ Superficial projection of the dermis

Lab

6 The Axial Skeleton

Overview of The Skeleton

While the lecture part of the course deals with bone cells and factors affecting bone growth, this lab primarily deals with bone structure and skeletal system anatomy. Typical adults have 206 bones in their skeleton, which is divided into two parts: the *axial skeleton* and the *appendicular skeleton*. In this lab you will study the axial skeleton and in the next you will study the appendicular skeleton. Since you will also be learning about major anatomical features of individual bones, you will need to learn the list of terms in **Table 6.1**. As you learn the names of specific bones of the skeleton, label them in **Figure 6-1** and the figures that follow.

Table 6.1 Bone Features

Term	Description
Condyle (KON-dīl; "knuckle")	Rounded knob that articulates with a depression in another bone
Crest	Narrow ridge for muscle attachment
Epicondyle (ep-i-KON-dīl)	Projection above a condyle for muscle attachment
Facet (FAS-et; "face")	Flat surface that articulates with a similar surface on another bone
Fissure (FISH-ur; "deep furrow")	Narrow groove through which vessels and nerves pass
Foramen (for-Ā-men; "opening")	Smooth-edged opening through which a nerve and/or vessel passes
Fossa (FOS-ah; "ditch")	Shallow depression in which another bone's process articulates
Fovea (FO-ve-ah; "pit")	Small depression for attachment of a ligament
Head	Rounded end of a long bone; connected to a narrow neck
Line	Narrow ridge along which a muscle attaches
Meatus (mē-Ā-tus; "passage")	Tubular canal in a bone
Neck	Constricted part of a long bone between the epiphysis and diaphysis
Notch	U-shaped depression between two projections
Process	Bony projection
Protuberance (prō-TŪ-ber-uns)	Rounded knob
Ramus (RĀ-mus; "branch")	Straight portion of a bone that bends at a right angle
Sinus (SĪ-nus)	Hollow cavity in a skull bone
Spine (spīn)	(**Spinous process**) Sharp projection
Sulcus (SUL-kus; "ditch")	Narrow groove
Tubercle (TŪ-ber-kul; "swelling")	Rounded projection
Tuberosity (tū-ber-OS-i-tē; "lump")	Roughened projection

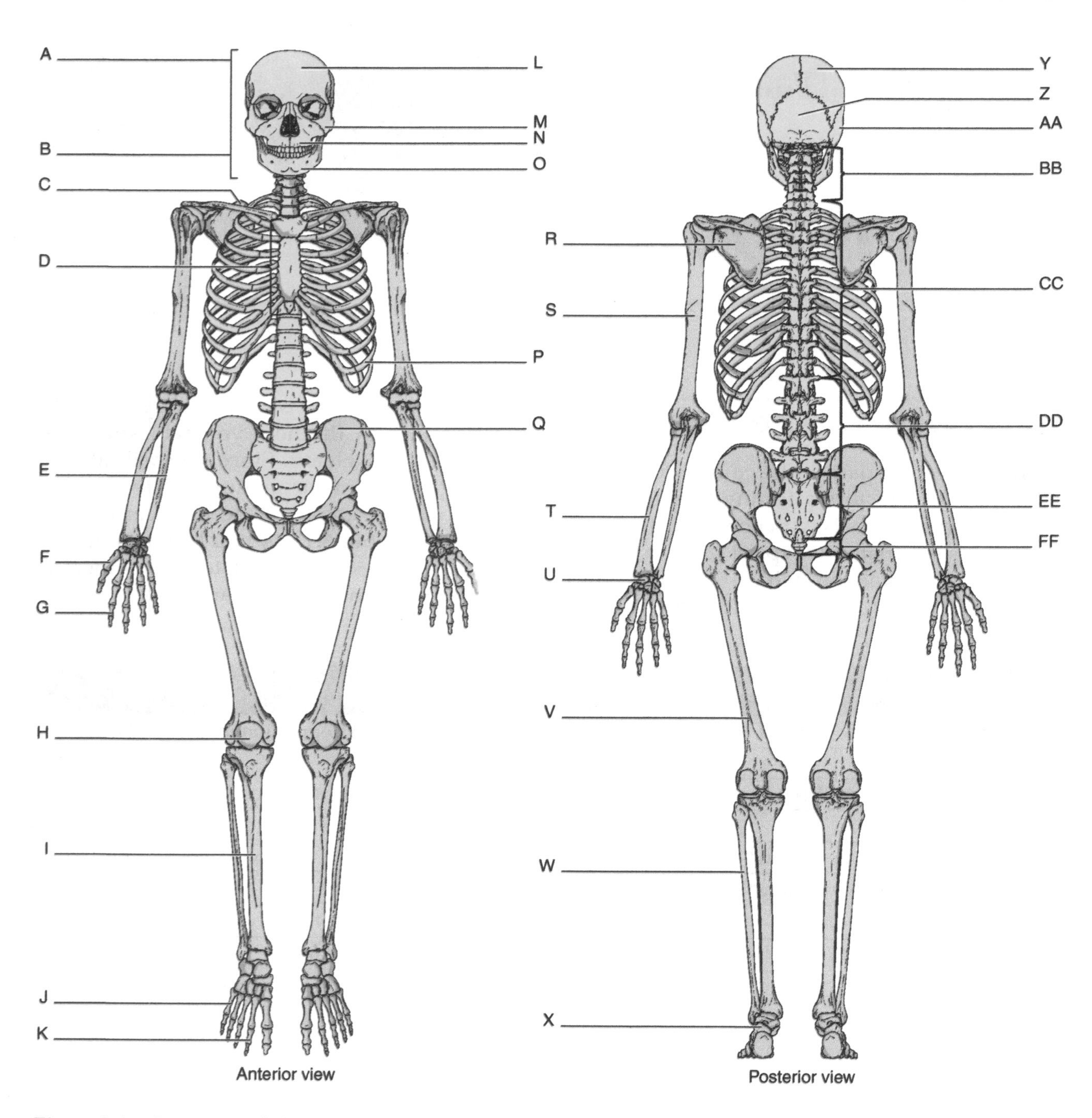

Figure 6-1 Overview of the skeleton.

Axial Skeleton

The **axial skeleton** (AX-ē-ul, "axle") forms the vertical axis of the skeleton to which the limb bones attach. As the name implies, the axial skeleton's orientation is much like that of a car's axle to which wheels attach. The typical axial skeleton has 80 bones, including 28 bones in the *skull*, 1 *hyoid bone* in the neck region, 26 bones in the *vertebral column* (spine), and 25 bones in the *thoracic cage*.

Skull

The **skull** contains 28 bones, including 6 tiny ear bones (3 in each ear) called *auditory ossicles* (OS-i-klz; "little bones"), which are used in hearing. We will identify the ossicles in a later chapter. Overall, the skull is divided into two groups of bones: *cranial bones* and *facial bones.*

Cranial Bones

The **cranium** (KRĀ-ne-um; "helmet") consists of 8 *cranial* bones (not including the ear ossicles) and it forms the large spherical part of the skull surrounding the brain. All cranial bones connect tightly to one another at immovable joints called *sutures* (SŪ-cherz; *sutur-*, seam). In some cases, small flat bones called *sutural* (SŪ-chur-ul), or *Wormian bones*, named after a person, may be found within a suture. **Tables 6.2** and **6.3** summarize the cranial bones and facial bones, respectively. After you become familiar with these bones and features, locate and label them on **Figures 6-2**, **6-3**, and **6-4.**

The calvaria and skull base The **calvaria** (kal-VĀR-ē-a; *calvar-*, "skull") or *skullcap* is the superior portion of the cranium consisting of parts of the frontal, parietal, occipital and temporal bones. The calvaria is often removed for

Table 6.2 Cranial Bones and Their Features

Basilar part of occipital bone (BĀ-si-lar; "base")	(**Basioccipital process**) Anterior part of occipital bone
Carotid canal (kar-OT-id; "put to sleep")	Passage in the skull base for internal carotid artery; Part of temporal bone
Clivus (KLĪ-vus; "slope")	Sloping internal portion of basilar part of occipital bone where the brain stem rests
Cribriform plate (KRI-bri-form; "sieve-like")	Flat groove on either side of the crista galli at superior portion of ethmoid bone; site of olfactory foramina; seen in cranial vault with the calvaria removed
Crista galli (kri-tuh-GAL-ē; "crested helmet")	Superior projection on the ethmoid; attachment site of brain covering
Ethmoid air cells (ETH-moyd: "sieve-like")	Numerous small sinuses in the ethmoid bone
Ethmoid bone (ETH-moyd; "sieve like")	Forms part of nasal cavity wall and base of cranial vault
External acoustic meatus (uh-KŪ-stik; "hear")	Passage in the temporal bone where sound waves pass to middle ear
External occipital protuberance	Knob on external posterior surface of occipital bone; attachment for nuchal ligament
Foramen lacerum (LAS-er-um; "torn")	Lateral to the basilar process and found between the sphenoid, temporal, and occipital bones; passage for an artery
Foramen magnum (MAG-num; "large")	Largest foramen in skull; part of the occipital bone; passage for spinal cord, nerves, vessels
Foramen ovale (ō-VAL-ē; "oval")	Lateral to the foramen lacerum part of sphenoid bone; passage for nerve
Foramen rotundum (rō-TUN-dum; "round")	Lateral to the sella turcica part of sphenoid bone; passage for nerve
Foramen spinosum (spī-NŌ-sum; "spine")	Slightly posterior and lateral to foramen ovale; adjacent to small spine on inferior base; part of sphenoid bone; passage for vessels
Frontal bone	Forms the forehead and brow ridges
Frontal sinus	Cavity that may or may not be present inside frontal bone
Glabella (gluh-BEL-a: ("smooth")	Smooth region of the frontal bone between the orbital ridges
Greater wing of the sphenoid	Lateral, flattened extension forming part of the orbit and temple
Hypoglossal canal ("hī-pō-GLOS-ul; "below tongue")	Opening superior to an occipital condyle; in occipital bone; passage for nerve

(Continued)

Table 6.2 Cranial Bones and Their Features *(Cont'd)*

Inferior orbital fissure (part of it)	Groove in the posterior, inferior orbit; passage for maxillary branch of trigeminal nerve and vessels to orbit
Internal acoustic meatus	On posterior side of temporal bone's petrous portion; Passage for nerves
Jugular foramen (JUG-ū-lar; "throat")	Large openings lateral to foramen magnum; between temporal and occipital bones; passage for vessels and nerves
Lesser wing of the sphenoid	Projection on superior portion seen in cranial vault; forms part of middle cranial fossa border
Mandibular fossa (man-DIB-ū-lar; "jaw")	Part of temporal bone that articulates with the condylar process of the mandible
Mastoid process (MAS-toyd; "breast")	On the inferior, posterior portion of temporal bone; attachment site for neck muscle
Middle nasal conchae (KON-kē; "shell")	Lateral extensions on either side of perpendicular plate
Occipital bone (ok-SIP-ih-tul; "back of head")	Forms back of skull
Occipital condyles	On external base of skull and adjacent to foramen magnum; articulate with the first cervical vertebra (atlas)
Olfactory foramina (OL-fak-tōr-ē; "smell")	Located in the cribriform plate of the ethmoid bone; passage for nerves
Optic canal (OP-tik; "eye")	Opening in the medial, posterior orbit; in the sphenoid bone; passage for optic nerve
Parietal bones (2)	Forms most of the skull's lateral walls
Perpendicular plate of the ethmoid	Vertical, inferior partition; forms most of nasal septum
Petrous portion (PET-rus; "rock")	Bony ridge of the temporal bone that contains inner ear components
Pterygoid process (TAIR-ih-goyd: "wing")	Inferior projection of the sphenoid; attachment site for muscles
Sella turcica (SEL-uh-TUR-si-ka; "Turkish saddle")	Depression on the superior surface of the sphenoid bone; holds pituitary gland
Sphenoid bone (SFĒ-noyd; "wedge-like");	Bird-shaped bone surrounded by other skull bones
Squamous part (SKWĀ-mus; "plate")	Smooth, superior half of the temporal bone
Styloid process (STĪ-loyd; "pillar")	Slender projections of the temporal bone; attachment site for a neck muscle
Stylomastoid foramen	Opening next to the styloid process; in the temporal bone; passage for facial nerve
Superior nasal conchae	Superior and posterior to the middle nasal conchae; part of the ethmoid bone
Superior orbital fissure	Groove in posterior, superior orbit; passage for nerves/vessels
Supraorbital foramen (SŪ-pruh; "above")	Small opening in the supraorbital margin of the frontal bone; sometimes it is just a notch; passage for vessels
Supraorbital margin (ridge)	(**Brow ridge**) Projection of the frontal bone that forms upper rim of orbit
Temporal bones (2) (TEM-por-ul; "time"):	Forms wall of the skull around the ear
Tympanic part (tim-PAN-ik; "drum")	Rough, inferior half of the temporal bone
Zygomatic bone (zī-gō-MAT-ik; ("yoke")	Forms the cheek bone; articulates with the zygomatic bone
Zygomatic process	Anterior projection of the temporal bone; articulates with temporal process of zygomatic bone; forms posterior portion of zygomatic arch

Table 6.3 Facial Bones and Their Features

Alveolar processes (al-VĒ-ō-lar; "sac")	V-shaped projections around alveolar sockets of the maxilla and mandible
Alveolar sockets	Cavities that hold teeth in the maxilla and mandible
Condylar process (KON-dih-lar)	(**Mandibular condyle**) Articulates with temporal bone's mandibular fossa
Coronoid process (KOR-ō-noyd)	Pointed projection on the mandible anterior to the mandibular notch; site of muscle attachment
Frontal processes	Articulate with frontal bone
Incisive fossa (in-SĪ-siv; "to cut")	Posterior to incisor (front) teeth; passage for nerve and vessel
Inferior nasal conchae (2):	Shell-like bones on the lateral walls of the nasal cavity; inferior to middle conchae; increases surface of nasal mucosa
Inferior orbital fissure	Passage for maxillary branch of trigeminal nerve and vessels to orbit
Infraorbital foramen	Passage for infraorbital nerve and artery
Infraorbital margin	Lower rim of orbit
Internasal suture	Articulation site between the two nasal bones
Lacrimal bones (2) (LAK-rih-mul; "tear")	In medial wall of the orbit; anterior to ethmoid and posterior to nasal bones
Mandible (MAN-dih-bul; "jaw")	The lower jaw
Mandibular angle	Region where the mandibular ramus connects to the mandibular body
Mandibular body	Straight portion containing alveolar sockets
Mandibular foramen	On medial surface of mandibular ramus; passage for nerve and vessels
Mandibular notch	U-shaped gap between the condylar and coronoid processes
Mandibular ramus	Region between the mandibular notch and mandibular notch and mandibular angle
Mandibular symphysis	Anterior, midline portion of mandible
Maxillae (2) (MAX-ih-lē; "jaw")	The upper jaw
Maxillary sinus	Cavity within each maxilla
Mental foramen	Lateral to mandibular symphysis; passage for nerves
Mental protuberance	Protruding anterior portion forming the chin
Nasal aperture	Opening of the nasal cavity
Nasal bones	Narrow bones that form the bridge of the nose
Nasal spine	Anterior projection of maxillae at inferior rim of nasal aperture
Nasolacrimal canal	Formed with lacrimal and maxilla bones; passage for tears into the nasal cavity
Palatine bones (2) (PAL-a-tīn; "roof of mouth")	Forms posterior portion of hard palate
Temporal process	Part of the zygomatic bone that articulates with the temporal bone's zygomatic process
Vomer (VŌ-mer; "plow")	Forms inferior portion of nasal septum
Zygomatic bones (zī-gō-MAT-ik; "yoke")	Rounded cheek bones inferior and slightly lateral to the orbits
Zygomatic process	Part of the temporal bone that articulates with zygomatic bone

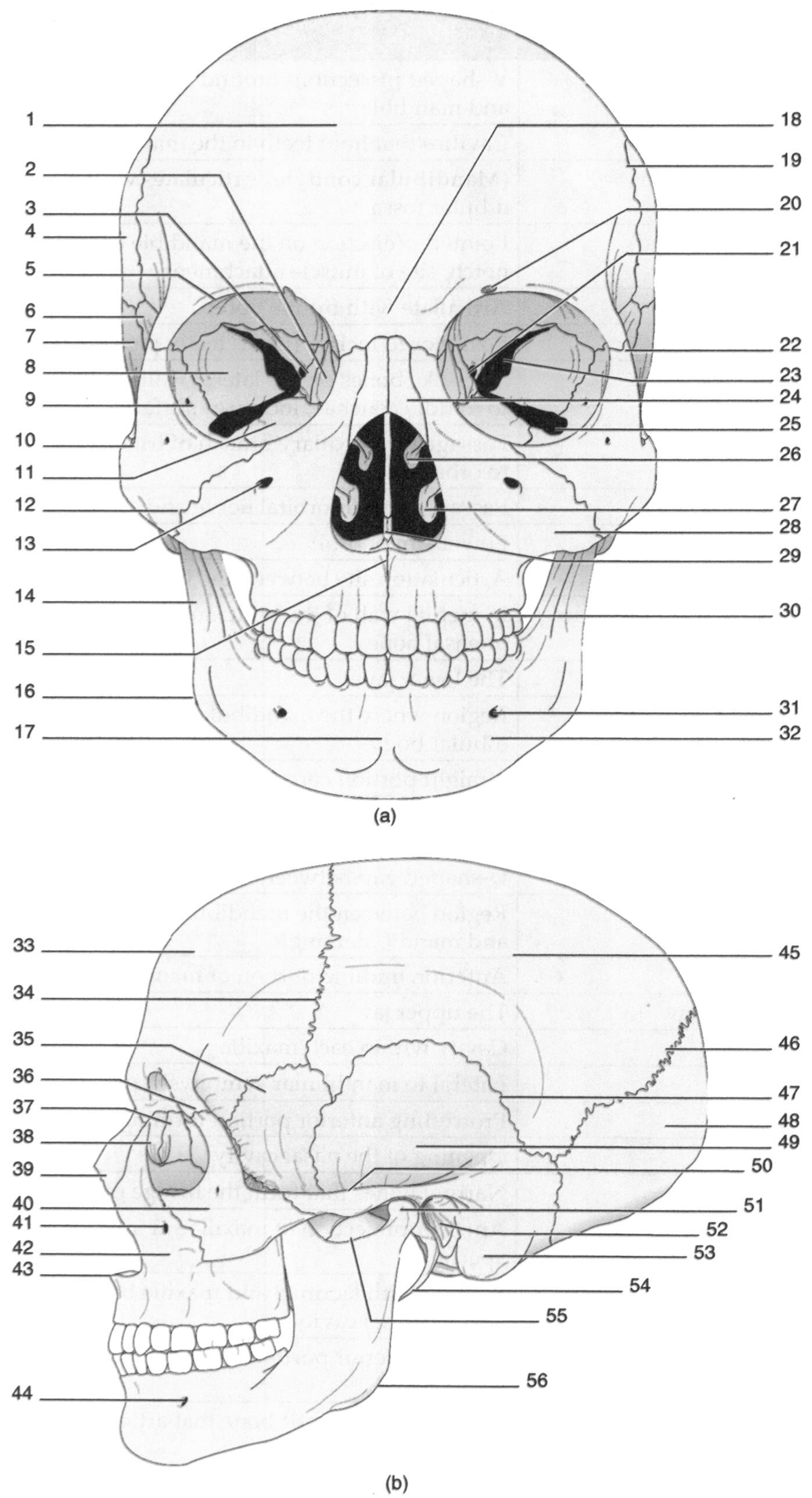

Figure 6-2 The skull (a) Anterior view (b) Lateral view.

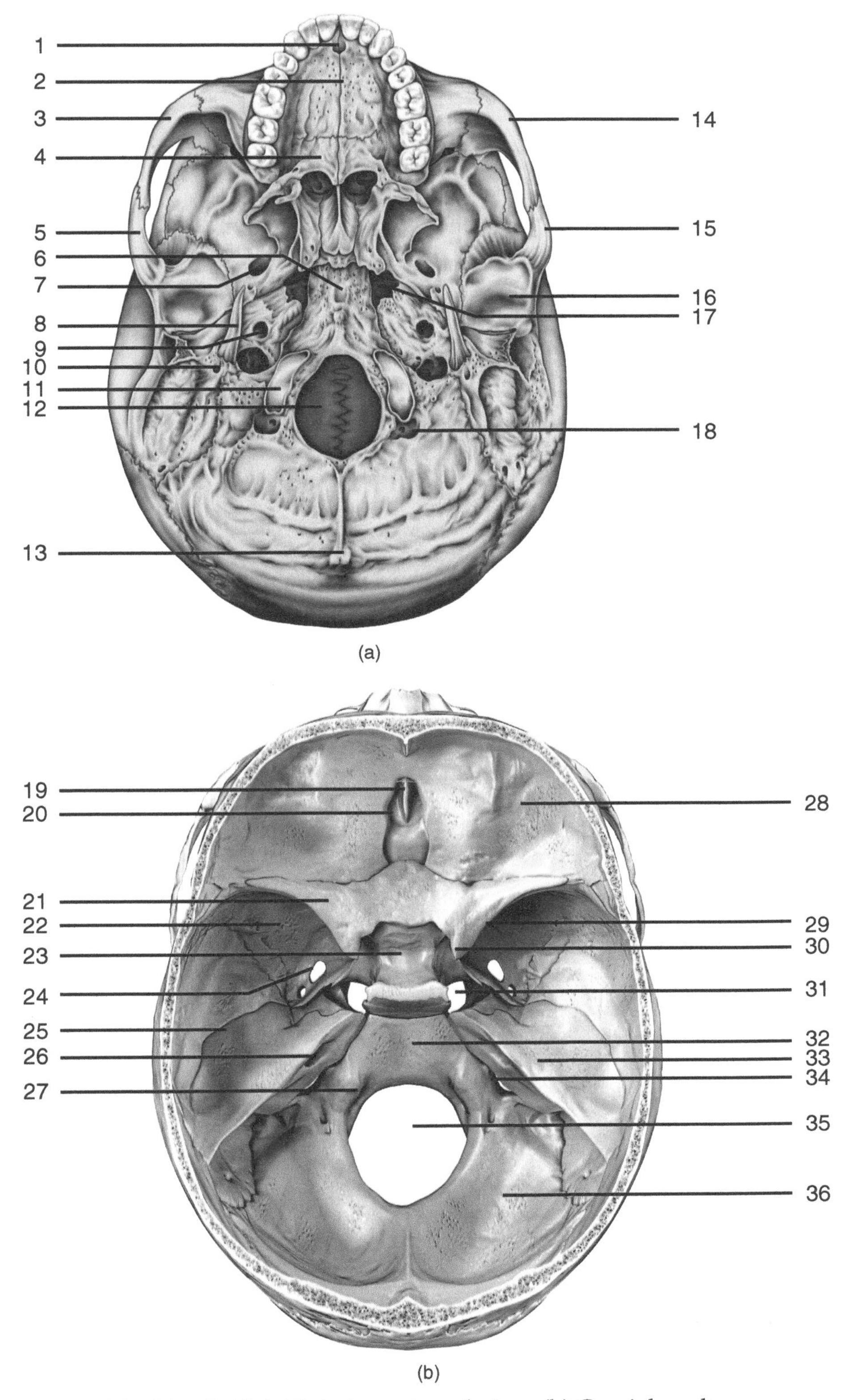

Figure 6-3 The skull: (a) Inferior, external view; (b) Cranial vault

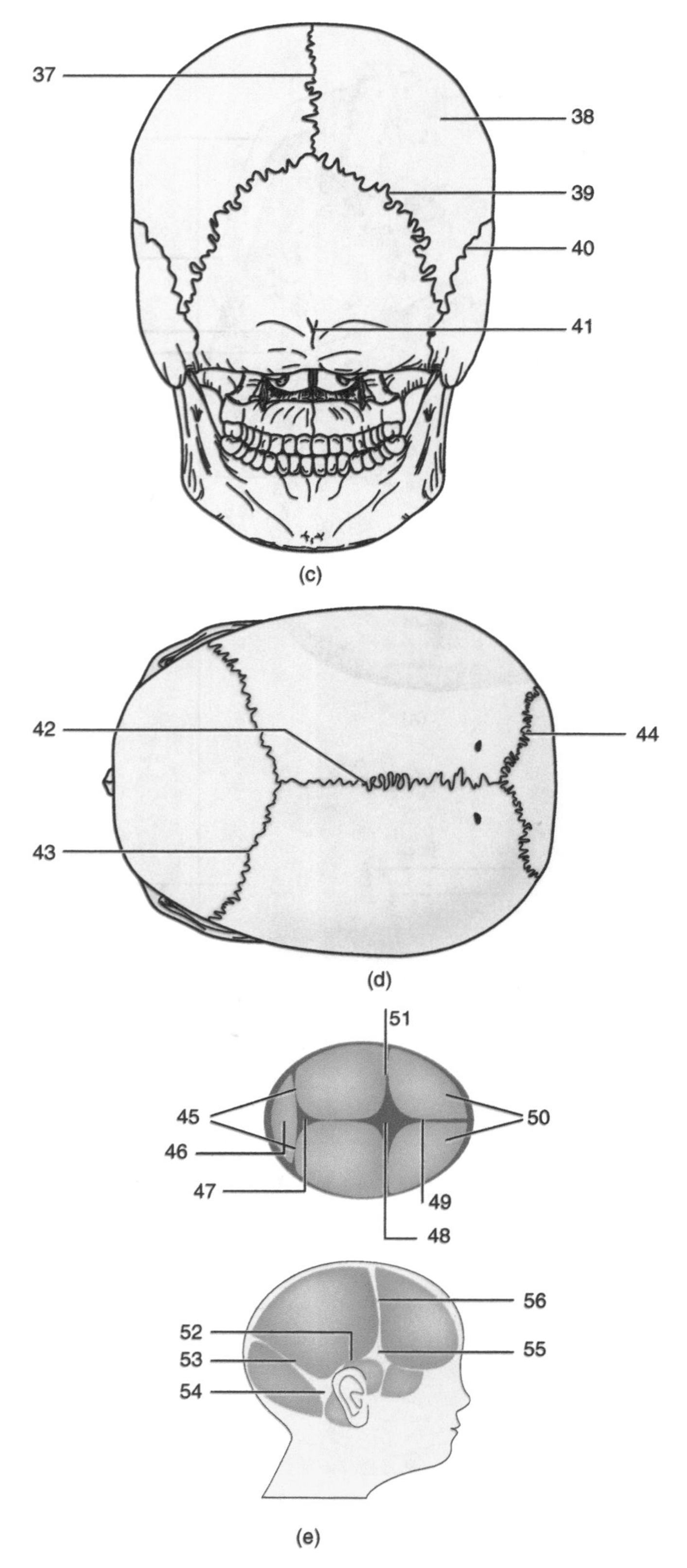

Figure 6-3 (*Continued*) (c) Posterior, external view; (d) Superior, external view; (e) Fontanels.

(a–d) Illustrations by Jamey Garbett. © 2003 Mark Nielsen. (e) © Grei/Shutterstock.com

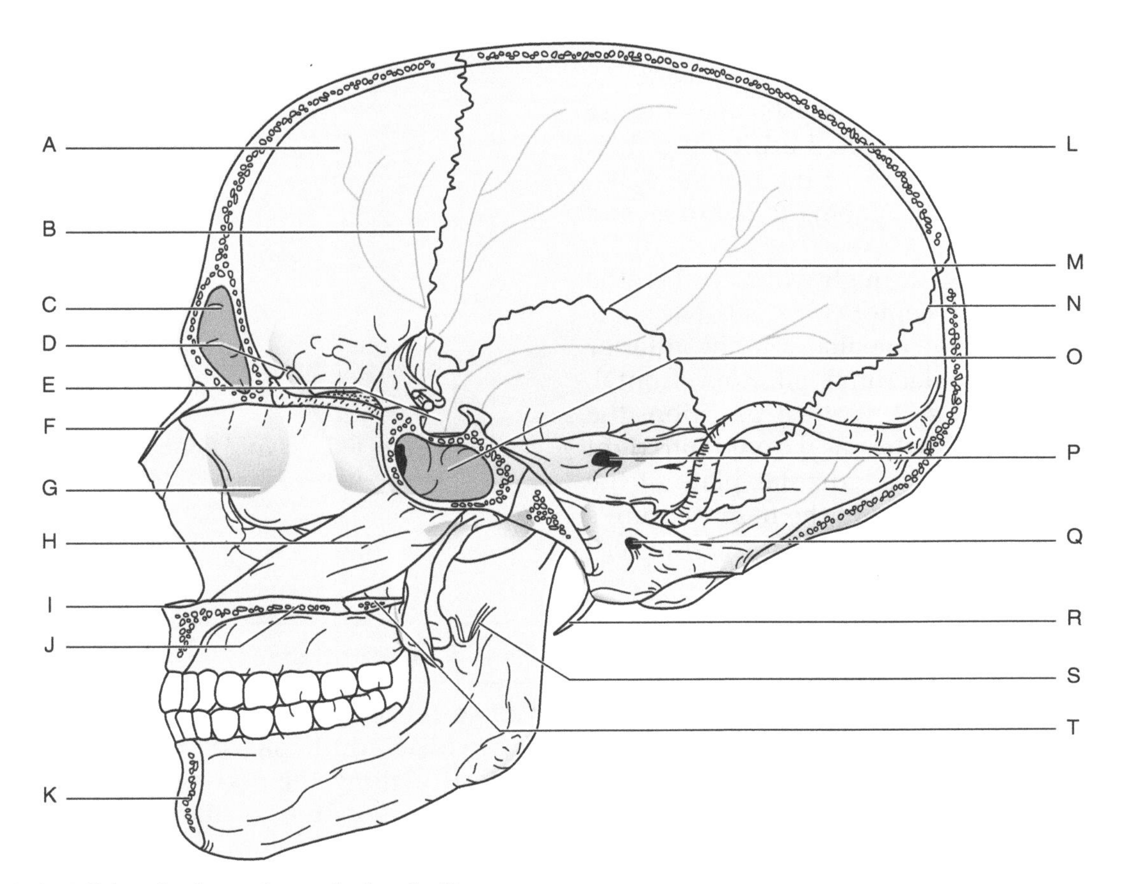

Figure 6-4 Midsagittal cut through the skull.
Illustration by Jamey Garbett. © 2003 Mark Nielsen.

brain surgery and autopsies. The **base** of the skull is its inferior surface consisting of portions of cranial and facial bones (except for the mandible); this is the part of the skull you see when the skull is turned upside-down. With the calvaria and brain removed, one can see the skull's internal base, or **cranial vault**, which consists only of cranial bones. The calvaria and internal base enclose the skull's largest cavity, the **cranial cavity**, which houses the brain.

Cranial fossae The skull's internal base consists of three large depressions, called *cranial fossae*. The **anterior cranial fossa** supports the frontal lobe of the brain and lies at the base of the frontal bone and a small portion of the ethmoid bone. Portions of the sphenoid and the right and left temporal bones form the **middle cranial fossa**, and it supports the temporal lobes of the brain. The occipital bone forms the **posterior cranial fossa** and it supports the brain's cerebellum.

Fontanels An infant is not born with fully developed cranial bones. Instead, in place of sutures, an infant's cranium has portions of fibrous membranes called **fontanels** (fon´-ta-NELZ) or "soft spots" that will eventually be replaced with bone material and sutures. There are six major fontanels present in an infant's skull at the time of birth. The **anterior fontanel** is the largest and is found where the coronal suture and the sagittal suture meet. The **posterior fontanel** is where the sagittal suture and the lambdoid suture intersect. A **sphenoidal fontanel** in each temple region is where the squamous suture and coronal suture intersect. A **mastoid fontanel** on both sides of the skull is where the lambdoid and squamous sutures meet. Label the fontanels in **Figure 6-3**.

Facial Bones

The **facial bones** include 14 bones that form the more flattened, anterior portion of the skull. Generally, these bones are smaller than cranial bones and form the borders of the facial cavities, which include the *orbits*, *nasal cavity*, and *paranasal sinuses*.

Orbits, Nasal Cavity, and Paranasal Sinuses

Outside of the cranial cavity, the skull contains several other notable cavities. The eyeballs are held within depressions called **orbits**, and their walls are formed by parts of the frontal, sphenoid, ethmoid, maxilla, zygomatic, lacrimal, and palatine bones.

The **nasal cavity**, through which we breathe and house sensory receptors for smell, has walls formed by parts of the maxilla, palatine, inferior nasal concha, nasal, lacrimal, ethmoid, frontal, and sphenoid bones. A vertical partition, the **nasal septum**, divides the nasal cavity into right and left sides. The perpendicular plate of the ethmoid bone forms most of the nasal septum's superior half, while the vomer forms the septum's inferior half.

Five skull bones contain internal cavities that border the nasal cavity; hence, they are collectively called **paranasal sinuses** (*para-*, "alongside"). These include the *frontal sinus, sphenoid sinus, ethmoid air cells,* and the *maxillary sinuses.*

Hyoid Bone

The only bone in the skeleton that does not articulate with another bone is the **hyoid bone** (HĪ-oyd, *hy-*, "U-shaped"). It is found in the anterior neck region, inferior to the mandible, and serves as an attachment for muscles that move the tongue and larynx (voice box). The prominent features are the **body**, which forms the anterior "base" of the "U" and two sets of projections: the **greater** and **lesser horns** (see **Figure 6-5**).

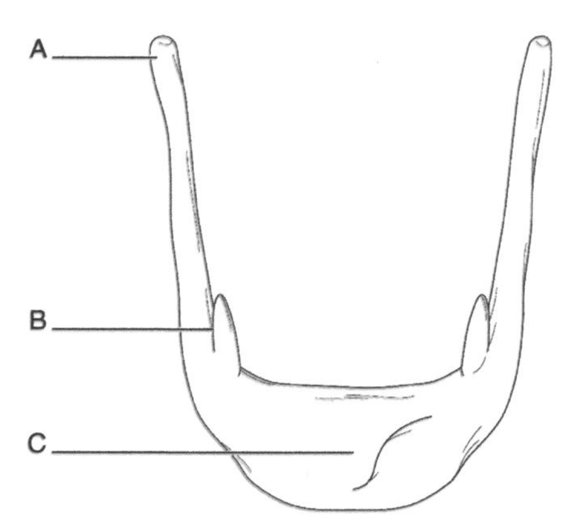

Figure 6-5 Hyoid bone.

Vertebral Column

The **vertebral column** (VER-te-bral), or **spine**, of a typical adult consists of 33 bones stacked vertically along the body's midline, from the base of the skull through the pelvis. It surrounds and protects the spinal cord, and it provides attachment sites for muscles that move the skull, back, shoulders, arms, hips, and thighs. The bones of the vertebral column are irregular in shape and are called **vertebrae** (VER-te-brā; *vertebr-*, "a joint"; singular is vertebra). From a lateral view, the spine of an adult has four curved regions, or *curvatures*, including the **cervical, thoracic, lumbar**, and **sacral curvatures**. Infants have only the thoracic and sacral curvatures. **Table 6.4** and **Figure 6-6** compare the different vertebrae in the vertebral column.

Table 6.4 Comparison of Vertebrae

Vertebrae	Cervical	Thoracic	Lumbar	Sacral	Coccygeal
Number	7	12	5	5	3-5
Location	Neck	Posterior thorax	Inferior back	Pelvis	Pelvis
Articulations	Occipital bone, C1-C7, T1	Ribs, C7, T1-T12, L1	T12, L2-L5; S1	L5, S1-S5, 1st coccygeal	S5, coccygeal vertebrae
Vertebral body	Small and oval (absent in C_1)	Heart shaped with rib facets	Large and oval	Fused in sacrum	Fused in coccyx
Vertebral foramen	Triangular	Circular	Diamond-shaped	Form sacral canal	NA
Transverse foramen	Present	Absent	Absent		

Table 6.4 Comparison of Vertebrae *(Cont'd)*

Spinous process	Short or absent; some split point downward	Long; point downward	Short; point outward	Form sacral crest	Absent
Transverse processes	Have transverse foramina and articular facets	No foramina; has articular and rib facets	No foramina; has articular facets	Fused; form **sacral foramina**	Absent, except on 1st vertebra

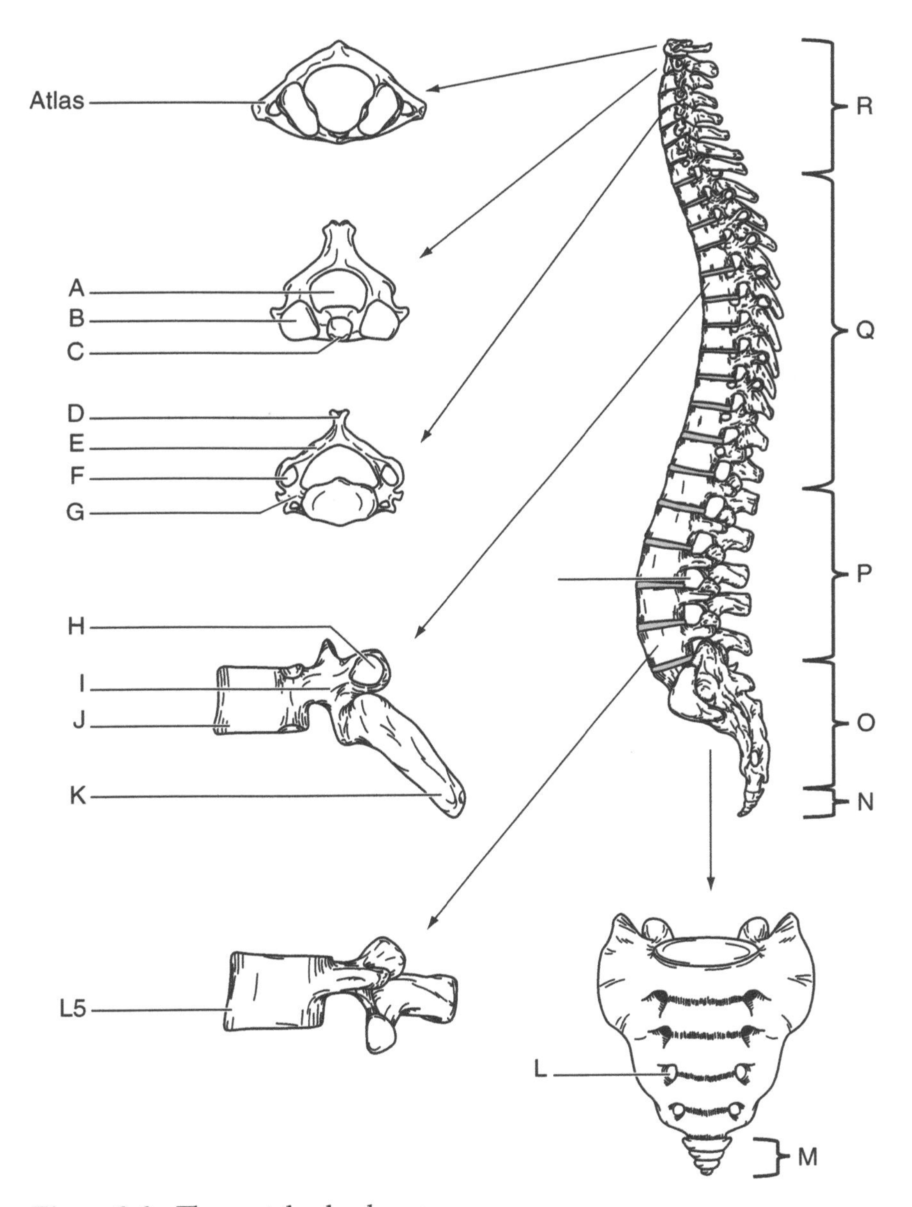

Figure 6-6 The vertebral column.

The Thoracic Cage

The part of the axial skeleton that forms the thorax is the **thoracic cage** (thō-RAS-ik), so named because most of its bones, the *ribs*, appear like the bars of a cage. In addition to protecting internal organs, the thoracic cage supports parts of the appendicular skeleton and provides attachment sites for muscles of the neck, torso, and arm. The thoracic cage forms three major parts: the *thoracic vertebrae* (covered earlier), the *ribs*, and the *sternum*. The features of the thoracic cage are provided in **Table 6.5** and **Figure 6-7**.

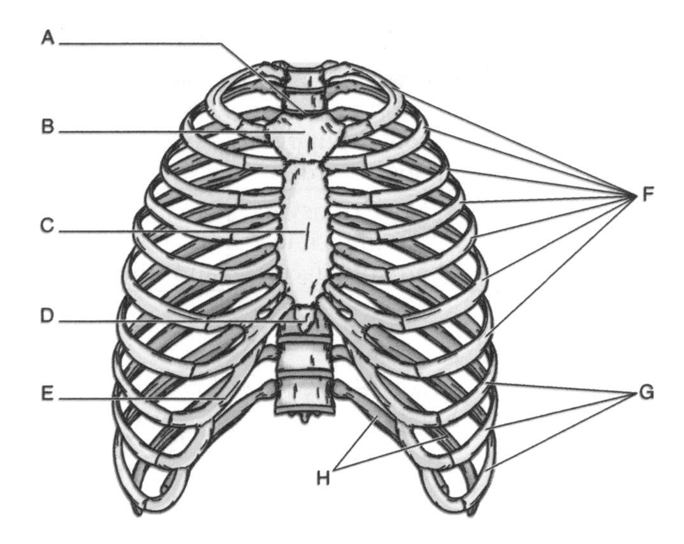

Figure 6-7 The thoracic cage.

Table 6.5 Components of the Thoracic Cage

Component	Features
Ribs (12 pairs)	Attach posteriorly to the thoracic vertebrae
Costal cartilage	Hyaline cartilage on anterior end of rib
Vertebrosternal ribs (VER-tuh-brō-STER-nul)	Pairs 1–7; true ribs; costal cartilages attach to sternum
Vertebrochondral ribs (VER-tuh-brō-KON-drul)	8, 9, 10; false ribs; costal cartilages attach to cartilages of more superior rib
Vertebral ribs (VER-tuh-brul)	11 and 12; floating false ribs; costal cartilages not attached
Sternum (STER-num; "breast plate")	Anterior center of the thoracic cage
Jugular notch (JUG-ū-lar; "throat")	Indentation on superior portion of manubrium
Manubrium (man-Ū-brē-um; "handle")	Superior part; articulates with clavicle and first pair of ribs
Body or **Gladiolus** (glad-ē-Ō-lus; "sword")	Long, middle part; attachment site for costal cartilages of true ribs (1–7)
Xiphoid process (ZIF-oyd; "sword")	Pointed, inferior end

Name ____________________ Course Number: __________ Lab Section ________

Lab 6: The Axial Skeleton Worksheet

Write out the names of the bones labeled in this lab's figures.

Figure 6-2.

1. ____________________
2. ____________________
3. ____________________
4. ____________________
5. ____________________
6. ____________________
7. ____________________
8. ____________________
9. ____________________
10. ____________________
11. ____________________
12. ____________________
13. ____________________
14. ____________________
15. ____________________
16. ____________________
17. ____________________
18. ____________________
19. ____________________
20. ____________________
21. ____________________
22. ____________________
23. ____________________
24. ____________________
25. ____________________
26. ____________________
27. ____________________
28. ____________________
29. ____________________
30. ____________________
31. ____________________
32. ____________________
33. ____________________
34. ____________________
35. ____________________
36. ____________________
37. ____________________
38. ____________________
39. ____________________
40. ____________________
41. ____________________
42. ____________________
43. ____________________
44. ____________________
45. ____________________
46. ____________________
47. ____________________
48. ____________________
49. ____________________
50. ____________________
51. ____________________
52. ____________________
53. ____________________
54. ____________________
55. ____________________
56. ____________________

Figure 6-3.

1. ____________________
2. ____________________
3. ____________________
4. ____________________
5. ____________________
6. ____________________
7. ____________________
8. ____________________
9. ____________________
10. ____________________

11. ____________________
12. ____________________
13. ____________________
14. ____________________
15. ____________________
16. ____________________
17. ____________________
18. ____________________
19. ____________________
20. ____________________
21. ____________________
22. ____________________
23. ____________________
24. ____________________
25. ____________________
26. ____________________
27. ____________________
28. ____________________
29. ____________________
30. ____________________
31. ____________________
32. ____________________
33. ____________________
34. ____________________
35. ____________________
36. ____________________
37. ____________________
38. ____________________
39. ____________________
40. ____________________
41. ____________________
42. ____________________
43. ____________________
44. ____________________
45. ____________________
46. ____________________
47. ____________________
48. ____________________
49. ____________________
50. ____________________
51. ____________________
52. ____________________
53. ____________________
54. ____________________
55. ____________________

Figure 6-4.

A. ____________________
B. ____________________
C. ____________________
D. ____________________
E. ____________________
F. ____________________
G. ____________________
H. ____________________
I. ____________________
J. ____________________
K. ____________________
L. ____________________
M. ____________________
N. ____________________
O. ____________________
P. ____________________
Q. ____________________
R. ____________________
S. ____________________
T. ____________________

Lab 7 The Appendicular Skeleton

The Appendicular Skeleton

The **appendicular skeleton** of a typical adult consists of 126 bones, including 4 in the *pectoral girdle*, 30 in each *upper limb*, 2 in the *pelvic girdle*, and 30 in each *lower limb*.

The Pectoral Girdle

The **pectoral girdle** (PEK-to-ral; *pector-*, breast), or **shoulder girdle**, connects the upper limbs to the thoracic cage of the axial skeleton, and it includes two *clavicles* (collarbones) and two *scapulae* (shoulder blades). Look at the pectoral girdle's features in **Table 7.1** and then label the items in **Figure 7-1**.

Table 7.1 Components of the PECTORAL GIRDLE (4 bones)

Component	Features
Clavicle	Collarbone; anterior portion of girdle; attaches sternum to scapula
Acromial end	Lateral end that articulates with the acromion of the scapula
Sternal end	Medial end that articulates with the manubrium of the sternum
Scapula	Shoulder blade; posterior portion of pectoral girdle
Acromion (uh-KRŌ-mē-on; "summit")	Tip of shoulder at lateral end of the spine; articulates with clavicle
Coracoid process (KOR-a-koyd; "beak")	Extends anteriorly and laterally; attachment site for muscles
Glenoid cavity (GLEN-oyd; "cavity")	Circular depression that articulates with head of humerus
Infraspinous fossa	Surface inferior to the spine
Lateral (axillary) border	Closest to the axillary region
Medial (vertebral) border	Closest to the vertebral column
Spine	Prominent ridge on posterior surface; ends laterally as the acromion
Subscapular fossa	Concave, anterior depression that rests on the posterior ribs' surfaces
Superior border	Horizontal, superior edge
Suprascapular notch (sū-pruh-SKAP-ū-lar)	Indentation on superior border through which a nerve passes
Supraspinous fossa (SŪ-pruh-SPĪ-nus)	Ditch-like depression on superior side of the spine

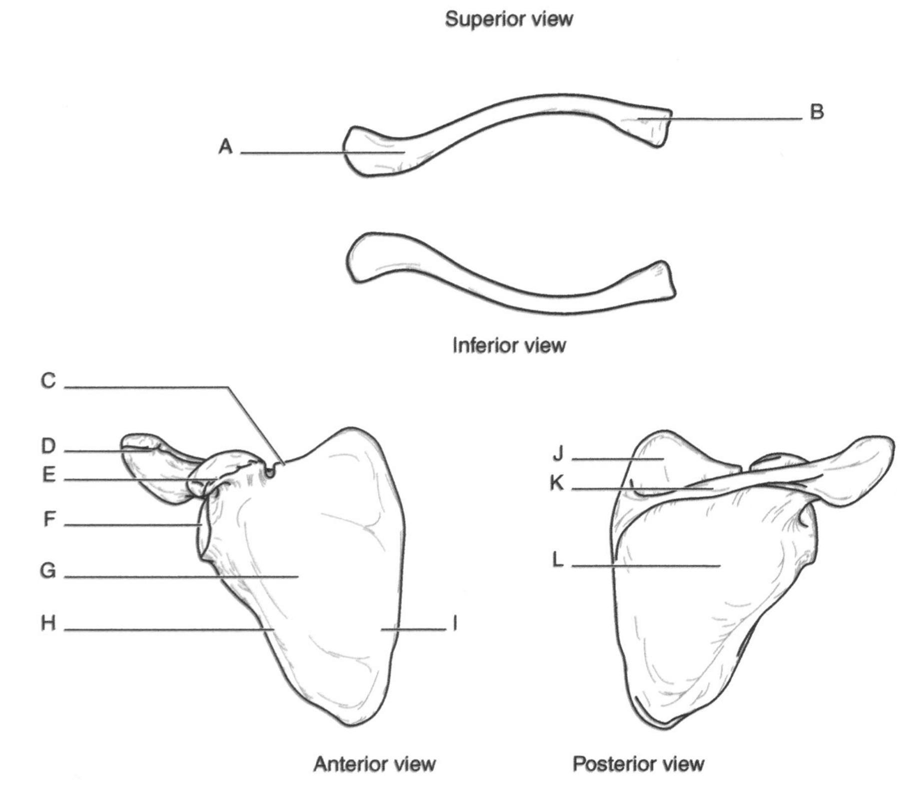

Figure 7-1 Clavicle (top) and scapula (bottom).

The Upper Limb

Each upper limb consists of 30 bones, including 1 bone in the arm, 2 bones in the forearm and 26 bones in the hand (including the wrist). All bones in the upper limb, except for the wrist bones, are *long bones;* whereas the wrist bones are *short bones.* The features of these bones are provided in **Table 7.2** and **Figure 7-2.**

Table 7.2 Components of the Upper Limb

Component	Features
HUMERUS (HŪ-mer-us; "shoulder")	Arm bone extending from the shoulder to the elbow
Head of humerus	Rounded proximal end that fits into scapula's glenoid fossa
Anatomical neck	Region where smooth head meets rougher surface of epiphysis
Greater tubercle	Lateral to anatomical neck; attachment site for arm muscle
Lesser tubercle	Inferior and medial to the greater tubercle; muscle attachment site
Intertubercular groove	Separates the tubercles; supports a muscle tendon
Surgical neck	Narrowing region between the head and diaphysis
Deltoid tuberosity	Roughened region halfway along diaphysis lateral side; site of muscle attachment
Radial groove	Shallow depression on posterior diaphysis; supports radial nerve
Trochlea	Condyle on medial, anterior, distal end; articulates with ulna
Capitulum (kuh-PIH-chū-lum; "head")	Round condyle on lateral, anterior, distal end; articulates with radius

Table 7.2 Components of the Upper Limb *(Cont'd)*	
Coronoid fossa	Superior to trochlea; receives coronoid process of ulna
Radial fossa	Superior to capitulum; receives head of radius when forearm bends
Olecranon fossa (ō-LEK-ruh-non)	Depression that receives the ulna's olecranon
Medial epicondyle	Rounded, medial projection at distal end; muscle attachment site
Lateral epicondyle	Rounded , lateral projection at distal end; muscle attachment site
ULNA (UL-nuh; "elbow")	Medial forearm bone that forms the elbow; in line with little finger
Olecranon process (ō-LEK-ruh-non; "elbow")	Rounded projection that forms the point of the elbow
Coronoid process	Pointed projection just distal to the olecranon process; fits into coronoid fossa of humerus when forearm is bent
Trochlear notch (TRŌ-klē-uh; "pulley")	U-shaped depression that articulates with the trochlea of the humerus
Radial notch	Shallow depression that articulates with the radial head
Head of ulna	Small, rounded, distal end of the ulna; can be palpated (felt) on the medial, posterior side of the wrist region
Styloid process	Short, pointed projection from the ulnar head
RADIUS (RĀ-dē-us; "ray")	Lateral forearm bone that is in line with the thumb
Head of radius	Disc-shaped, narrow, proximal end; articulates with the capitulum
Radial tuberosity	Projection just distal to the radial head; attachment site for muscle
Styloid process	Pointed projection on the lateral, distal end
Ulnar notch	Depression on the lateral, distal end that receives the ulnar head.
CARPALS (KAR-pulz; "wrist")	Wrist bones; includes 8 bones in each wrist
Scaphoid (SKAF-oyd; "boat-like")	Largest carpal; most lateral and proximal; articulates with radius
Lunate (LŪ-năt; "moon")	Medial to scaphoid; articulates with radius
Triquetrum (trī-KEH-trum; "3-cornered")	Medial to lunate; articulates with the ulna
Pisiform (PĒ -zih-form; "pea-like")	Medial, proximal carpal; palpated as a small knob near end of ulna
Trapezium (truh-PĒ-zē-um; "table")	Lateral, distal carpal; articulates with the scaphoid and 1st metacarpal
Trapezoid (TRAP-ih-zoyd; "4-sided")	Medial to the trapezium
Capitate (KAP-ih-tăt; "head")	Medial to the trapezoid
Hamate (HAM-ăt; "hooked")	Most medial, distal carpal; has a hook on its anterior surface
METACARPALS (*meta-*, "beyond")	Hand bones; numbered 1-5 from lateral to medial
Base	Proximal end
Head	Rounded distal end
Body	Between the base and head
PHALANGES (fuh-LAN-jēz; "line of soldiers")	Long bones of the digits (fingers)
Proximal phalanx (FĀ-langks)	Articulates with a metacarpal
Middle phalanx	Distal to the proximal phalanx (absent in the thumb—digit #1)
Distal phalanx	Distal to the middle phalanx (or proximal phalanx in digit #1)

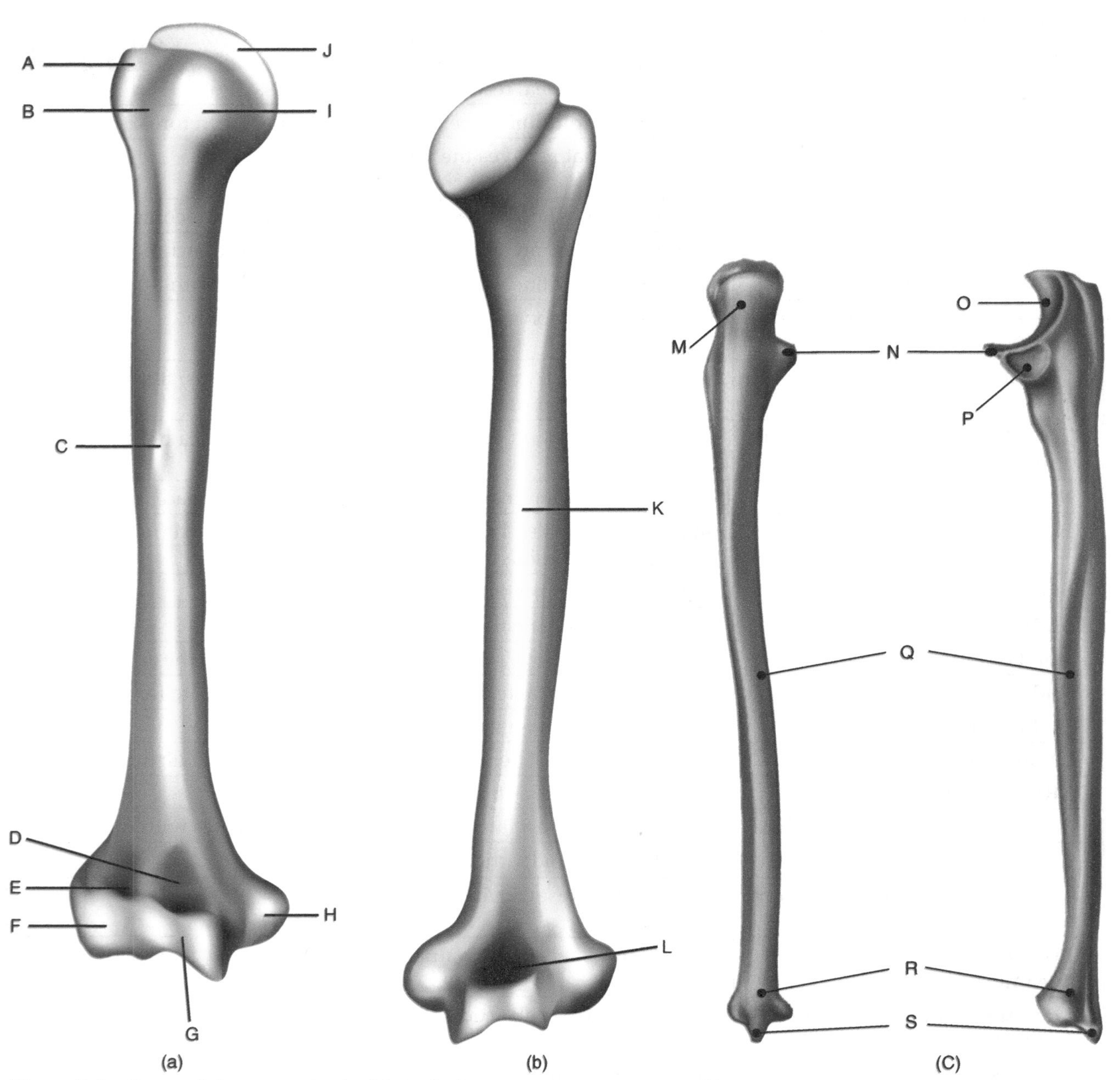

Figure 7-2 Bone of the upper limb. (a) Right humerus, anterior view; (b) Posterior view; (c) Ulna, posterior and lateral views;

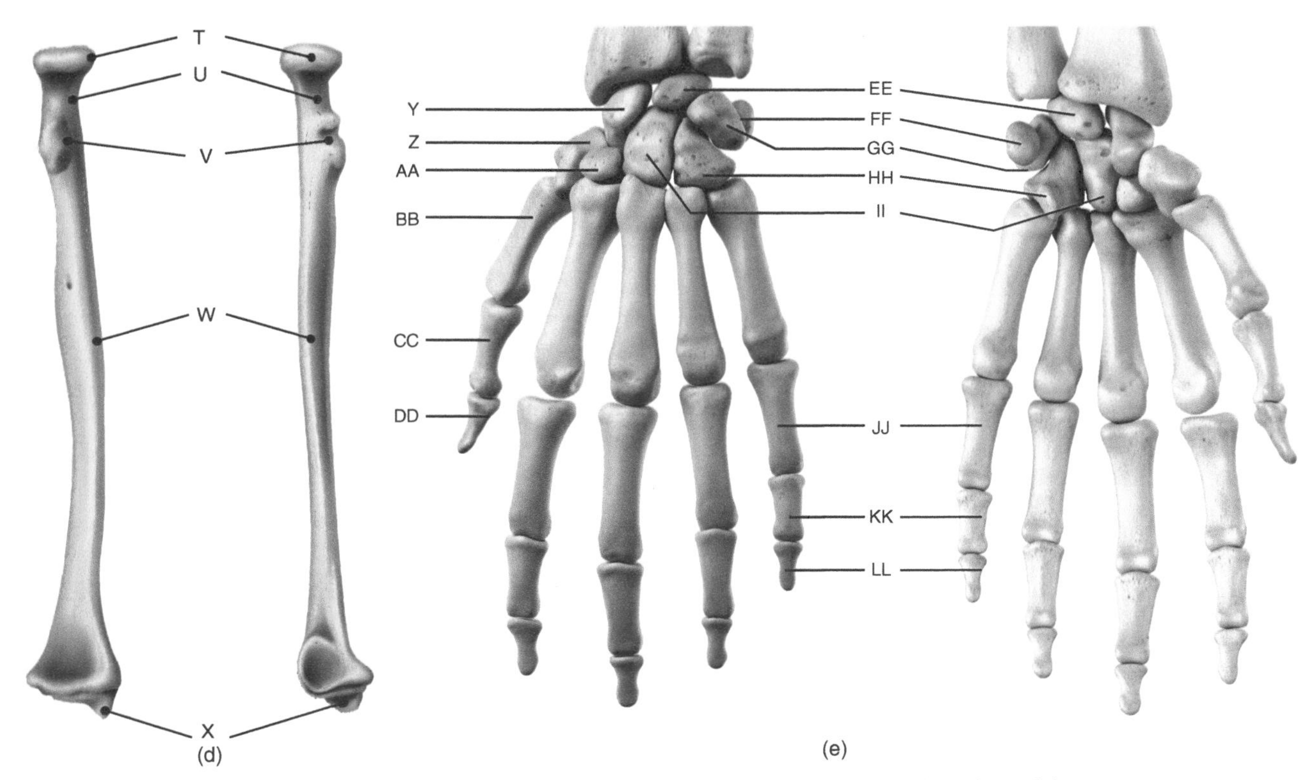

Figure 7-2 (*continued*) (d) Radius, posterior and lateral views; (e) Right wrist, hand, and fingers; posterior view on left and anterior view on right.

The Pelvic Girdle And Pelvis

Whereas the pectoral girdle joins the upper limbs to the axial skeleton, the **pelvic girdle** (PEL-vik) joins the lower limbs to the axial skeleton. The pelvic girdle is not synonymous with *pelvis*, but is a part of it. The **pelvis** (PEL-vis; "basin") is the bowl-shaped ring of bone that includes the *pelvic girdle*, *sacrum*, and *coccyx*. Thus, the pelvis consists of axial and appendicular skeleton components. The pelvic girdle lies between the sacrum and femurs (thighbones), and it consists of two **hipbones**, or **coxal bones** (KOK-sal; *coxa-*, "hip"). Each hipbone in an adult consists of three fused bones, but all of these are separate bones in an infant. The coxal bones lie on opposite sides of the body's midline and articulate with the opposite hipbone, anteriorly (a joint called the *pubic symphysis*), and with the sacrum, posteriorly. Since there are some notable structural differences between the pelvis of a male and a female, it is important to identify them here. The major features of the pelvic girdle and the differences between male and female pelves (PEL-vēz) are shown in **Table 7.3** and **7.4**, and in **Figure 7-3**.

Table 7.3 Components of the Pelvic Girdle and Pelvis

Component	Features
COXAL BONES	Hipbones (one on each side) that form the pelvic girdle; each coxal bone contains 3 fused bones in the adult
Acetabulum (as-i-TAB-ū-lum; "vinegar cup")	Cup-like socket that articulates with the head of the femur; formed from parts of the hipbone's ilium, ischium, and pubis
Obturator foramen (OB-tur-ā-tor: "stopped up")	Large opening formed by curved extensions of the ischium and pubis; so-named due to being covered by a web of ligaments

(*continued*)

Table 7.3 Components of the Pelvic Girdle and Pelvis *(Cont'd)*

Ilium (IL-ē-um; "flank")	Largest part of hipbone; forms superior half of girdle
Ala (Ā-la; "wing")	Resembles wings of a butterfly
Iliac crest	Thick ridge on superior portion of ala
Iliac fossa	Large depression on anterior side of ala; site of muscle attachment
Anterior superior iliac spine	Most anterior projection of hipbone; anterior end of iliac crest
Anterior inferior iliac spine	Lies inferior and slightly medial to the anterior superior iliac spine
Posterior superior iliac spine	Most posterior projection on hipbone; posterior end of iliac crest
Posterior inferior iliac spine	Slightly inferior and anterior to the posterior superior iliac spine
Greater sciatic notch	Immediately inferior and posterior to posterior inferior iliac spine; passage for the body's large nerve (*sciatic nerve*)
Ischium (IS-kē-um; "hip")	Second largest part of hipbone; forms posterior, inferior half
Ischial spine	Sharp projection from posterior rim of ischium
Ischial tuberosity	Most noticeable feature of ischium; site of muscle attachment
Lesser sciatic notch	Small indentation between ischial tuberosity and ischial spine; passage for nerves and blood vessels
Ramus	Extends anteriorly from the ischial tuberosity to the pubis
Pubis (PŪ-bis; "genital")	Smallest part of hipbone; forms anterior, inferior half
Pubic arch	V-shaped angle formed where right and left pubis meet
PELVIS	Includes the coxal bones, sacrum, and coccyx
False pelvis	Wide, shallow space extending between the iliac crest to the pelvic brim
Pelvic brim	Rim of bone that forms the superior, oval-shaped opening (or pelvic inlet) of the true pelvis
True pelvis	Narrow, deep space inferior to the pelvic brim; holds pelvic organs
Pelvic inlet	Larger, superior opening to the true pelvis; formed by pelvic brim
Pelvic outlet	Smaller, inferior opening of the true pelvis

Table 7.4 Comparison of Male and Female Pelvis

Feature	Female	Male
Acetabulum	Smaller diameter	Larger diameter
Bone thickness	Not as thick	Thicker
Coccyx	More in line with the sacrum	More noticeable curving
Greater sciatic notch	Greater diameter	Smaller diameter
Iliac crest	Curves less anteriorly	Curves more anteriorly
Iliac fossa	Not as deep	More depressed
Ilium	Oriented more in the coronal plane	Narrower wings
Ischial spines	Relative distance between them is greater	Relative distance between them is less
Ischial tuberosities	Relatively long	Relatively short

Table 7.4 Comparison of Male and Female Pelvis *(Cont'd)*

Pelvic inlet	Relatively wide and oval shaped	Relatively narrow and more triangular shaped
Pelvic outlet	Relatively wider	Relatively narrow
Pubic arch	Greater than 90°	Less than 90°
Sacrum	Relatively wide and straighter	Relatively narrower and more curved

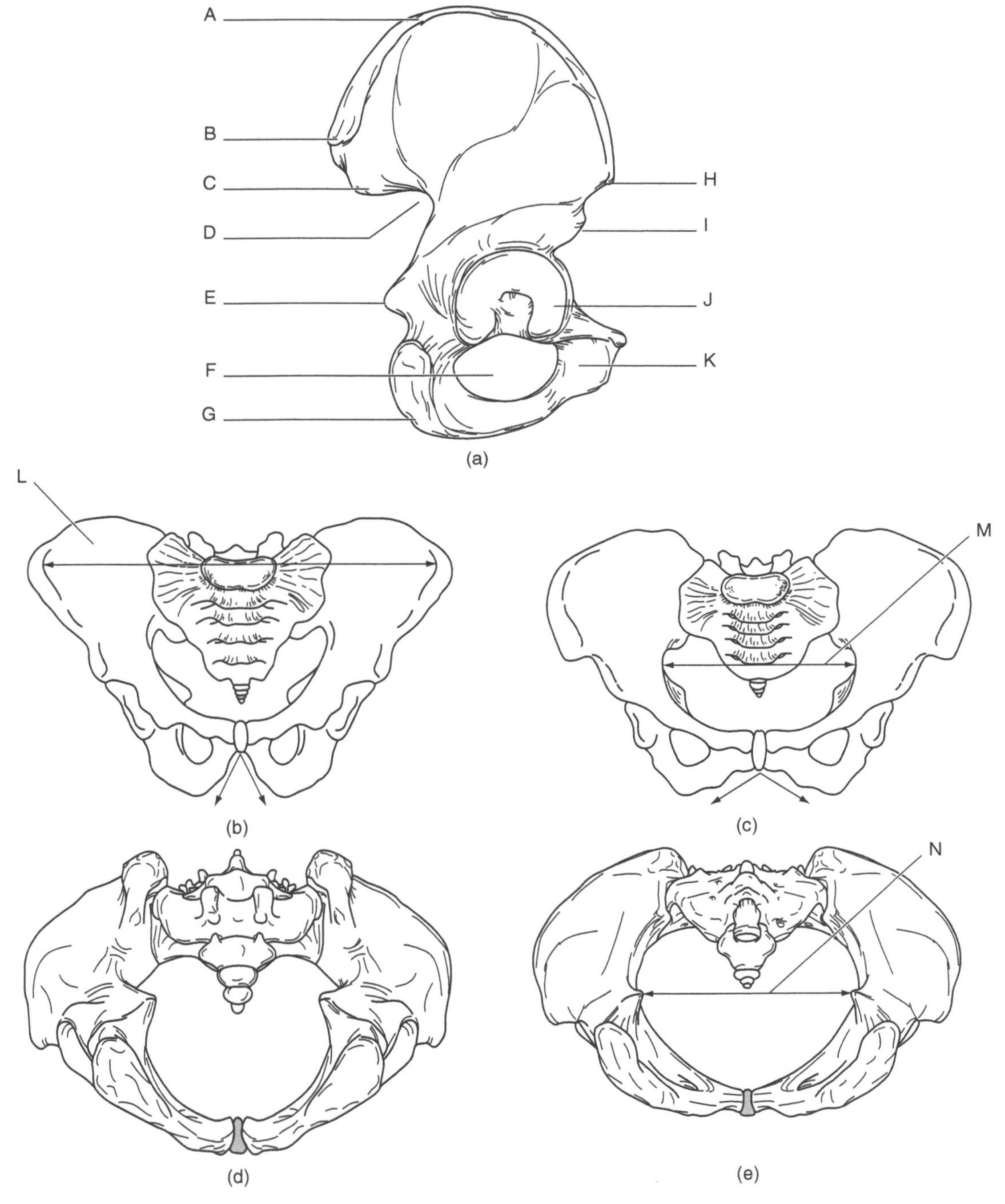

Figure 7-3 The pelvis. (a) Right os coxa, lateral view; (b) Male pelvis, anterior view; (c) Female pelvis, anterior view; (d) Male pelvis, inferior view; (e) Female pelvis, inferior view

The Lower Limb

The arrangement of bones in the lower limb is similar to that in the upper limb, with larger bones being proximal to the axial skeleton and smaller bones being distal. In addition, each lower limb consists of 30 bones, the same as in each upper limb. Each lower limb consists of 1 bone in the thigh, 2 bones in the leg, 1 large sesamoid bone in the knee, and 25 bones in the foot (including the ankle). Although the foot contains one less bone than the hand, the sesamoid bone in the knee makes up the difference in number. **Table 7.5** and **Figure 7-4** reveal features of the lower limb.

Table 7.5 Components of the Lower Limb

Component	Features
FEMUR	Thighbone
Head of femur	Ball-shaped proximal end that fits into acetabulum of pelvis
Fovea capitis	Small pit in the femoral head; attachment site of a ligament
Neck of femur	Constricted region connecting the head to the remainder of epiphysis
Greater trochanter	Large process lateral and superior to neck; muscle attachment site
Lesser trochanter	Smaller process on medial side of proximal epiphysis
Gluteal tuberosity	Roughened line on superior, posterior shaft; attachment site for muscle
Linea aspera	Long ridge from gluteal tuberosity; attachment site for muscle
Lateral condyle	On distal, lateral end; articulates with lateral condyle of tibia
Medial condyle	On distal, medial end; articulates with the medial condyle of tibia
Lateral epicondyle	Rounded process on lateral edge of lateral condyle
Medial epicondyle	Rounded process on medial edge of medial condyle
Intercondylar fossa	Depression between the condyles; holds ligaments that stabilize knee
Patellar surface	Smooth, shallow surface on which patella articulates
PATELLA (puh-TELL-uh; "little dish")	Knee cap
TIBIA (TIB-ē-uh; "flute")	Large, medial bone in leg; supports body weight
Lateral condyle	Concave proximal, lateral end; articulates with lateral condyle of femur
Medial condyle	Concave proximal, medial end; articulates with medial condyle of femur
Intercondylar eminence	Double ridge separating the lateral and medial condyles
Tibial tuberosity	Rough raised region just inferior and anterior to the condyles; attachment site for ligament leading from the patella
Medial malleolus (muh-LĒ-o-lus; "hammer")	Large process on medial, distal end
Fibular notch	Shallow depression on distal, lateral end; articulates with fibula
FIBULA (FIB-ū-luh; "fastener")	Smaller, lateral bone in leg; does not support body weight
Head of fibula	Proximal epiphysis; articulates with facet on tibia's lateral condyle
Lateral malleolus	Process projecting inferiorly from the distal epiphysis
TARSALS (TAR-sulz; "flat surfaces")	Ankle bones
Calcaneus (kal-KĀ-nē-us; "heel")	Largest and most posterior tarsal
Talus (TĀ-lus; "ankle")	Second largest tarsal; on superior surface of the calcaneus
Navicular (nuh-VIK-ū-lar; "boat")	Curved like a boat; articulates with the talus and cuneiform bones

Table 7.5 Components of the Lower Limb *(Cont'd)*

Cuboid (KŪ-boyd; "cube-shaped")	Most lateral tarsal; articulates with calcaneus, lateral cuneiform, navicular, and 4th and 5th metatarsals
Lateral cuneiform (kū-NĒ-ih-form; "wedge-like")	Medial to the cuboid and lateral to the intermediate cuneiform
Intermediate cuneiform	Between the lateral and medial cuneiforms
Medial cuneiform	Medial to the intermediate cuneiform
METATARSALS (1, 2, 3, 4, 5)	Foot bones located between the tarsals and proximal phalanges of the digits; metatarsal #1 is most medial (in line with big toe) and #5 is most lateral (in line with little toe)
Phalanges of Digits	Bones of toes
Proximal phalanx	Articulates with a metatarsal and a middle phalanx (except in hallux—big toe—where it articulates with a distal phalanx)
Middle phalanx	Located between a proximal and distal phalanx; not present in the big toe
Distal phalanx	Articulates with a middle phalanx (except in big toe, where it articulates with a proximal phalanx)

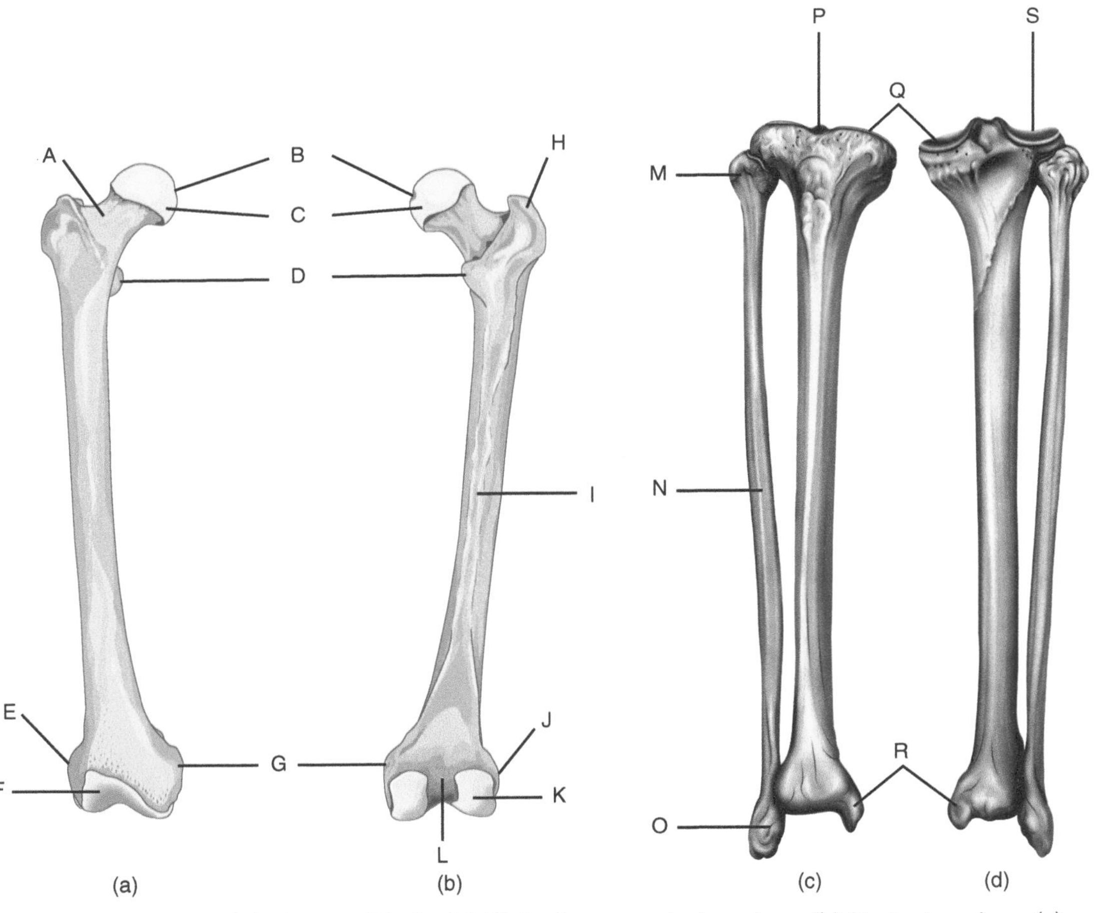

Figure 7-4 Bones of the Lower Limb. (a) Right femur, anterior view; (b) Posterior view; (c) Tibia and fibula, anterior view; (d) Posterior view;

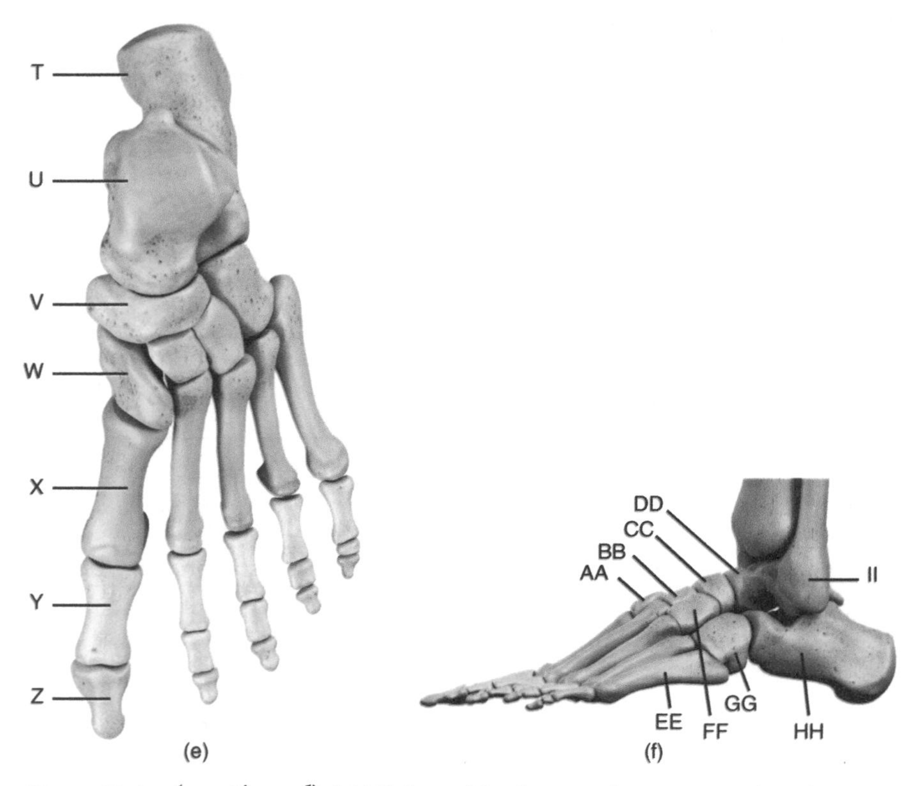

Figure 7-4 (*continued*) (e) Right ankle, foot, and toes, superior view.

Name ________________ Course Number: ________ Lab Section ______

Lab 7: The Appendicular Skeleton Worksheet

Write out the names of the bones labeled in this lab's figures.

Figure 7-1.

A. ________________
B. ________________
C. ________________
D. ________________
E. ________________
F. ________________
G. ________________
H. ________________
I. ________________
J. ________________
K. ________________
L. ________________

Figure 7-2.

A. ________________
B. ________________
C. ________________
D. ________________
E. ________________
F. ________________
G. ________________
H. ________________
I. ________________
J. ________________
K. ________________
L. ________________
M. ________________
N. ________________
O. ________________
P. ________________
Q. ________________
R. ________________
S. ________________
T. ________________
U. ________________
V. ________________
W. ________________
X. ________________
Y. ________________
Z. ________________
AA. ________________
BB. ________________
CC. ________________
DD. ________________
EE. ________________
FF. ________________
GG. ________________
HH. ________________
II. ________________
JJ. ________________
KK. ________________
LL. ________________

Figure 7-3.

A. ________________
B. ________________
C. ________________
D. ________________
E. ________________
F. ________________
G. ________________
H. ________________
I. ________________
J. ________________
K. ________________
L. ________________
M. ________________
N. ________________

Figure 7-4.

A. ________________
B. ________________
C. ________________

D. ______________________
E. ______________________
F. ______________________
G. ______________________
H. ______________________
I. ______________________
J. ______________________
K. ______________________
L. ______________________
M. ______________________
N. ______________________
O. ______________________
P. ______________________
Q. ______________________
R. ______________________
S. ______________________
T. ______________________
U. ______________________
V. ______________________
W. ______________________
X. ______________________
Y. ______________________
Z. ______________________
AA. ______________________
BB. ______________________
CC. ______________________
DD. ______________________
EE. ______________________
FF. ______________________
GG. ______________________
HH. ______________________
II. ______________________

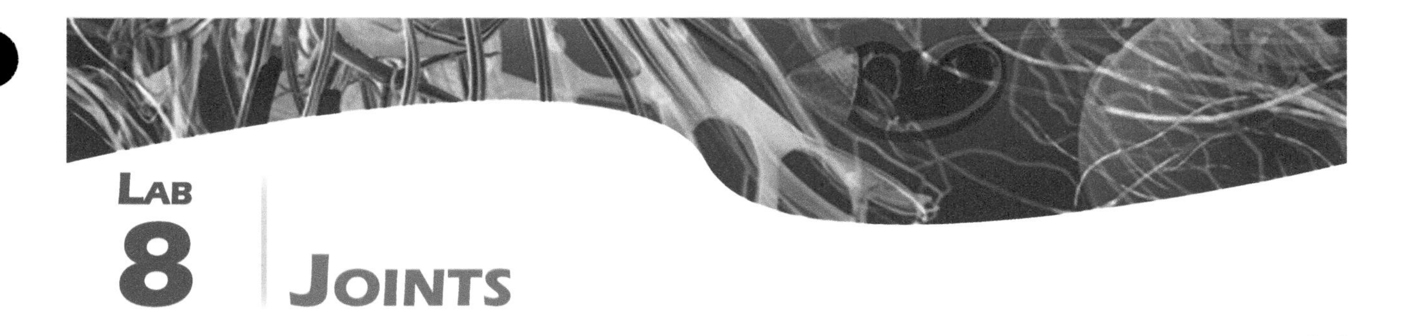

LAB 8 JOINTS

A **joint**, or **articulation** (ar-tik-u-Ā-L-shun), is a site where two bones come together. In most cases, the name of a joint is based on the names of bones (or their components) that interact at the joint. You should be familiar with bone names from previous labs, and this is the key to naming joints. Therefore, before proceeding to the next section that deals with joint classification, fill in the right column in **Table 8.1** and then place appropriate numbers in the boxes in **Figure 8-1**.

Table 8.1 Major Joints of the Body

Figure 8-1 No.	Joint	Articulating Bones
1	Acromioclavicular	
2	Atlantoaxial	
3	Atlanto-occipital	
4	Carpometacarpal	
5	Coxal	
6	Dento-alveolar	
7	Femoropatellar	
8	Glenohumeral	
9	Humeroradial	
10	Humeroulnar	
11	Intercarpal	
12	Interphalangeal in lower limb	
13	Interphalangeal in upper limb	
14	Intertarsal	
15	Intervertebral	
16	Manubriosternal	
17	Metacarpophalangeal	
18	Metatarsophalangeal	
19	Pubic symphysis	

(Continued)

Table 8.1 Major Joints of the Body *(Cont'd)*		
20	**Radiocarpal**	
21	**Radioulnar (distal)**	
22	**Radioulnar (proximal)**	
23	**Sacroiliac**	
24	**Sternoclavicular**	
25	**Sternocostal**	
26	**Suture**	
27	**Talocrural**	
28	**Tarsometatarsal**	
29	**Temporomandibular**	
30	**Tibiofibular (distal)**	
31	**Tibiofibular (proximal)**	
32	**Tibiofemoral**	
33	**Vertebrocostal**	
34	**Xiphisternal**	

Classification of Joints

Any joint can be classified according to two criteria: (1) *function* (amount of movement possible) and (2) *structure* (connecting material). Structurally, a joint is *fibrous, cartilaginous*, or *synovial*. Functionally, a joint is *synarthrotic, amphiarthrotic*, or *diarthrotic*. In addition, both structural and functional classifications have subdivisions. **Table 8.2** summarizes the classification system for joints. After reviewing the table and skeletal models, classify the joints in **Figure 8-2**.

Mode of Articulation at Synovial Joints

Since synovial joints allow more movement than other types of joints, we can classify them based on the type of body movement possible. **Table 8.3** summarizes the different types of movements possible at specific synovial joints and **Figure 8-3** depicts some of them. When you become familiar with these movements, complete the exercise that follows.

Actions Depicted in Figure 8-3

As you determine the specific action for each number in Figure 8-3, write that action in the blank by the appropriate number. Be sure that you include the name of the part highlighted in red; e.g., arm, forearm, trunk, thigh, leg, etc.

1. ______________________
2. ______________________
3. ______________________
4. ______________________
5. ______________________
6. ______________________
7. ______________________
8. ______________________
9. ______________________
10. ______________________

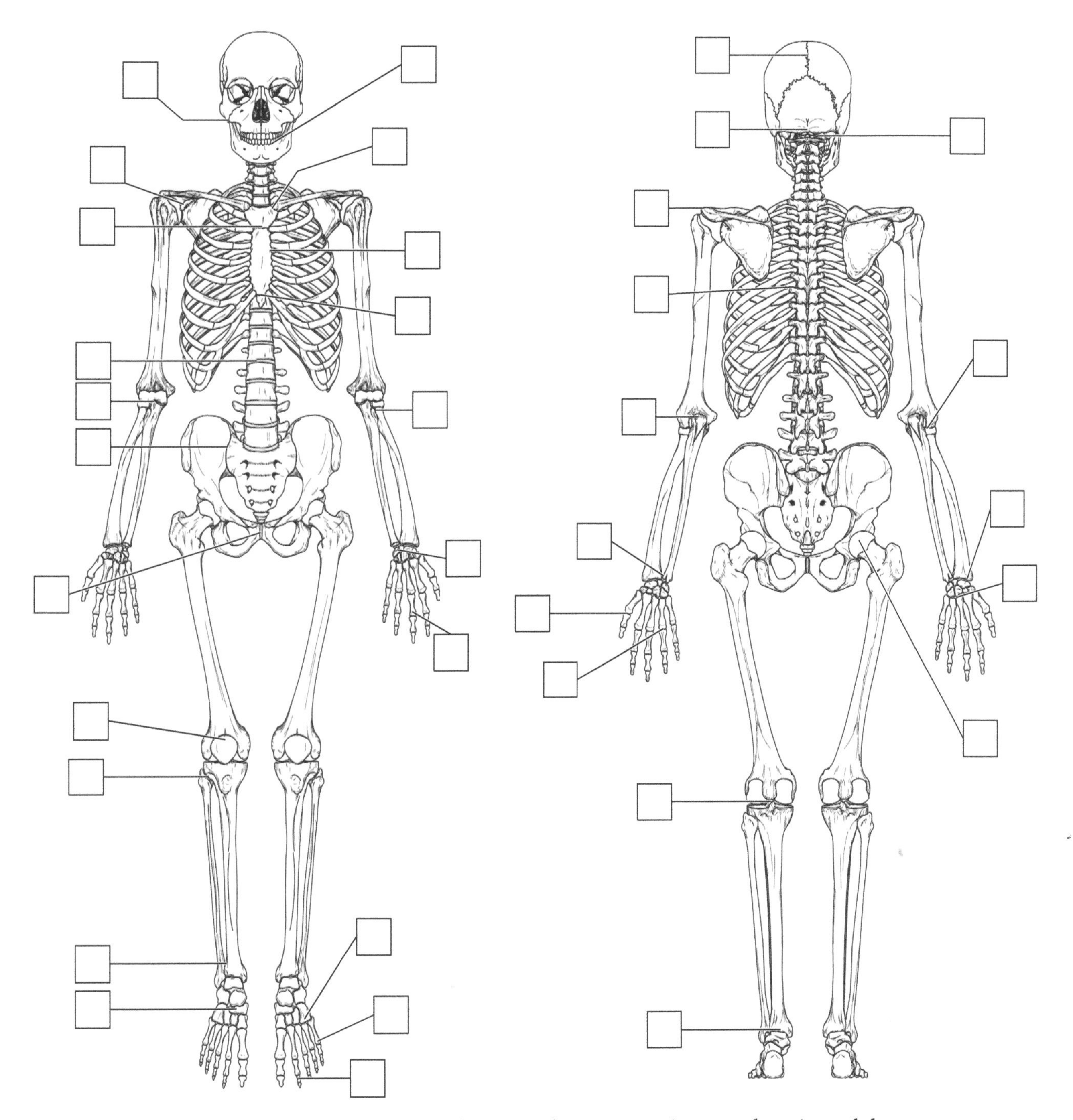

Figure 8-1 Major joints of the body. Use Table 8.1 to place appropriate numbers in each box.

11. ____________________
12. ____________________
13. ____________________
14. ____________________
15. ____________________
16. ____________________
17. ____________________
18. ____________________
19. ____________________
20. ____________________
21. ____________________
22. ____________________
23. ____________________
24. ____________________
25. ____________________
26. ____________________
27. ____________________
28. ____________________
29. ____________________
30. ____________________
31. ____________________
32. ____________________
33. ____________________
34. ____________________
35. ____________________
36. ____________________
37. ____________________
38. ____________________
39. ____________________
40. ____________________
41. ____________________
42. ____________________
43. ____________________
44. ____________________

Table 8.2 Major Classification of Joints

Classification	Characteristics
BASED ON FUNCTION	
Synarthrosis (sin-ar-THRŌ-sis)	Relatively immovable
Amphiarthrosis (am-fē-ar-THRŌ-sis; "both")	Partially movable
Diarthrosis (dī-ar-THRŌ-sis)	Freely movable
BASED ON STRUCTURE	
Fibrous	Thick collagen fibers; synarthrotic or amphiarthrotic
Suture (SŪ-chur; "seam")	Short *sutural ligaments*; synarthrotic
Syndesmosis (sin-dez-MŌ-sis; "binding")	Long fibers between parallel bones; amphiarthrotic
Gomphosis (gom-FŌ-sis; "bolt")	Short *periodontal ligaments*; synarthrotic
Cartilaginous	Hyaline or fibrocartilage; synarthrotic or amphiarthrotic
Synchondrosis (sin-kon-DRŌ-sis ("together cartilage")	Hyaline cartilage; synarthrotic
Symphysis (SIM-fih-sis; "together")	Fibrocartilage pad; amphiarthrotic

Synovial (sin-Ō-vē-ul; "with egg")	Joint enclosed by a synovial capsule; diarthrotic
Plane joint	Flat surfaces slide back-and-forth
Hinge joint	Convex and concave surfaces
Pivot joint	Knobby process fits through a "ring" formed by a shallow depression and ligament
Condylar (KON-dih-lar; "knuckle")	**Ellipsoid**; a convex surface fits into an oval-shaped fossa
Saddle	Two convex parts fit into concave part on opposing bone
Ball-and-Socket	Rounded knob fits into cuplike fossa

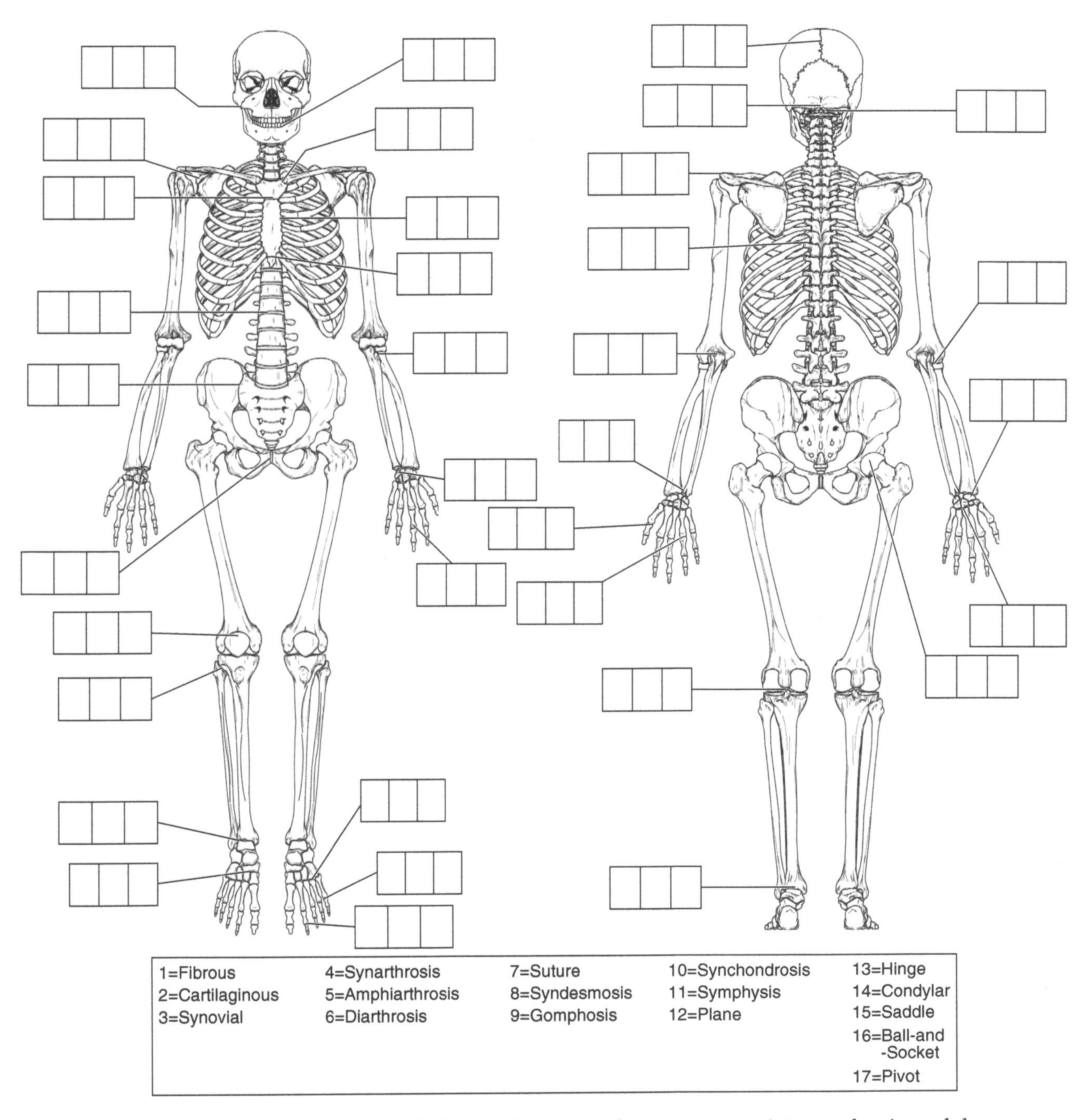

Figure 8-2 Classification of joints. Use the key at the top to place an appropriate number in each box.

Table 8.3 Bone Movements at Synovial Joints

Movement	Direction
GLIDING	Only one plane
ANGULAR	Changes the angle between two bones
Abduction (ab-DUK-shun; "take away")	Away from a midline
Adduction (ad-DUK-shun; "bring toward")	Toward a midline
Circumduction (ser-kum-DUK-shum; "around")	End of a bone makes a circular motion
Extension (eks-TEN-shun; "stretch")	Increases the angle between bones
Flexion (FLEK-shun; "bend")	Decreases the angle between bones
Hyperextension	Extends the end of a bone beyond its anatomical position
ROTATION	Around the bone's longitudinal axis
Lateral rotation	Anterior bone surface moves away from a midline
Left rotation	Head or thorax turns to the left
Medial rotation	Anterior bone surface moves toward a midline
Right rotation	Head or thorax turns to the right
SPECIAL	Unique to specific joints
Depression (dē-PRESH-un; "push down")	Inferiorly along the sagittal plane
Dorsiflexion (dor-sih-FLEK-shun; "bend up")	Anterior portion of foot moves superiorly
Elevation (el-ih-VĀ-shun; "lift up")	Superiorly along the sagittal plane
Eversion (ē-VER-shun; "turn outward")	Turning the sole of the foot away from the midline
Excursion (eks-KUR-shun; "run out")	Shifting the mandible back and forth from side-to-side
Inversion (in-VER-shun; "turn inward")	Turning the sole of the foot towards the midline
Lateral flexion	Tilting the vertebral column right or left
Opposition (op-ō-ZISH-un; "go against")	Thumb moves toward the little finger
Plantar flexion	Posterior portion of foot moves superiorly
Pronation (prō-NĀ-shun; "bend forward")	Palm turns posteriorly from anatomical position
Protraction (prō-TRAK-shun; "draw forth")	Anteriorly along a sagittal plane
Reposition (rē-pō-ZISH-un; "replace")	Thumb moves from palm back to anatomical position
Retraction (re-TRAK-shun; "draw back")	Posteriorly along a sagittal plane
Supination (sū-pih-NĀ-shun; "turn backward")	Palm turns anteriorly back to anatomical position

Anatomy of the Knee Joint

Since the knee joint usually gets a significant amount of attention in the medical profession, it is deemed important to consider it in more detail. At the knee joint, articulation occurs between the femur and tibia and between the femur and patella. Superiorly, the *quadriceps tendon* attaches a large thigh muscle to the patella. This tendon extends inferiorly from the patella as the **patellar ligament** and inserts on the tibial tuberosity.

Two extracapsular ligaments on each side of the knee joint help stabilize the joint. The **tibial collateral ligament** connects the medial condyle of the femur to the medial condyle of the tibia and prevents medial displacement of the tibia. The **fibular collateral ligament** connects the lateral condyle of the femur to the head of the fibula and prevents lateral displacement of the tibia.

Two intracapsular ligaments prevent excessive front-to-back movements of the tibia. The **anterior cruciate ligament** (KRŪ-shē-āt;

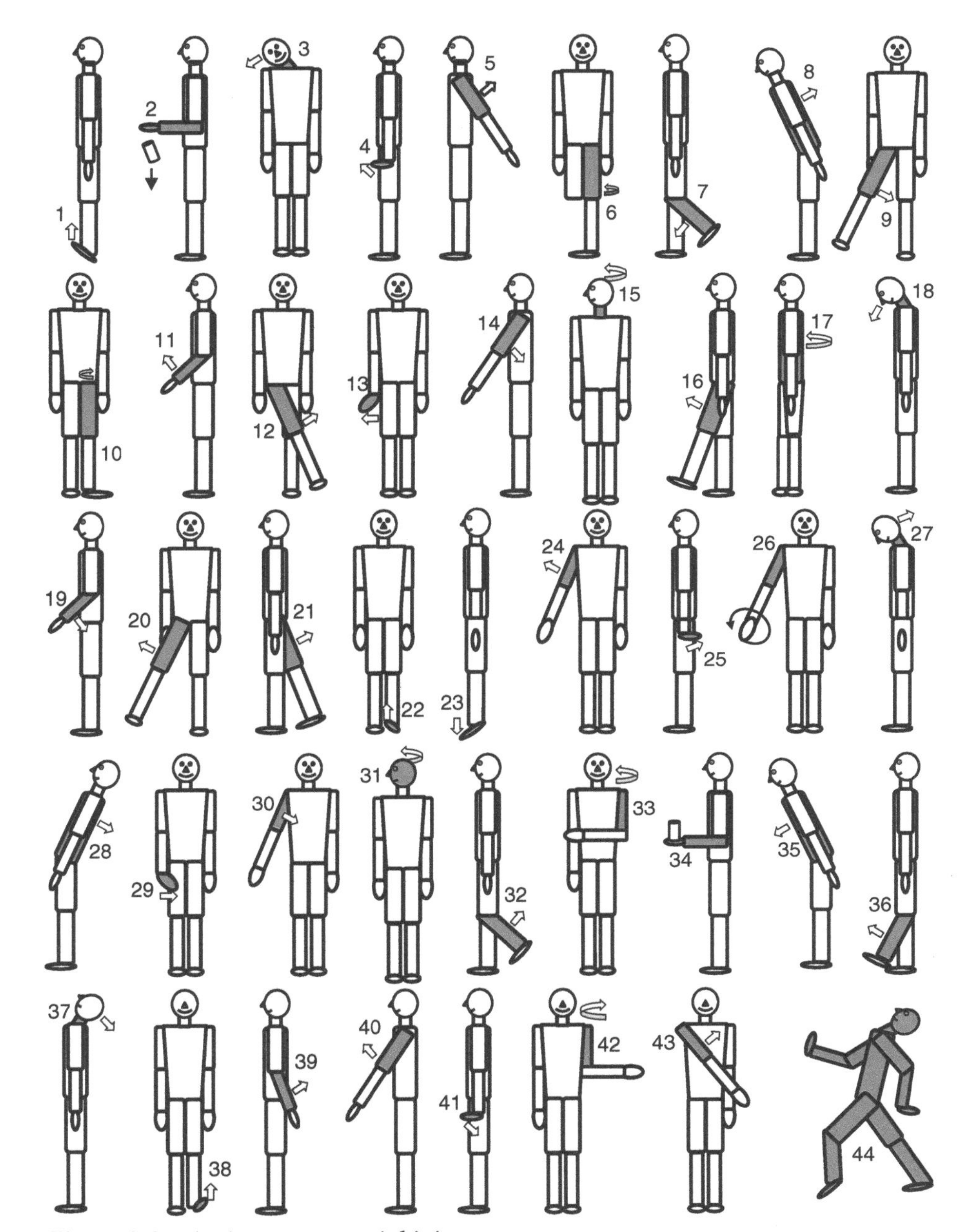

Figure 8-3 Actions at synovial joints.
Source: Mark Taylor

"cross"), or **ACL**, extends from the posterior side of the femur to the anterior portion of the tibia and prevents hyperextension of the tibia. The **posterior cruciate ligament (PCL)** extends from the anterior side of the femur and inserts on the posterior side of the tibia and it prevents excessive back-and-forth sliding of the tibia.

Also inside the knee joint are several fibrocartilaginous discs called *menisci* (mih-NIS-sī; "crescents"), or *articular discs* that help distribute weight across the joint and act as shock absorbers. The **lateral meniscus** (men-IS-kus) is a pad of fibrocartilage that rests on the lateral condyle of the tibia, while the **medial meniscus** rests on the medial condyle.

Several fluid-filled sacs, called **bursae** (BER-sē, "purse") help reduce friction between the bones in the knee and the overlying skin. Their names are based on their location relative to the patella: the **suprapatellar bursa** is superior to it; the **prepatellar bursa** is anterior to it, and the **infrapatellar bursa** is inferior to it. An **infrapatellar fat pad** lies at the anterior surface of the knee joint. It cushions the femoral condyles and helps prevent them from sliding forward. Label the structures in **Figure 8-4**.

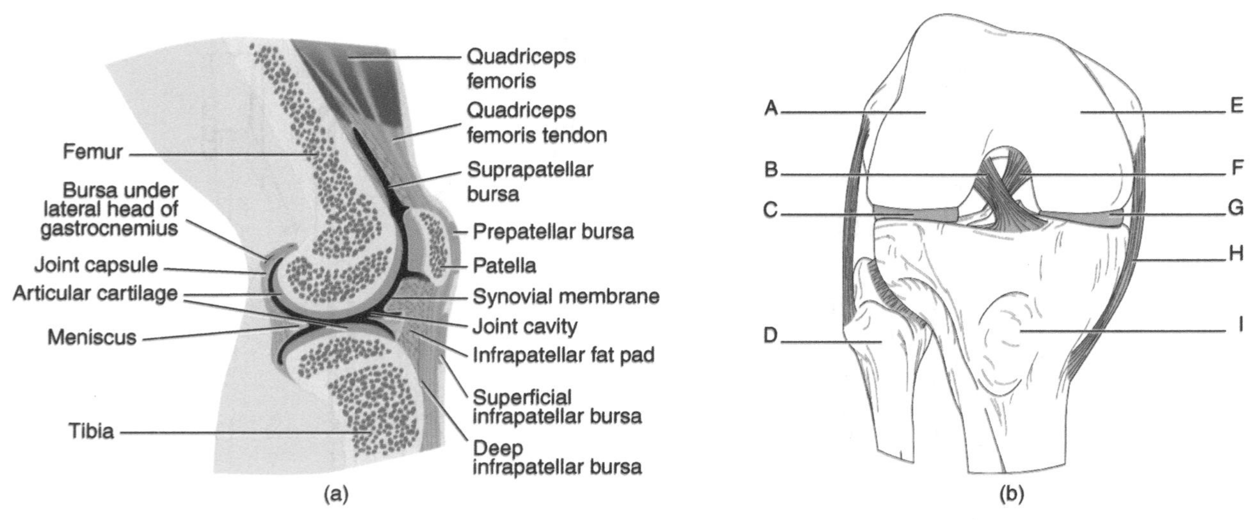

Figure 8-4 Structure of the knee joint (a) Midsagittal section. (b) Anterior view of right knee.

Summary of Major Joints

To complete your study of joints, look at Table 8.4, which provides information about major joints in the body. You knowledge of the skeletal system will allow you to locate these joints on model skeletons in the lab. Be prepared to answer questions related to the five categories listed at the top of this table.

Table 8.4 Classification and Actions of Selected Joints

Joint	Structure	Function	Major Actions
Acromioclavicular†	Plane	Diarthrosis	Scapula rotation
Atlantoaxial	Pivot/plane	Diarthrosis	Head rotation
Atlanto-occipital	Condylar	Diarthrosis	Head flexion, extension, lateral flexion, circumduction
Carpometacarpal 1	Saddle	Diarthrosis	1st metacarpal flexion, extension, abduction, adduction, circumduction, opposition
Carpometacarpal 2–5	Plane	Diarthrosis	Metacarpal gliding
Coxal	Ball-and-socket	Diarthrosis	Thigh flexion, extension, abduction, adduction, circumduction, rotation
Dento-alveolar	Gomphosis	Synarthrosis	None
Femoropatellar	Plane	Diarthrosis	Patella gliding
Glenohumeral	Ball-and-socket	Diarthrosis	Arm flexion, extension, abduction, adduction, circumduction

Table 8.4 Classification and Actions of Selected Joints *(Cont'd)*

Humeroradial	Hinge/pivot	Diarthrosis	Radius flexion, extension, rotation
Humeroulnar	Hinge	Diarthrosis	Ulnar flexion, extension
Intercarpal	Plane	Diarthrosis	Carpal gliding
Interphalangeal-finger	Hinge	Diarthrosis	Finger flexion, extension
Interphalangeal-toe	Hinge	Diarthrosis	Toe flexion, extension
Intertarsal	Plane	Diarthrosis	Inversion, eversion
Intervertebral discs	Symphysis	Amphiarthrosis	Vertebra flexion, extension, rotation
Intervertebral facets	Plane	Diarthrosis	Vertebrae gliding
Manubriosternal[††]	Symphysis	Amphiarthrosis	Slight movement
Metacarpophalangeal	Condylar	Diarthrosis	Finger flexion, extension, abduction, adduction, circumduction
Metatarsophalangeal	Condylar	Diarthrosis	Toe flexion, extension, abduction, adduction
Pubic symphysis	Symphysis	Amphiarthrosis	Slight movement
Radiocarpal	Condylar	Diarthrosis	Hand flexion, extension, abduction, adduction, circumduction
Radioulnar-distal[†]	Pivot	Diarthrosis	Hand pronation, supination
Radioulnar-proximal	Pivot	Diarthrosis	Radius rotation
Sacroiliac	Plane	Diarthrosis	Slight gliding
Sternoclavicular[†]	Saddle	Diarthrosis	Clavicle protraction, retraction, elevation, depression
Sternocostal 1	Synchondrosis	Synarthrosis	None
Sternocostal 2–7	Plane	Diarthrosis	Rib gliding
Suture	Fibrous	synarthrosis	None
Talocrural (ankle)	Hinge	Diarthrosis	Foot dorsiflexion, plantar flexion
Tarsometatarsal	Plane	Diarthrosis	Metatarsal gliding
Temporomandibular[†]	Hinge/plane	Diarthrosis	Elevation, depression, protraction, retraction, excursion
Tibiofemoral[†]	Hinge	Diarthrosis	Leg flexion, extension, slight rotation
Tibiofibular-distal	Syndesmosis	Amphiarthrosis	Fibula rotation
Tibiofibular-proximal	Plane	Diarthrosis	Fibula gliding
Vertebrocostal	Plane	Diarthrosis	Rib gliding
Xiphisternal[††]	Symphysis	Amphiarthrosis	Slight movement

[†] Joint contains a meniscus.
[††] Joint usually becomes a synostosis later in life.

Name ______________________ Course Number: ________ Lab Section ______

Lab 8: Joints Worksheet

As you determine the specific action for each number in **Figure 8-3**, write that action in the blank by the appropriate number. Be sure to include the name of the part highlighted in red. Some examples are provided.

1. Foot dorsiflexion
2. ____________
3. ____________
4. ____________
5. ____________
6. ____________
7. ____________
8. Trunk (Torso) extension
9. ____________
10. ____________
11. ____________
12. ____________
13. ____________
14. Arm extension
15. ____________
16. ____________
17. ____________
18. ____________
19. ____________
20. Thigh abduction
21. ____________
22. ____________
23. ____________
24. ____________
25. ____________
26. Arm circumduction
27. ____________
28. ____________
29. ____________
30. ____________
31. Head rotation
32. Leg flexion
33. ____________
34. ____________
35. ____________
36. Leg hyperextension
37. ____________
38. ____________
39. ____________
40. ____________
41. ____________
42. ____________
43. ____________
44. Variety of motions

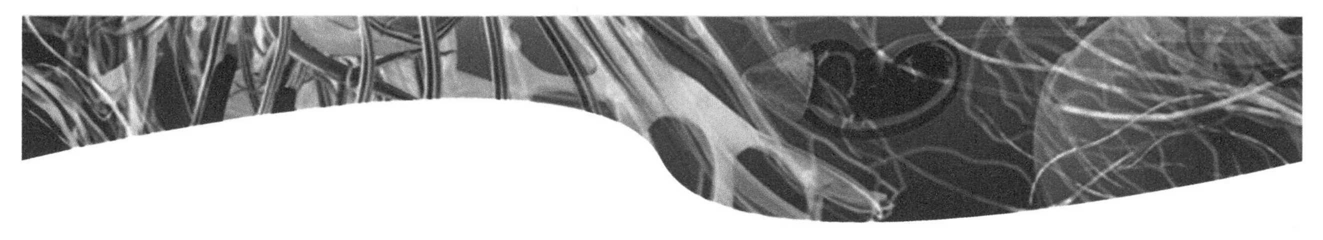

Lab 9 The Muscular System: Muscles of the Head and Torso

The muscular system includes *skeletal muscles,* which contain skeletal, or voluntary, muscle tissue. The majority of this lab deals with identifying skeletal muscles and learning their actions. However, it is also important to understand the anatomy and physiology of individual skeletal muscle cells. Before proceeding to the next section, look at **Table 9.1** and then label the skeletal muscle components in **Figure 9-1**.

Muscle Attachment Sites

There are approximately 700 skeletal muscles in the body and most have tendons that attach to the skeletal system. These muscles can move these components because one of the tendons remains relatively stationary during contraction. This site is the muscle's **origin**; whereas, the movable attachment site is the **insertion**. Your lab instructor will specify the muscles for which you need to know their origin and insertion.

Primary Role of a Skeletal Muscle

While you learn specific actions of skeletal muscles, it is important to understand the primary role the muscles play when they contract. A skeletal muscle may play one or more of the following roles:

- A **prime mover**, or **agonist** (AG-o-nist), is primarily responsible for a specific movement, such as flexion or extension.
- An **antagonist** (an-TAG-o-nist) opposes the action of an agonist. Shortening an agonist causes its antagonist to stretch.
- A **synergist** (SIN-er-jist) works together (*syn-*, together; *erg-*, work) with a prime mover to produce a certain movement.
- A **fixator** (FIK-sā-ter) is a synergist that stabilizes or "fixes" a joint (holds it steady) while a prime mover moves another joint. While large superficial back muscles move the arm to row a boat, deep back muscles keep the vertebral column straight.

Over thousands of years, a few skeletal muscles have lost important roles in the body. A **vestigial** (ves-TIH-jul; "trace") muscle is a remnant that may have had a more significant function in past human history but now is no longer considered necessary. Some examples of vestigial muscles include the *auricularis* in the head, the *palmaris longus* in the forearm, and the *plantaris* in the leg.

Features Associated with Muscles

Several features are associated with skeletal muscles. A **tendon** is a strong, slender band of dense, regular connective tissue that anchors

Table 9.1 Components of Skeletal Muscles and Skeletal Myofibers	
A band	Dark band in skeletal/cardiac myofiber; overlap of myofilaments
Actin	Contractile, filamentous protein and major component of thin filament
Endomysium	Dense irregular connective tissue covering individual myofibers
Epimysium	Dense irregular connective tissue covering an entire muscle
Fascicle	Bundle of myofibers within a muscle
H zone	Light-colored band in the center of an A band; contains only thick filaments
I band	Light-colored band around a Z disc; contains only thin filaments and titin
M line	In middle of A band; consists of protein *myomesin* that keeps thick filaments in line
Myofiber	Muscle cell
Myofibril	Cylindrical organelle within skeletal myofibers; composed of sarcomeres
Myofilament	Filamentous proteins that interact to produce muscle movement
Myosin	Contractile protein of thick myofilaments; has cross bridges that attach to actin
Perimysium	Dense irregular connective tissue covering a fascicle
Sarcolemma	Muscle cell membrane
Sarcomere	Functional unit of skeletal myofiber; between two Z discs; make up myofibril
Sarcoplasm	Muscle cell cytoplasm
Sarcoplasmic reticulum	(SR); Endoplasmic reticulum in a myofiber; stores Ca^{2+} for muscle contraction
Thin filament	Made of actin, tropomyosin, and troponin
Titin	Protein that attaches myosin to Z disc; largest polypeptide known
Transverse tubule	(T tubule); Invagination of a muscle cell's membrane; pathway for impulse to SR
Tropomyosin	Regulatory protein that covers myosin-binding sites on actin to prevent contraction
Troponin	Regulatory protein attached to tropomyosin; when bound to Ca^{2+}, troponin moves tropomyosin off myosin-binding sites so myosin cross bridges can attach to actin
Z disc	Protein disc that separates sarcomeres; anchoring site for actin and titin
Zone of overlap	Region on either side of H zone where thin and thick filaments overlap

a skeletal muscle to a bone. (Recall, a *ligament* is similar in structure but attaches a bone to a bone.) Some connective tissue bands are not cylindrical but are flattened out, more like a ribbon. An example is the *linea alba* (LIN-ē-a AL-ba; *linea*, "line"; *alb-*, "white") that runs along the midline of the abdomen to separate the right and left *rectus abdominis* muscles. Three shorter, horizontal bands of connective tissue called *tendinous intersections* divide each rectus abdominis into four rectangular segments.

An **aponeurosis** (a-po-nur-O-sis; "from sinew") is a broad, flat tendon sheath attached to a flat muscle. One example is the *epicranial aponeurosis* (ep-ih-KRĀ-nē-ul; "above skull"), also called the *galea aponeurotica*: gal-Ē-uh ā-pō-nūr-ō-tih-ka; "helmet"), on top of the skull. Another example is the *aponeurosis of the external oblique*, which extends from the external oblique muscle to the linea alba and inguinal ligament.

Fascia (FASH-uh; "band") is a fibrous sheet that encloses a group of muscles. **Deep fascia** (FASH-uh; "band") separates muscles from one another. When separating muscles, the deep fascia pulls apart like cotton candy and looks like a spiderweb. Recall, *superficial fascia* is adipose tissue that comprises the hypodermis (subcutaneous tissue). The **thoracolumbar** (or **lumbodorsal**) **fascia** is a fibrous sheet in the lower back that anchors the latissimus dorsi muscle to vertebrae.

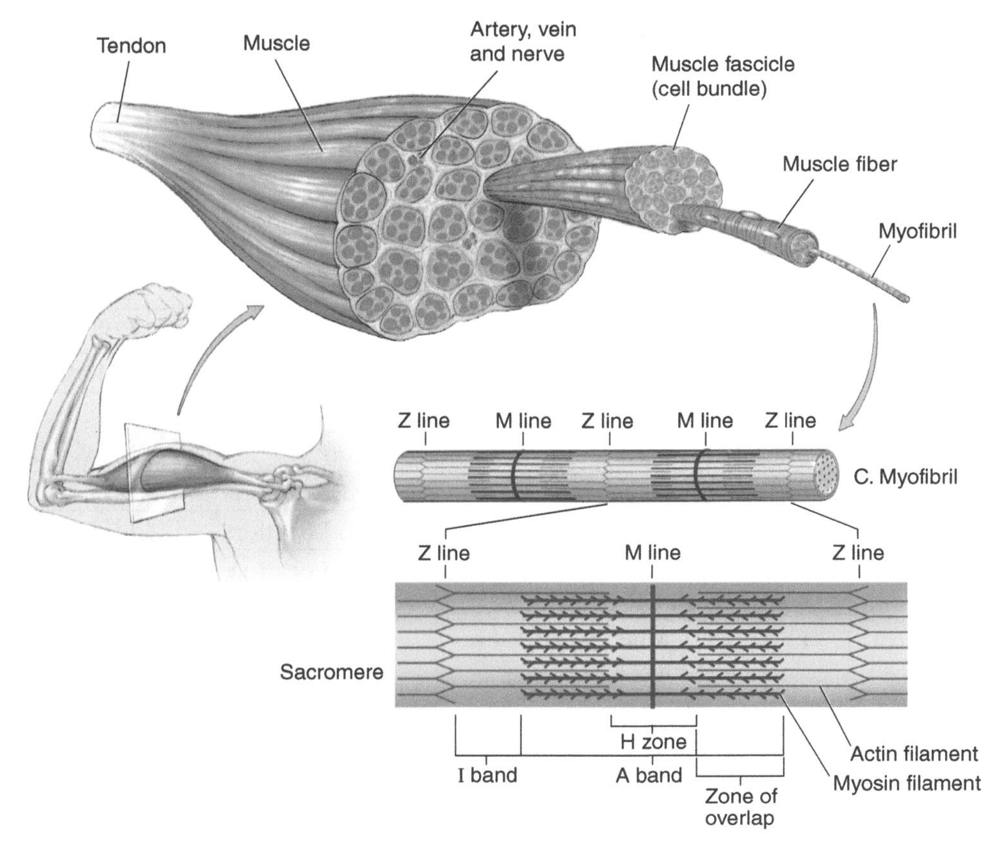

Figure 9-1 Components of a skeletal muscle

Muscle Names

A muscle's name may relate to the muscle's *location, shape, size, orientation, number of origins*, or *action*. See **Table 9.2** for this and more information about muscle names. Then review the muscles in **Tables 9.3** and **9.4** and identify them in **Figures 9.2 and 9.3.**

Dissections

Your lab instructor may provide a pre-dissected fetal pig or cat on which you are to identify selected skeletal muscles. A handout for the information you need to know will be provided.

Electromyography

Electromyography is the study of electrical changes that occur on muscle cells. A muscle contraction is a *mechanical* event and it results in the development of tension. However, before a muscle can contract and produce tension, it must experience an electrical change called *depolarization*, which is immediately followed by a *repolarization*. The depolarization and repolarization of muscle cells can be detected by special equipment and then recorded on a graph called an **electromyogram**. In contrast, a graph that shows the amount of tension that a contracting muscle is exerting is called a **myogram**. The lab instructor may present information about how to record data for an EMG and/or myogram. Handouts will be provided for this exercise.

Table 9.2 Names of Selected Head and Torso Muscles

Muscle	Pronunciation	Basis of Name	Literal meaning
Auricularis	aw-rik-ū-LAR-is	Location	Ear
Buccinator	BUK-si-nā-tor	Location	Cheek
Corrugator supercilii	KOR-uh-gā-tor sū-per-SIL-ē-ī	Shape/Location	Wrinkle above eyelid
Deltoid	DEL-toyd	Shape	Delta (diamond) shaped
Depressor anguli	dē-PRESS-or ANG-ū-lī	Action/Location	Pull down angle
Depressor labii	dē-PRESS-or LĀ-bē-ī	Action/Location	Pull down lip
External oblique	External ō-BLĒK	Location/Orientation	Diagonal
Iliocostalis	il-ē-ō-kos-TAL-is	Location	Flank rib
Latissimus dorsi	la-TIS-i-mus DOR-sī	Shape/Location	Broad back
Levator anguli	LĒ-vā-tor ANG-ū-lī	Action/Location	Lift up angle
Levator labii superioris	LĒ-vā-tor LĀ-bē-īsu-PĒR-ē-OR-is	Action/Location	Lifts top lip
Levator labii superioris alaeque nasi	LĒ-vā-tor LĀ-bē-ī sū-PĒR-ē-OR-is al-ā-kā NA-sī	Action/Location	Lift lip above wing of nose
Levator scapulae	LĒ-vā-tor SKAP-ū-lē	Action/Location	Lift up shoulder blade
Longissimus	long-JIS-ih-mus	Shape	Length
Longus colli	LONG-us KOL-ī	Shape/Location	Long neck
Masseter	MAS-ih-ter	Action	Chew
Mentalis	men-TAL-is	Location	Chin
Multifidus	mul-TIF-ih-dus	Shape	Many divisions
Nasalis	nā-SAL-is	Location	Nose
Occipitofrontalis	ok-SIP-i-tō-frun-TAL-is	Location	Back of head and front
Orbicularis oculi	or-bik-ū-LĀR-is OK-ū-lī	Location	Around eye
Orbicularis oris	or-bik-ū-LĀR-is OR-is	Location	Around lip
Pectoralis	pek-tor-AL-is	Location/Size	Chest
Platysma	pluh-TIZ-muh	Shape	Flat plate
Procerus	prō-SĒ-rus	Shape	Slender
Pterygoid	TĀR-ih-goyd	Location	Wing-like
Quadratus lumborum	qwa-DRĀ-tus lum-BOR-um	Shape/Location	4-sides; lumbar region
Rectus abdominis	REK-tus ab-DOM-ih-nus	Shape/Location	Straight abdomen
Rhomboid	ROM-boyd	Shape	Parallel sides
Risorius	rī-SOR-ē-us	Action	Laugh
Scalene	SKĀ-lēn	Shape	Unequal sides
Semispinalis	sim-ē-spī-NAL-is	Shape/Location	Half spine
Serratus	ser-Ā-tus	Shape/Location	Jagged front
Spinalis	spī-NAL-is	Location	Spine (vertebral column)
Splenius	SPLĒN-ē-uz	Shape	Bandage
Sternocleidomastoid	ster-nō-klī-dō-MAS-toyd	Location	Sternum, clavicle, mastoid

Table 9.2 Names of Selected Head and Torso Muscles *(Cont'd)*

Subclavius	sub-KLĀ-vē-us	Location	Below clavicle
Subscapularis	sub-skap-ū-LĀR-is	Location	Under scapula
Temporalis	tem-por-AL-is	Location	Time
Temporoparietalis	Tem-por-ō-puh-rī-ih-TAL-is	Location	Time wall
Transversus abdominis	trans-VER-sus ab-DOM-ih-nus	Orientation/Location	Across abdomen
Trapezius	tra-PĒ-zē-us	Shape	Trapezoid
Zygomaticus	zī-gō-MAT-ih-kus	Location	Zygomatic bone

Table 9.3 Muscles of the Head

Muscle	Origin	Insertion	Major Action
MUSCLES THAT MOVE THE SCALP			
Auricularis*	Scalp skin	Skin around ear	Moves ear
Corrugator supercilii (a deep muscle)	Frontal bone	Skin of brow	Wrinkles brow
Occipitofrontalis (includes frontalis and occipitalis)	Epicranial aponeurosis	Scalp skin	Raises eyebrows; wrinkles forehead
Procerus	Nasal bone	Frontalis muscle	Lowers eyebrows; wrinkles forehead
Temporoparietalis	Scalp skin	Epicranial aponeurosis	Tightens scalp
MUSCLES THAT MOVE THE EYELIDS AND NOSE			
Levator labii superioris alaeque nasi	Maxilla	Nose, upper lip	Flares nostril
Levator palpebrae	Orbit	Upper eyelid	Raises eyelid
Nasalis	Maxilla	Bridge of nose	Flares nostril
Orbicularis oculi	Orbit	Eyelids	Closes eye
MUSCLES THAT MOVE THE CHEEKS, LIPS, AND MOUTH			
Buccinator	Mandible, maxilla	Orbicularis oris	Compresses cheek
Depressor anguli	Mandible	Mouth angle	Lowers mouth angle
Depressor labii	Mandible	Lower lip	Curls lower lip
Levator anguli	Maxilla	Mouth angle	Raises mouth angle
Levator labii superioris	Maxilla	Upper lip	Curls upper lip
Levator labii superioris alaeque nasi	Maxilla	Nose, upper lip	Raises upper lip
Mentalis	Mandible	Skin of chin	Raises lower lip; wrinkles chin
Orbicularis oris	Muscles around mouth	Lips	Closes/protrudes lips
Risorius	Cheek fascia	Mouth	Pulls mouth angle
Zygomaticus major	Zygomatic bone	Mouth	Raises mouth angle
Zygomaticus minor	Zygomatic bone	Upper lip	Elevates upper lip

(Continued)

Table 9.3 Muscles of the Head *(Cont'd)*

MUSCLES THAT MOVE THE MANDIBLE			
Masseter	Zygomatic arch	Mandible	Raises jaw
Platysma	Skin of chest	Mandible	Lowers jaw; tenses neck skin
Pterygoids (deep)	Sphenoid bone	Mandible	Raises jaw, lateral excursion
Temporalis (deep)	Temporal bone	Mandible	Raises jaw

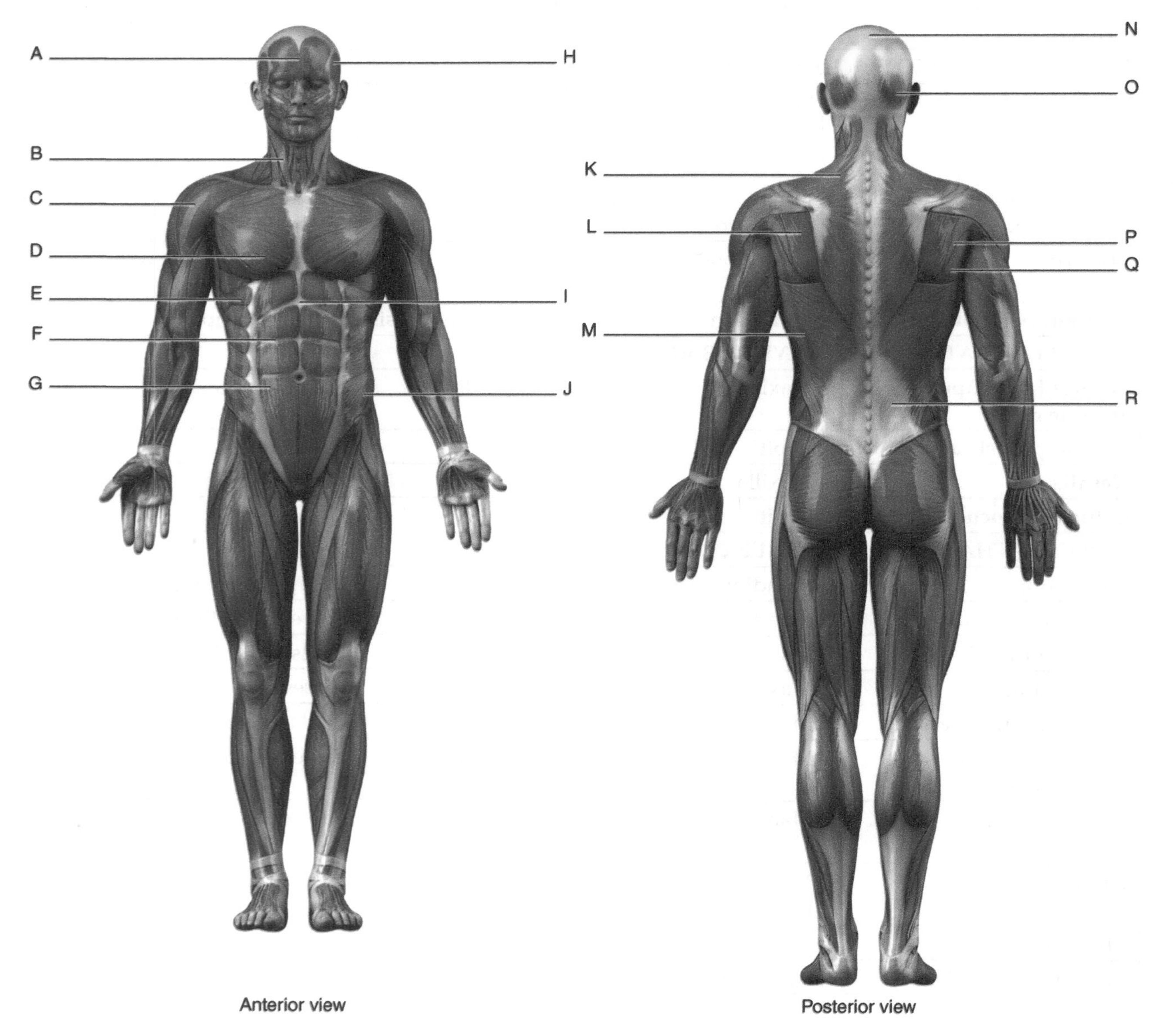

Figure 9-2 Selected muscles of head/torso.

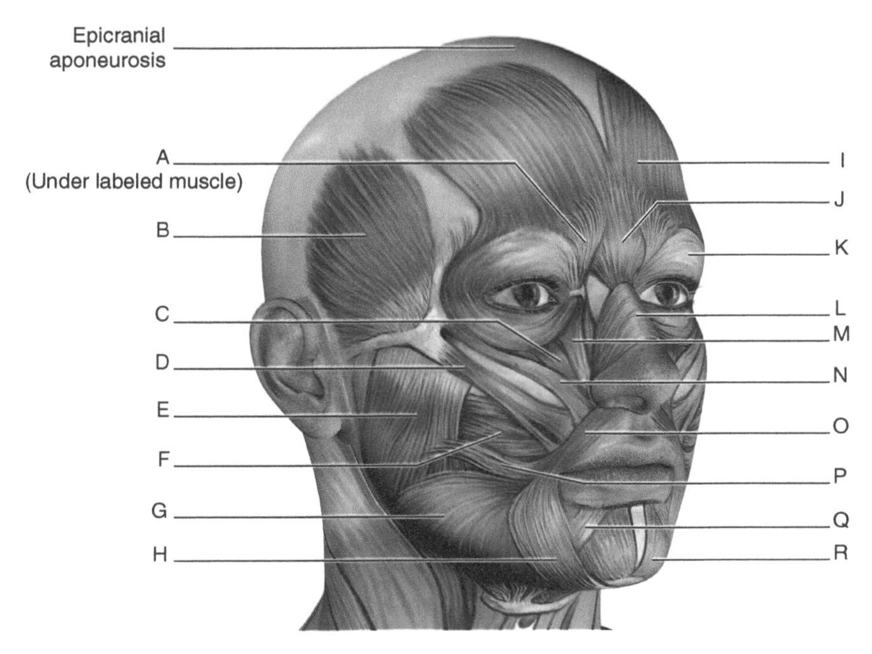

Figure 9-3 Muscles of the head.

Table 9.4 Muscles of the Neck and Torso

Muscle	Origin	Insertion	Major Action
MUSCLES THAT MOVE THE HEAD AND BACK			
Erector spinae group			
Iliocostalis group	Ribs/ilium	Vertebrae, ribs	Flexes/extends neck/back
Longissimus group	Vertebrae	Temporal bone/vertebrae/ ribs	Flexes/extends neck/back
Spinalis group	Vertebrae	Occipital bone/ vertebrae	Extends neck/back
Longus colli*	Vertebrae	Atlas	Flexes neck
Multifidus	Vertebrae	Vertebrae	Back lateral flexion/ extension
Quadratus lumborum	Ilium	Vertebrae	Laterally flexes/extends spine
Scalene group	Cervical vertebrae	Ribs	Flexes/rotates neck; raises ribs
Semispinalis group	Vertebrae	Occipital bone/ vertebrae	Rotates/extends neck/ back
Serratus posterior	Vertebrae	Ribs	Extends/rotates back
Splenius group	Nuchal ligament, vertebrae	Temporal bone/ vertebrae	Rotates/extends neck
Sternocleidomastoid	Sternum/clavicle	Temporal bone	Rotates/flexes neck
Trapezius	Occipital bone/nuchal ligament/vertebrae	Clavicle/scapula	Rotates/raises/lowers/ abducts scapula; tilts/ extends neck

(Continued)

Table 9.4 Muscles of the Neck and Torso *(Cont'd)*

MUSCLES THAT MOVE THE ABDOMEN AND BACK			
External oblique	Ribs	Ilium/linea alba	Tenses abdomen; flexes/ rotates back
Internal oblique	Ilium/inguinal liga- ment/ thoracolumbar fascia	Ribs/linea alba	Tenses abdomen; flexes/ rotates back
Rectus abdominis	Pubis	Xiphoid process/ribs	Flexes trunk; compresses abdomen
Transversus abdominis	Ilium/thoracolumbar fascia/ribs	Linea alba/pubis	Tenses abdomen
MUSCLES THAT CHANGE THORACIC VOLUME			
Diaphragm	Sternum/ribs	Central tendon	Increases thoracic volume
External intercostals	Ribs	Ribs	Elevates rib cage; increases thoracic volume
Internal intercostals	Ribs	Ribs	Depresses rib cage; decreases thoracic volume
MUSCLES THAT MOVE THE PECTORAL GIRDLE			
Levator scapulae	Cervical vertebrae	Scapula	Raises/rotates scapula
Pectoralis minor	Scapula	Ribs	Protracts scapula, raises ribs
Rhomboid group	Vertebrae	Scapula	Adducts/rotates scapula
Serratus anterior	Ribs	Scapula	Protracts/rotates scapula
Subclavius	1st rib	Clavicle	Lower/protracts clavicle
Trapezius	Occipital bone/nuchal ligament/vertebrae	Clavicle/scapula	Rotates/raises/lowers/ abducts scapula; tilts/ extends neck
TORSO MUSCLES THAT MOVE THE ARM			
Coracobrachialis	Scapula coracoid process	Humerus-shaft	Adduct/flex arm
Deltoid	Clavicle/scapula	Deltoid tuberosity	Abduct/flex/extend arm
Latissimus dorsi	Thoracolumbar fascia	Humerus intertubular sulcus	Extend/adduct/med. rotate arm
Pectoralis major	Clavicle/sternum/ribs	Intertubercular groove	Adduct/medially rotate arm
Rotator Cuff Muscles (4)			
Infraspinatus	Infraspinous fossa	Greater tubercle	Laterally rotate arm
Supraspinatus	Supraspinous fossa	Greater tubercle	Abducts arm
Subscapularis	Subscapular fossa	Lesser tubercle	Medially rotate arm
Teres minor	Scapula lateral border	Greater tubercle	Laterally rotate arm
Teres major	Scapula lateral border	Lesser tubercle	Adduct/medially rotate arm
TORSO MUSCLES THAT MOVE THE THIGH			
Iliopsoas			
Iliacus	Ilium/sacrum	Lesser trochanter	Flexes thigh/trunk
Psoas major	Vertebrae	Lesser trochanter	Flexes thigh/trunk

*Often injured in accidents that cause whiplash

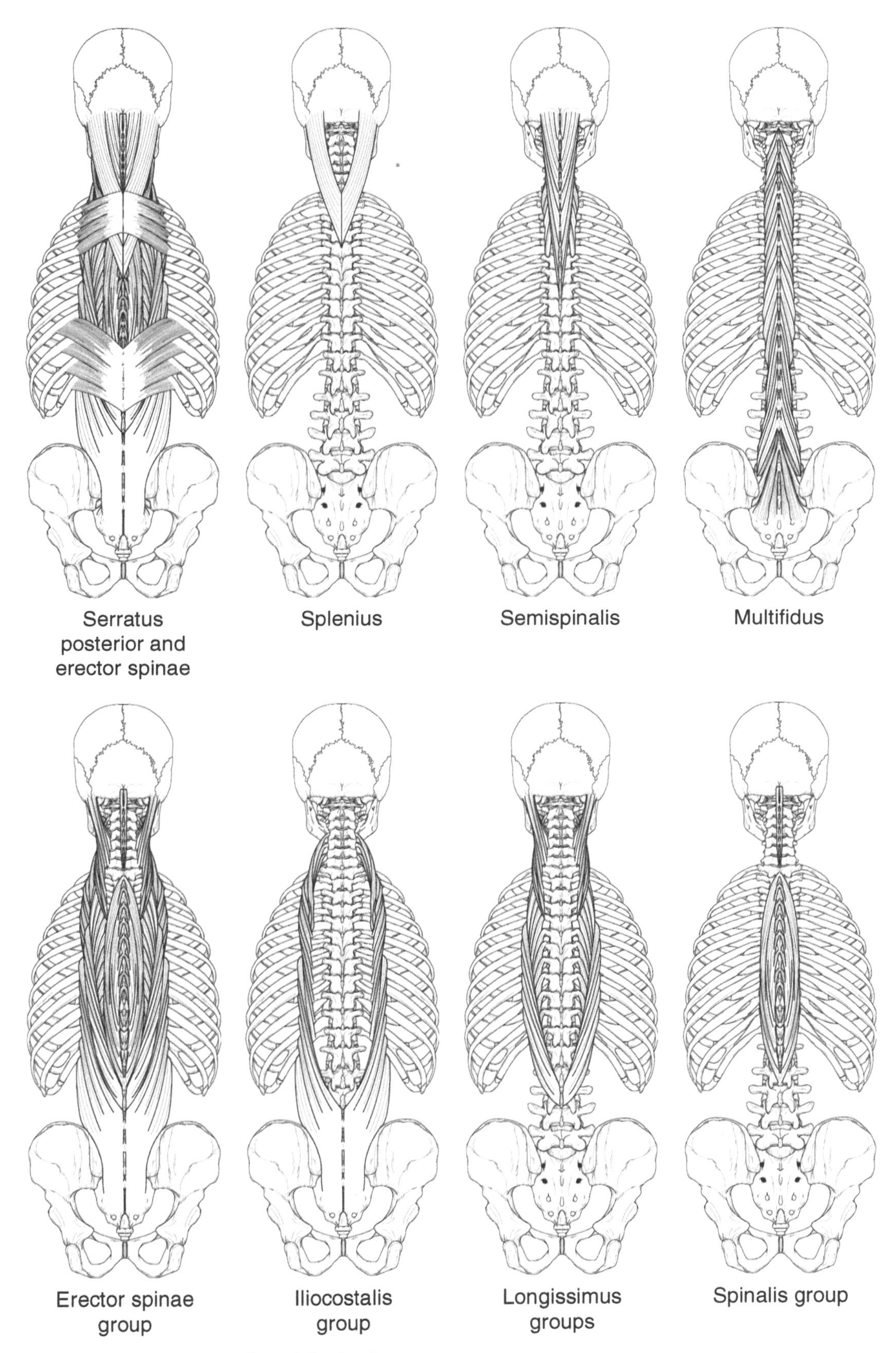

Figure 9-4 Deep muscles of the back.

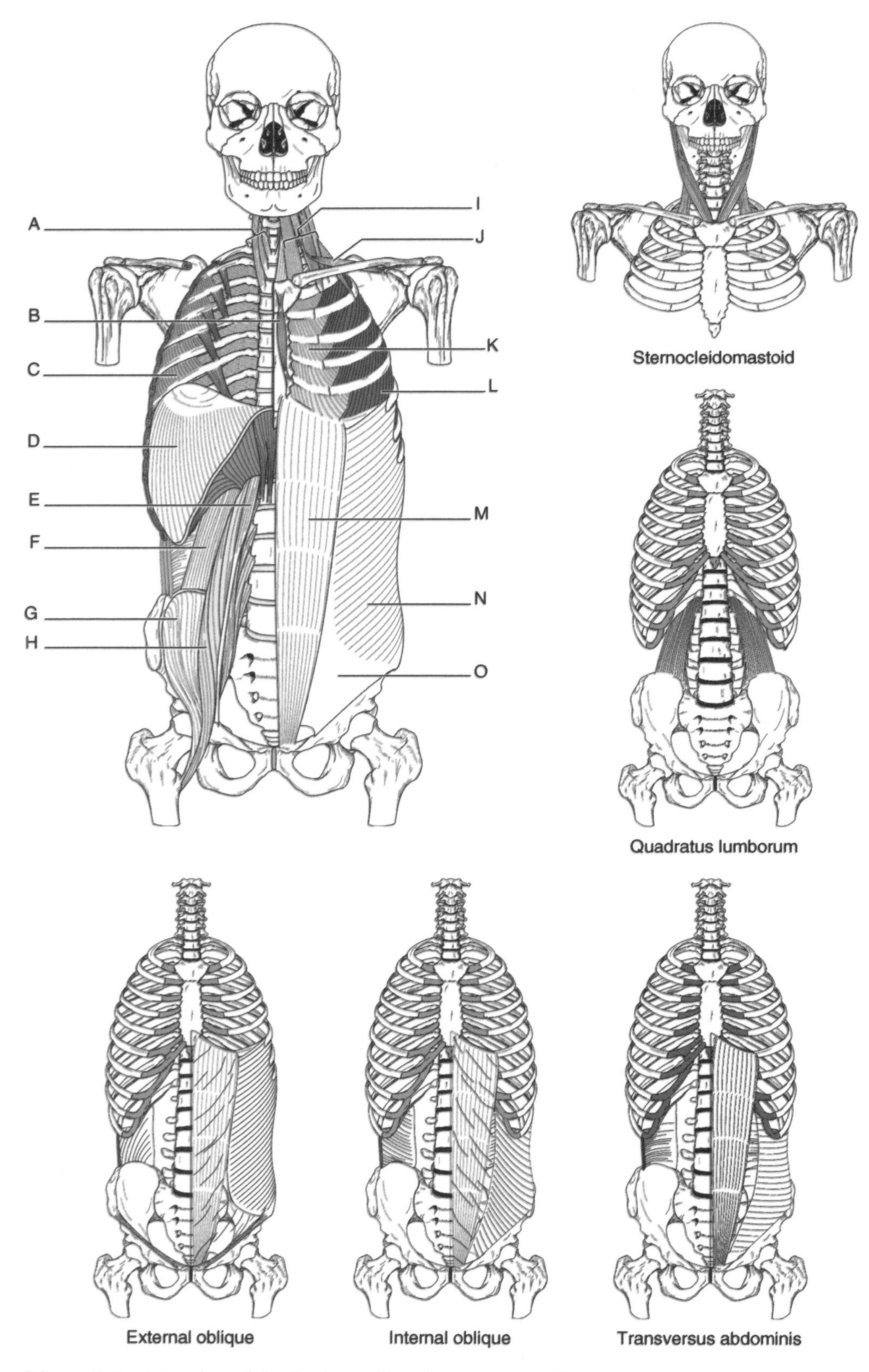

Figure 9-5 Muscles of the body wall and anterior neck.

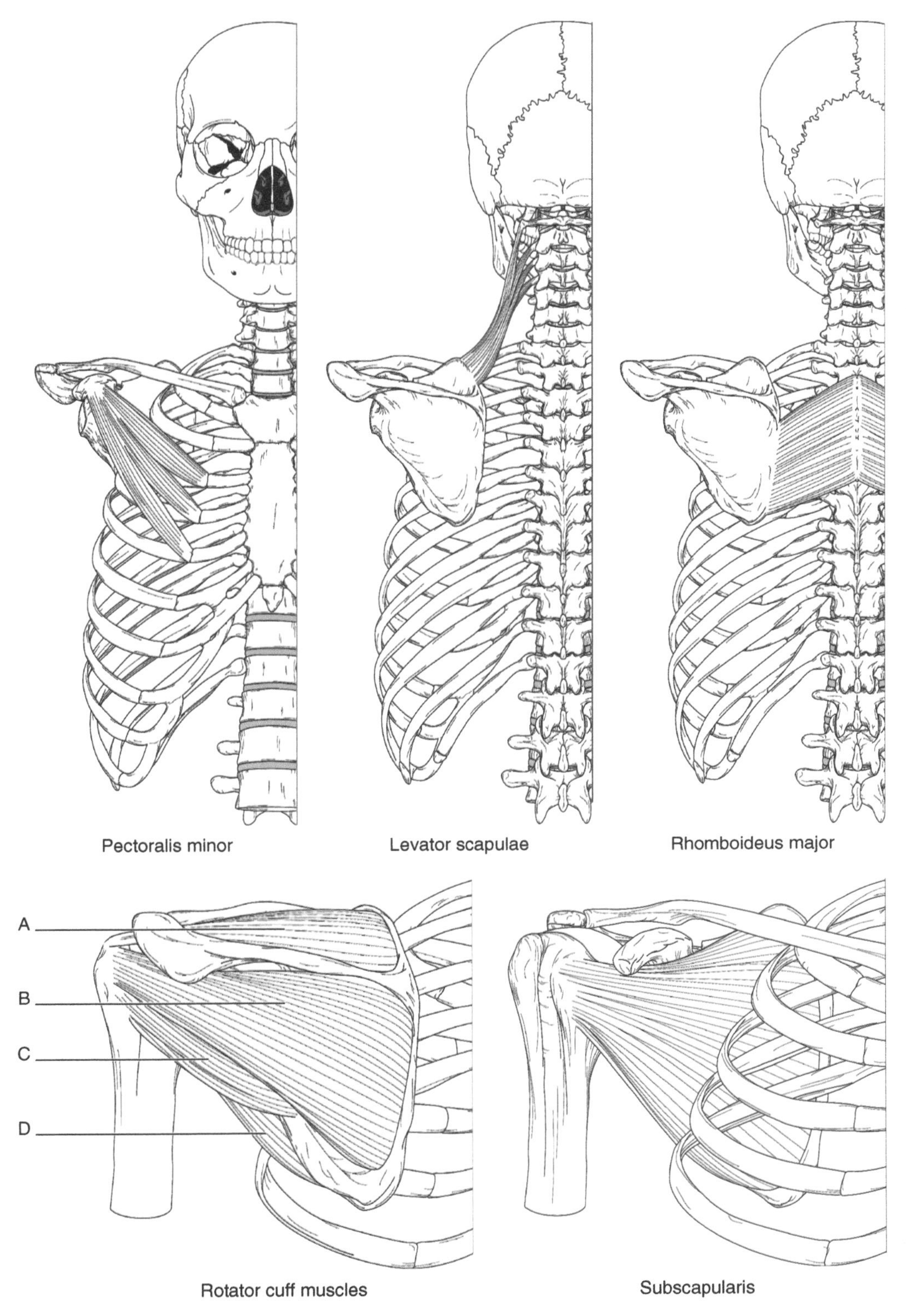

Figure 9-6 Torso muscles that move the scapula and arm.

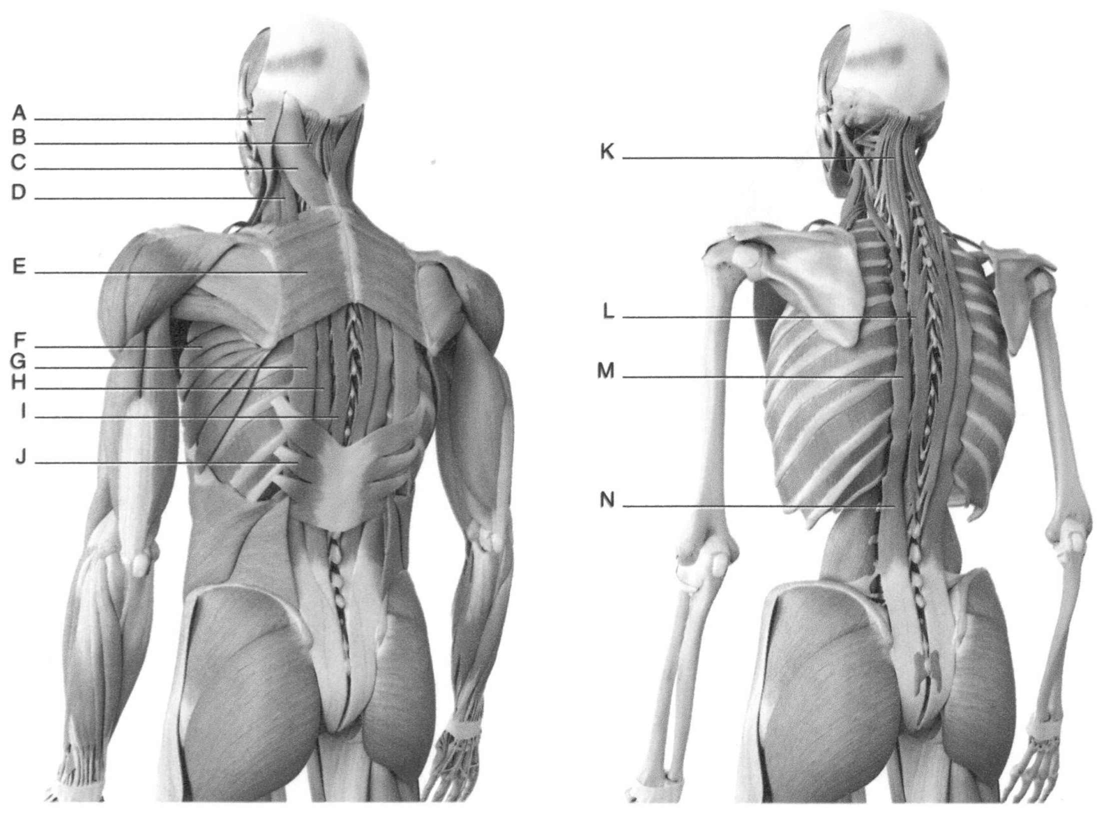

Figure 9-7 Superficial and deep back muscles.

Name ____________________ Course Number: ________ Lab Section ______

Lab 9: Muscles of the Head and Torso Worksheet

Write out the names of the muscles labeled in the selected figures.

Figure 9-2.

A. ____________________
B. ____________________
C. ____________________
D. ____________________
E. ____________________
F. ____________________
G. ____________________
H. ____________________
I. ____________________
J. ____________________
K. ____________________
L. ____________________
M. ____________________
N. ____________________
O. ____________________
P. ____________________
Q. ____________________
R. ____________________

Figure 9-3.

A. ____________________
B. ____________________
C. ____________________
D. ____________________
E. ____________________
F. ____________________
G. ____________________
H. ____________________
I. ____________________
J. ____________________
K. ____________________
L. ____________________
M. ____________________
N. ____________________
O. ____________________
P. ____________________
Q. ____________________
R. ____________________

Figure 9-6.

A. ____________________
B. ____________________
C. ____________________
D. ____________________

Figure 9-7.

A. ____________________
B. ____________________
C. ____________________
D. ____________________
E. ____________________
F. ____________________
G. ____________________
H. ____________________
I. ____________________
J. ____________________
K. ____________________
L. ____________________
M. ____________________
N. ____________________

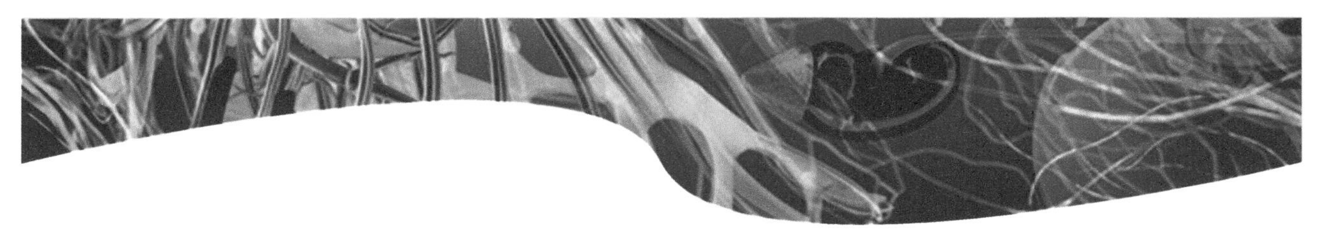

Lab 10 The Muscular System: Muscles of the Limbs and Hip Region

Overview of Limb and Hip Muscles

While the previous lab presented muscles anchored to the axial skeleton and pectoral girdle, this lab presents muscles anchored in the upper and lower appendages and selected muscles of the pelvic girdle. Since a number of different muscles have similar actions, but a single muscle may also have a variety of actions, the tables in this lab are in a slightly different format from those seen earlier. For example, tables that help you locate muscles will provide only muscle's origin and insertion. Additional tables will then list the actions of those muscles.

Muscle Names

The last lab provided a table that included information about pronunciation and literal meanings of names for muscles in the head and torso. **Table 10.1** provides similar information for names of muscles in the limbs and hip. **Figure 10-1** provides an overview of superficial muscles of the limbs and hip region.

Table 10.1 Names of Selected Limb and Hip Muscles

Muscle	Pronunciation	Basis of Name	Literal meaning
Abductor	ab-DUK-tor	Action	Take away
Adductor	ad-DUK-tor	Action	Bring back
Anconeus	an-KŌ-nē-us	Location	Elbow
Biceps	BĪ-seps	Origins	Two heads
Brachi-	brā-kē	Location	Arm
Brevis	BREV-is	Shape	Short
Carpi	KAR-pē	Location	Wrist
Digit-	DIJ-jit	Location	Digit (finger or toe)
Extensor	eks-TEN-sor	Action	Straighten
Externus	eks-TER-nus	Location	Outer

(Continued)

Table 10.1 Names of Selected Limb and Hip Muscles *(Cont'd)*

Fascia	FASH-ē	Shape	Band
Femoris	FEM-or-is	Location	Thigh
Flexor	FLEK-sor	Action	Bend
Gastrocnemius	gas-trok-NĒ-mē-us	Shape/Location	Belly/calf of leg
Gemellus	jih-MEL-us	Shape	Little twin
Gluteus	GLŪ-tē-us	Location	Buttock
Gracilis	GRAS-ih-lis	Shape	Slender
Iliacus	il-ē-Ā-kus	Location	Ilium
Indicis	EN-dih-kis	Location	Index finger
Intermedius	in-ter-MĒ-dē-us	Location	Between
Interossei	en-ter-OS-ē-Ī	Location	Between bones
Latae	LĀ-tē	Location	Side
Lateralis	lat-er-AL-is	Location	Side
Longus	LONG-us	Shape	Long
Lumbrical	LUM-brih-kulz	Shape	Earthworms
Magnus	MAG-nus	Size	Large
Maximus	MAX-ih-mus	Size	Large
Medi-	MĒ-dēi	Location	Between
Membranous	MEM-bruh-nus	Shape	Membrane
Minim-	MEN-em	Size	Little
Obturator	OB-ter-ā-tor	Action/Location	Stop up/outside
Opponens	ō-PŌ-nenz	Action	Oppose
Palmaris	pal-MĀR-is	Location	Palm
Pectineus	pek-TEN-ē-us	Shape	Comb
Piriformis	pēr-ih-FOR-mis	Shape	Pear-like
Plantaris	plan-TĀR-is	Location	Foot
Pollicis	POL-ih-sis	Location	Thumb
Popliteus	pop-LIH-tē-us	Location	Behind knee
Pronator	PRŌ-nā-tor	Action	Prone-down
Psoas	SŌ-us	Location	Loin (lower back)
Quadratus	kwod_RĀ-tus	Shape	Four sides
Radialis	rā-dē-AL-is	Location	Radius
Rectus	REK-tus	Action	Straighten
Sartorius	sar-TOR-ē-us	Action	Tailor (crossing legs)
Semi-	sem-ē	Shape	Half
Supinator	SŪ-pih-nā-tor	Action	Supine (up)
Tendinosus	TEN-din-Ō-sis	Shape	Sinew (tendon)
Tensor	TEN-sor	Action	Tighten
Teres	TAĪR-ēz	Action	Prone (down)
Triceps	TRĪ-seps	Origins	Three heads
Ulnaris	ul-NĀR-is	Location	Ulna
Vastus	VAS-tus	Size	Large

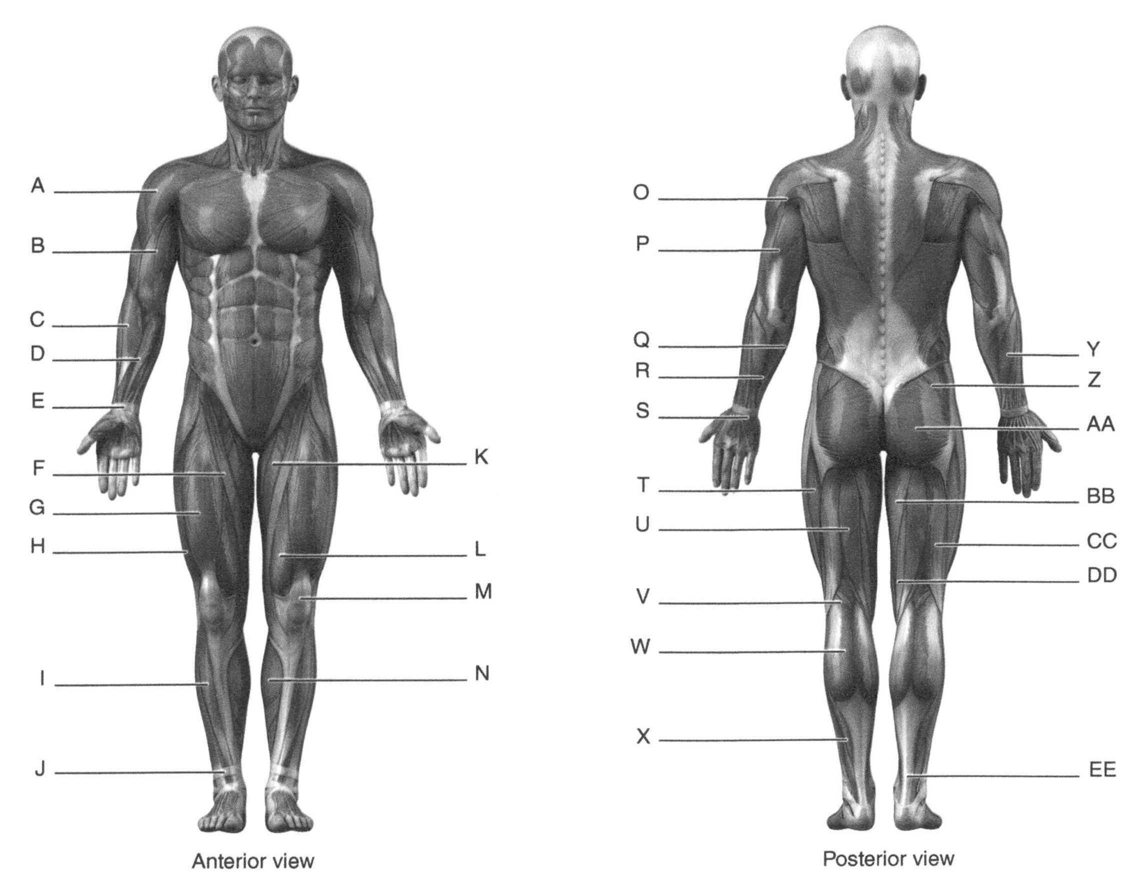

Figure 10-1 Selected muscles of the limbs and hip.

Muscles of the Upper Limb

The upper limb includes the **brachium** (*arm*—from the shoulder to the elbow), **antebrachium** (*forearm*—from the elbow to the wrist) and the **hand**, which includes the *carpus* (wrist), *hand*, and *fingers*. In the previous lab, you learned about certain torso muscles that move the arm. These include the *deltoid*, *latissimus dorsi*, and the *rotator cuff muscles*. Look back at the previous lab if you need to review the location of torso muscles that move the arm.

Muscles in the upper limb may have actions on different parts of the upper limb. For example, several muscles in the arm can move the arm and the forearm. Most muscles in the forearm, however, move only the hand and fingers. Since these forearm muscles are not located in the hand or fingers, but simply have tendons reaching into those regions, we refer to them as *extrinsic* (eks-TREN-sik; "outside") hand muscles. Muscles located entirely within the hand and move different parts of the hand itself, including the fingers, are *intrinsic* (in-TREN-sik; "inside") hand muscles. Intrinsic muscles that form a bulge between the thumb and wrist are **thenar muscles** (*thenar*, palm of the hand). Intrinsic muscles that form the bulge between the little finger and wrist are **hypothenar muscles**.

Additional features associated with upper limb

There are several unique features associated with muscles of the upper limb that deserve mentioning here.

The **anatomical snuffbox** is a depression on the posterior, lateral part of the wrist seen on the pronated hand while abducting the thumb. It is between the forearm's *adductor pollicis* and *extensor pollicis* tendons, both of which lead out to the thumb.

A **retinaculum** (ret-in-AK-ū-lum; "band") is a flat band encircling the wrist or ankle and it holds muscle tendons of the arm or leg in place when the muscles contract. A *flexor retinaculum* holds in the tendons of flexor muscles, whereas, an *extensor retinaculum* holds in tendons of extensor muscles.

A **vestigial muscle** in the forearm is the *palmaris longus*. It may be present in both forearms, only one forearm, or not present in either forearm. Its absence does not significantly affect grip strength. To see if you have this muscle, place your fingers under the edge of the table with the palm up. Press upward. If you have a palmaris longus muscle, then you will see its narrow tendon rise upward at the midline of the wrist. If it is not present, then you will see only tendons elevated away from the forearm's midline.

On the next few pages, **Tables 10.2** through **10.5** provide information about attachment sites and actions of upper limb muscles. After the tables, you will find figures that illustrate these muscles. When you are able to recognize these muscles, label them in **Figures 10-2, 10-3, and 10-4.**

Table 10.2 Muscles That Move the ARM

Muscle	Origin	Insertion
Biceps brachii	Coracoid process/glenoid rim	Radial tuberosity
Coracobrachialis	Scapula coracoid process	Humerus-shaft
Deltoid	Clavicle/scapula	Deltoid tuberosity
Infraspinatus	Infraspinous fossa	Greater tubercle
Latissimus dorsi	Thoracolumbar fascia	Intertubular sulcus
Pectoralis major	Clavicle/sternum/ribs	Intertubercular groove
Subscapularis	Subscapular fossa	Lesser tubercle
Supraspinatus	Supraspinous fossa	Greater tubercle
Teres major	Scapula lateral border	Lesser tubercle
Teres minor	Scapula lateral border	Greater tubercle
Triceps brachii	Glenoid fossa rim/ humerus shaft	Olecranon process

	Arm Flexion	**Arm Extension**	**Arm Abduction**	**Arm Adduction**	**Arm Medial Rotation**	**Arm Lateral rotation**
Biceps brachii	✓					
Coracobrachialis	✓			✓		
Deltoid	✓	✓	✓		✓	✓
Infraspinatus			✓			✓
Latissimus dorsi		✓		✓	✓	
Pectoralis major	✓			✓	✓	
Subscapularis					✓	
Supraspinatus			✓			
Teres major		✓		✓	✓	
Teres minor				✓		✓
Triceps brachii		✓		✓		

Table 10.3 Muscles That Move the FOREARM

	Origin	Insertion
Anconeus	Lateral epicondyle	Olecranon process
Biceps brachii	Coracoid process/glenoid fossa rim	Radial tuberosity

Table 10.3 Muscles That Move the FOREARM *(Cont'd)*

Brachialis	Humeral shaft		Ulna	
Brachioradialis	Lateral epicondyle		Radius styloid process	
Pronator quadratus	Ulnar shaft		Distal radius	
Pronator teres	Medial epicondyle		Radius	
Supinator	Lateral epicondyle		Below radial tuberosity	
Triceps brachii	Glenoid fossa rim/ humeral shaft		Olecranon process	
	Forearm Flexion	**Forearm Extension**	**Forearm Pronation**	**Forearm Supination**
Anconeus		✓		
Biceps brachii	✓			✓
Brachialis	✓			
Brachioradialis	✓			
Pronator quadratus			✓	
Pronator teres	✓		✓	
Supinator				✓
Triceps brachii		✓		

Table 10.4 Forearm Muscles That Move the HAND at the Wrist

Muscle	**Origin**		**Insertion**	
Abductor pollicis longus	Proximal radius and ulna		1st metacarpal	
Extensor carpi radialis	Humerus lateral epicondyle		2nd/3rd metacarpals	
Extensor carpi ulnaris	Humerus lateral epicondyle		5th metacarpal	
Extensor digitorum	Humerus lateral epicondyle		Phalanges	
Flexor carpi radialis	Humerus lateral epicondyle		2nd/3rd metacarpals	
Flexor carpi ulnaris	Humerus lateral epicondyle		Carpals, 5th metacarpal	
Palmaris longus*	Humerus medial epicondyle		Retinaculum, palmar aponeurosis	
Pronator quadratus	Humerus medial epicondyle		Radial shaft	
Pronator teres	Medial epicondyle/ coronoid process		Radial shaft	
Supinator	Humerus lateral condyle/ulna		Radial shaft	
	Hand Flexion	**Hand Extension**	**Hand Abduction**	**Hand Adduction**
Abductor pollicis longus			✓	
Extensor carpi radialis		✓	✓	
Extensor carpi ulnaris		✓		✓
Extensor digitorum		✓		
Flexor carpi radialis	✓		✓	
Flexor carpi ulnaris	✓			✓
Flexor digitorum muscles	✓			
Palmaris longus	✓			

Table 10.5 Muscles That Move the FINGERS

Muscle	Origin	Insertion
Abductor digiti minimi	Carpals, flexor carpi tendon	Proximal phalanx
Abductor pollicis brevis	Carpals, flexor retinaculum	Proximal phalanx
Abductor pollicis longus*	Radius, ulna	1st metacarpal
Adductor pollicis	Carpals, metacarpals	Proximal phalanx
Dorsal Interossei	Metacarpals	Proximal phalanges
Extensor digiti minimi*	Humerus	Proximal phalanx
Extensor digitorum superficialis*	Humerus	Phalanges
Extensor indicis*	Ulna	Extensor digitorum tendon
Extensor pollicis brevis*	Radius & ulna	Proximal phalanx of digit 1
Extensor pollicis longus*	Radius & ulna	Distal phalanx of digit 1
Flexor digiti minimi brevis	Carpals, flexor retinaculum	Proximal phalanx
Flexor digitorum*	Humerus, radius, ulna	Phalanges
Flexor pollicis brevis	Carpals, flexor retinaculum	Proximal phalanx
Flexor pollicis longus*	Radius	Phalanx
Lumbricals	Flexor digitorum tendon	Extensor digitorum tendon
Opponens digiti minimi	Carpals, flexor retinaculum	5th metacarpal
Opponens pollicis	Carpals, flexor retinaculum	1st metacarpal
Palmar Interossei	Metacarpals	Proximal phalanges

Muscle	Flexion	Extension	Abduction	Adduction	Opposition
Abductor digiti minimi			✓		
Abductor pollicis muscles					
Adductor pollicis				✓	✓
Dorsal interossei**	✓	✓	✓		
Extensor digiti minimi		✓			
Extensor digitorum superficialis		✓			
Extensor indicis		✓			
Extensor pollicis muscles		✓			
Flexor digiti minimi brevis	✓				
Flexor digitorum muscles	✓				
Flexor pollicis muscles	✓				
Lumbricals	✓	✓			
Opponens digiti minimi					✓
Opponens pollicis					✓
Palmar interossei*	✓	✓		✓	

*An extrinsic hand muscle (located in forearm)

**Flexion at metacarpophalangeal joints and extension at interphalangeal joints

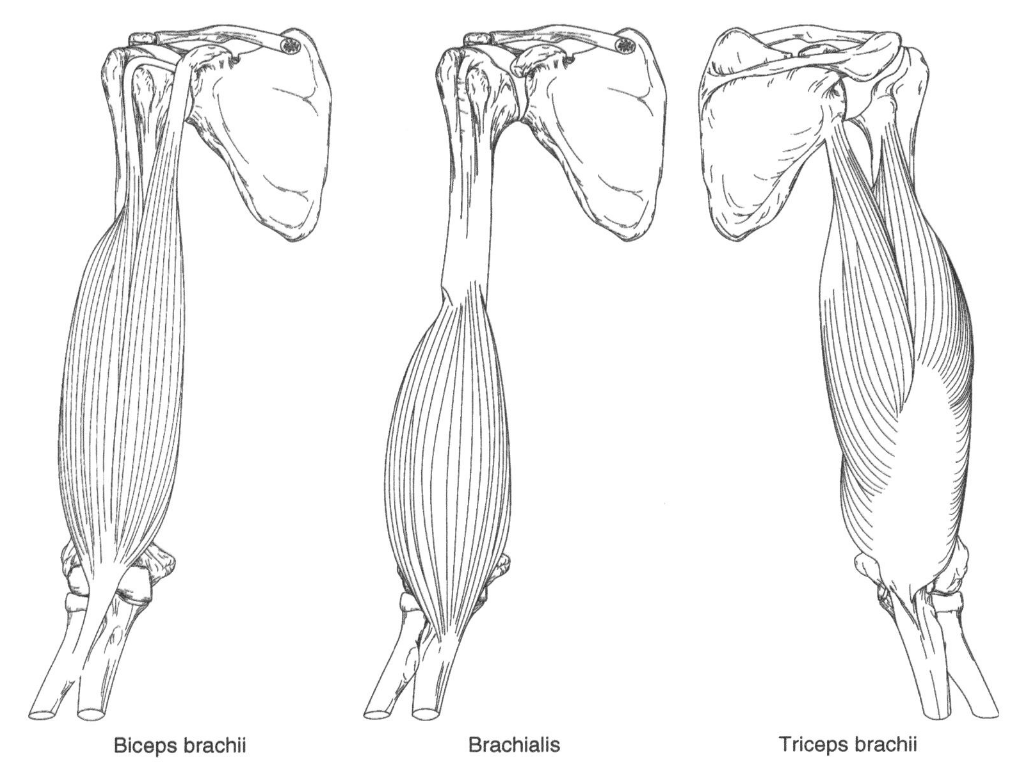

Figure 10-2 Muscles of the arm.

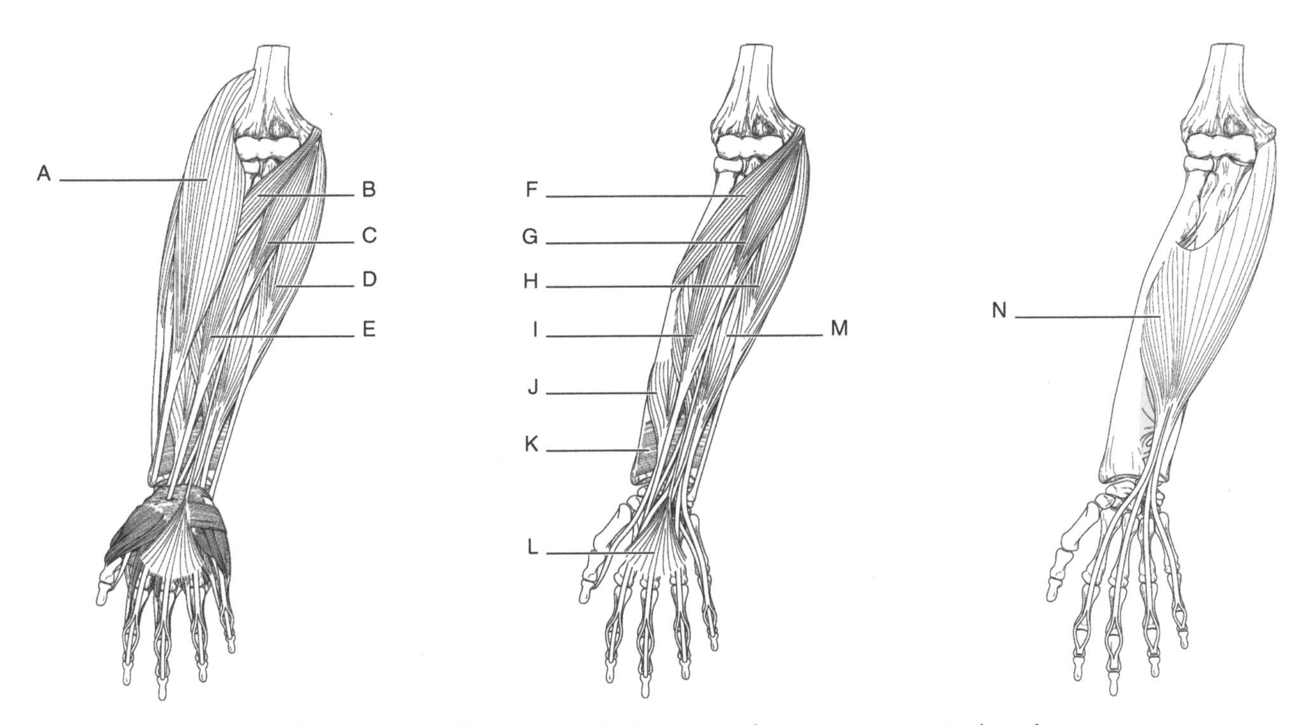

Figure 10-3 Muscles of the forearm. Top row: anterior views; bottom row: posterior views.

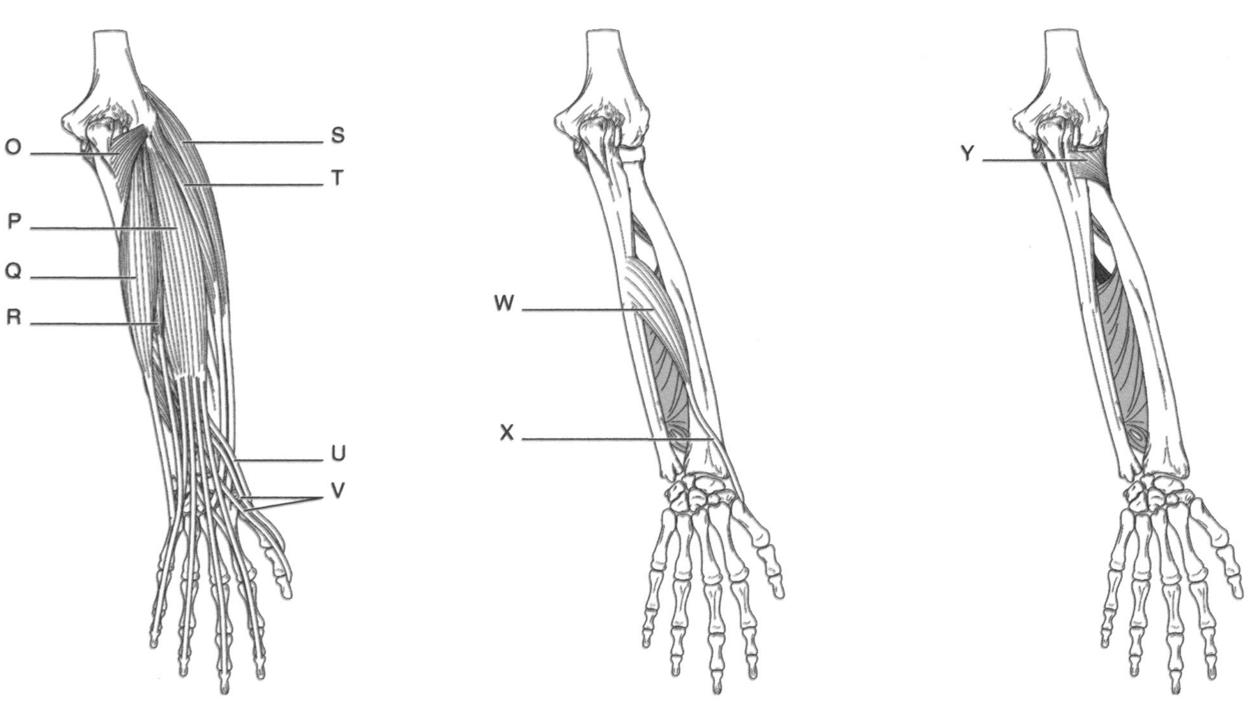

Figure 10-3 Continued

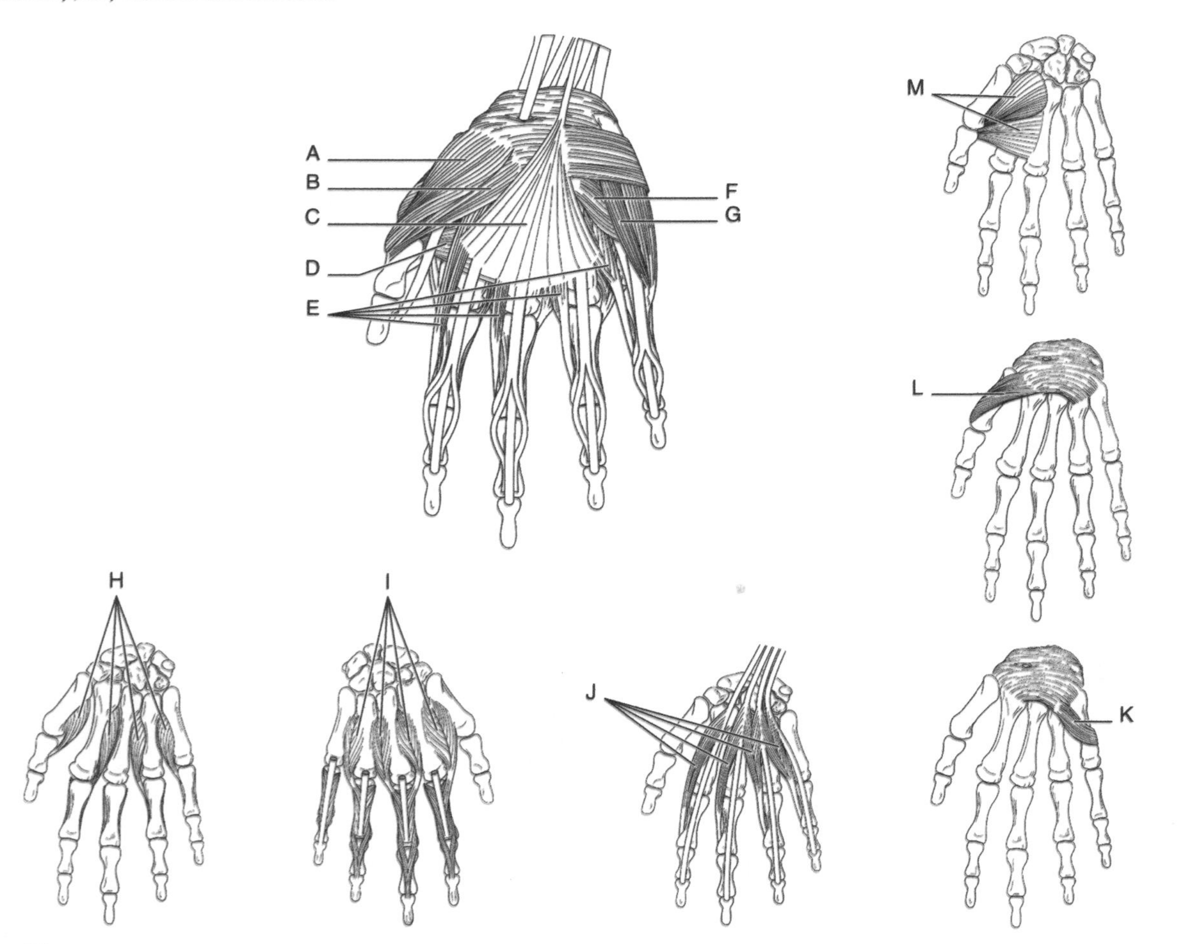

Figure 10-4 Intrinsic muscles of the hand.

Muscles of the Hip and Lower Limb

In this last section, you will learn about muscles of the hip (pelvis) and lower limb. Muscles of the hip and inguinal regions may move the thigh or support and have actions only in the pelvic region. At this time, you are not responsible for learning about muscles that have actions on pelvic structures since we will present those muscles in a later lab. However, you need to know all information presented in the following tables and figures.

Additional features associated with lower limb

The **fascia lata** is a broad band of fibrous connective tissue that encloses the thigh muscles and it is most visible on the lateral side of the thigh where it is called the **iliotibial** tract. This appears as a gray or white band along the outer part of the thigh.

A vestigial muscle in the leg is the *plantaris*. It may be present in both legs, only one leg, or not present in either leg. Often, a segment of its tendon is used to repair other damaged ligaments, such as those in the knee.

Tables 10.6 through **10.9** provide information about the location of muscles in the hip (pelvis), thigh, leg, and foot.

Table 10.6 Muscles That Move the THIGH

Muscle	Origin	Insertion
Adductor brevis	Pubis	Linea aspera
Adductor longus	Pubis	Linea aspera
Adductor magnus	Pubis/ischial tuberosity	Linea aspera
Biceps femoris	Ischial tuberosity/linea aspera	Tibia lateral condyle
Gemellus group	Ischial spine/tuberosity	Greater trochanter
Gluteus maximus	Iliotibial tract	Femur gluteal tuberosity
Gluteus medius	Iliac crest	Greater trochanter
Gluteus minimus	Ilium	Greater trochanter
Gracilis	Pubis	Tibia
Iliacus	Iliac fossa	Below lesser trochanter
Obturator group	Obturator foramen margins	Greater trochanter
Pectineus	Pubis	Near lesser trochanter
Piriformis	Sacrum	Greater trochanter
Popliteus	Femur lateral condyle	Tibia shaft
Psoas major	T12-L5 vertebrae	Lesser trochanter
Quadratus femoris	Ischial tuberosity	Between trochanters
Rectus femoris	Ant.inf. iliac spine	Patella
Sartorius	Ant. sup. iliac spine	Tibia
Semimembranosus	Ischial tuberosity	Tibia medial condyle
Semitendinosus	Ischial tuberosity	Tibia
Tensor fascia latae	Iliac crest/ant. sup. iliac spine	Iliotibial tract

(Continued)

Table 10.6 Muscles That Move the THIGH *(Cont'd)*

Muscle	Flexion	Extension	Abduction	Adduction	Lateral Rotation	Medial Rotation
Adductor muscles	✓			✓		✓
Biceps femoris		✓			✓	
Gemellus muscles					✓	
Gluteus maximus		✓			✓	
Gluteus medius			✓			✓
Gluteus minimus			✓			✓
Gracilis				✓		✓
Iliacus	✓					
Obturator muscles					✓	
Pectineus	✓			✓		✓
Piriformis			✓		✓	
Psoas major	✓					
Quadratus femoris					✓	
Rectus femoris	✓					
Sartorius	✓		✓		✓	
Semimembranosus		✓				
Semitendinosus		✓				
Tensor fasciae latae	✓		✓			✓

Table 10.7 Muscles that Move the LEG

Muscle	Origin	Insertion
Biceps femoris*	Ischial tuberosity/linea aspera	Fibula, tibia
Gastrocnemius	Femur	Calcaneus
Gracilis	Pubis	Tibia
Plantaris	Femur	Calcaneus
Popliteus	Femur lateral condyle	Tibia shaft
Rectus femoris**	Ilium	Tibia
Sartorius	Ant. sup. iliac spine	Tibia
Semimembranosus*	Ischial tuberosity	Tibia
Semitendinosus*	Ischial tuberosity	Tibia
Vastus intermedius**	Linea aspera	Tibial tuberosity
Vastus lateralis**	Linea aspera	Tibial tuberosity
Vastus medialis**	Linea aspera	Tibial tuberosity

Table 10.7 Muscles that Move the LEG *(Cont'd)*

Muscle	Leg Flexion	Leg Extension
Biceps femoris	✓	
Gastrocnemius	✓	
Gracilis	✓	
Plantaris	✓	
Popliteus	✓	
Rectus femoris		✓
Sartorius	✓	
Semimembranosus	✓	
Semitendinosus	✓	
Vastus intermedius		✓
Vastus lateralis		✓
Vastus medialis		✓

*One of the hamstring muscles
**One of the quadricep muscles

Table 10.8 Leg Muscles That Move the FOOT at the Ankle

Muscle	Origin	Insertion
Extensor digitorum longus	Tibia,/fibula	Distal phalanges
Fibularis group	Tibia/fibula	Metatarsals, tarsal
Gastrocnemius	Femur	Calcaneus
Plantaris	Femur lateral condyle/fibula	Calcaneus
Soleus	Fibula	Calcaneus
Tibialis anterior	Tibia	1st metatarsal; tarsal
Tibialis posterior	Tibia,/fibula	Tarsals, metatarsals

Muscle	Plantar Flexion	Foot Dorsiflexion	Foot Inversion	Foot Eversion
Extensor digitorum		✓		
Extensor hallucis		✓	✓	
Fibularis	✓			✓
Flexor digitorum	✓		✓	
Flexor hallucis	✓		✓	
Gastrocnemius	✓			
Plantaris	✓			
Soleus	✓			
Tibialis anterior		✓	✓	
Tibialis posterior	✓		✓	

Table 10.9 Muscles that move the TOES

Muscle	Origin	Insertion
Abductor digiti minimi	Calcaneus	Proximal phalanx
Abductor hallucis	Calcaneus	Proximal phalanx
Adductor hallucis	Toe ligaments	Proximal phalanx
Dorsal interossei	Metatarsals	Proximal phalanges
Extensor digitorum brevis	Calcaneus	Proximal phalanges
Extensor digitorum longus*	Proximal tibia & fibula	Middle/distal phalanges
Extensor hallucis brevis	Calcaneus	Proximal phalanx
Extensor hallucis longus*	Fibula	Distal phalanx
Flexor digiti minimi brevis	5th metatarsal	Proximal phalanx
Flexor digitorum brevis	Calcaneus	Middle phalanx
Flexor digitorum longus*	Tibia	Distal phalanges
Flexor hallucis brevis	Tarsals	Proximal phalanx
Flexor hallucis longus*	Fibula	Distal phalanx
Lumbricals	Flexor digitorum tendon	Proximal phalanx
Plantar interossei	Metatarsals	Proximal phalanges
Quadratus plantae	Calcaneus	Flexor digitorum

Muscle	Flexion	Extension	Abduction	Adduction
Abductor digiti minimi	✓		✓	
Abductor hallucis	✓		✓	
Adductor hallucis	✓			✓
Dorsal interossei	✓	✓	✓	
Extensor digitorum muscles		✓		
Extensor hallucis muscles		✓		
Flexor digiti minimi brevis	✓			
Flexor digitorum muscles	✓			
Flexor hallucis	✓			
Lumbricals	✓	✓		
Plantar interossei	✓	✓		✓

*An extrinsic foot muscle (located in leg)

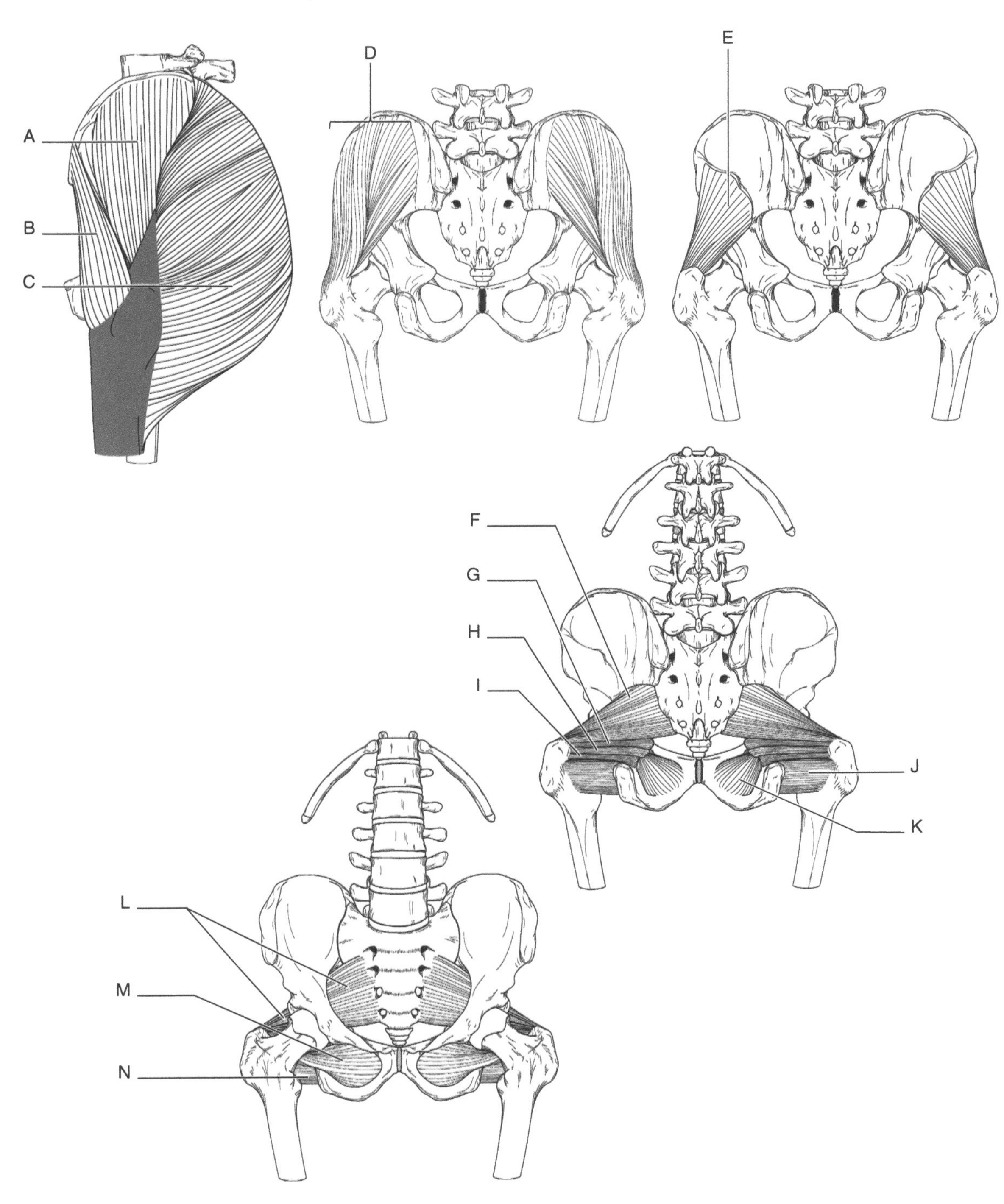

Figure 10-5 Selected muscles of the hip region.

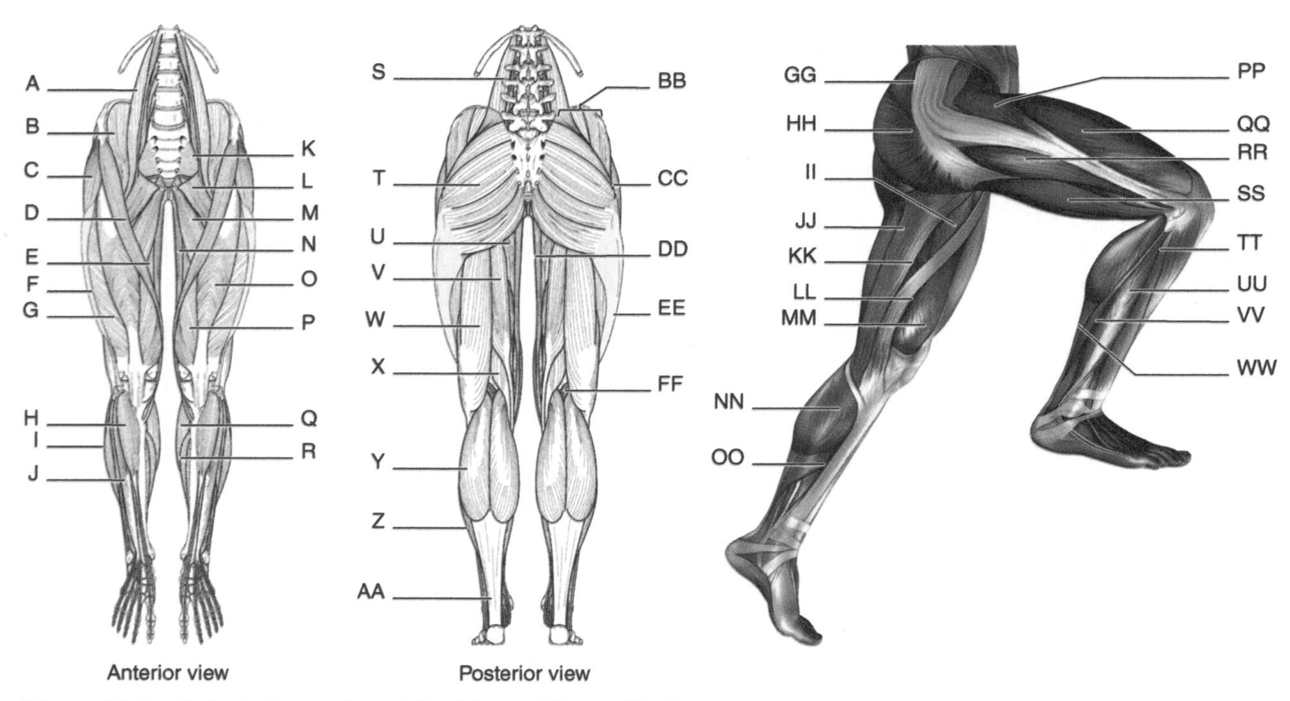

Figure 10-6 Selected muscles of the hip and lower limb.

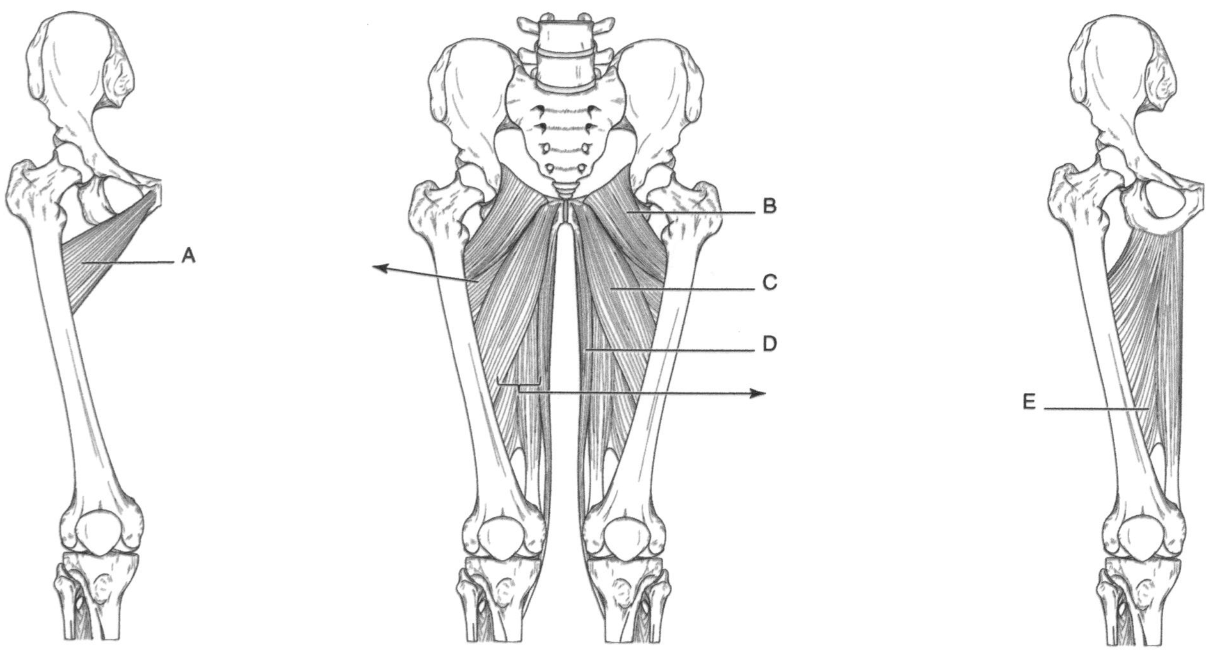

Figure 10-7 Selected muscles of the thigh.

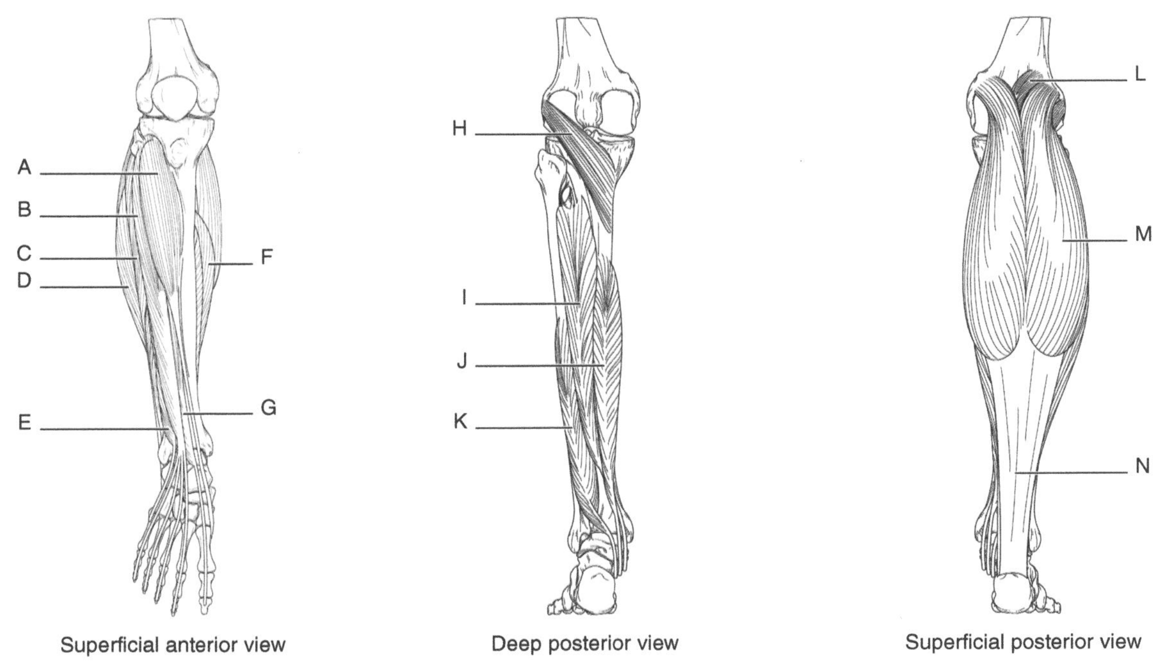

Figure 10-8 Selected muscles of the leg.

A
B
C
D
E
F
G
H

First layer
(Plantar view)

Dorsal view

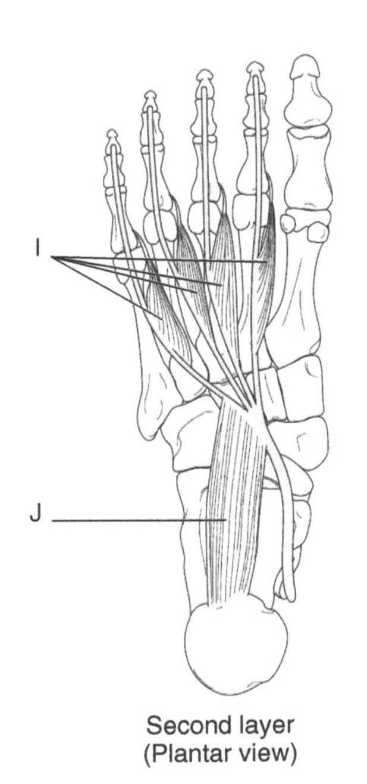

Figure 10-9 Intrinsic foot muscles.

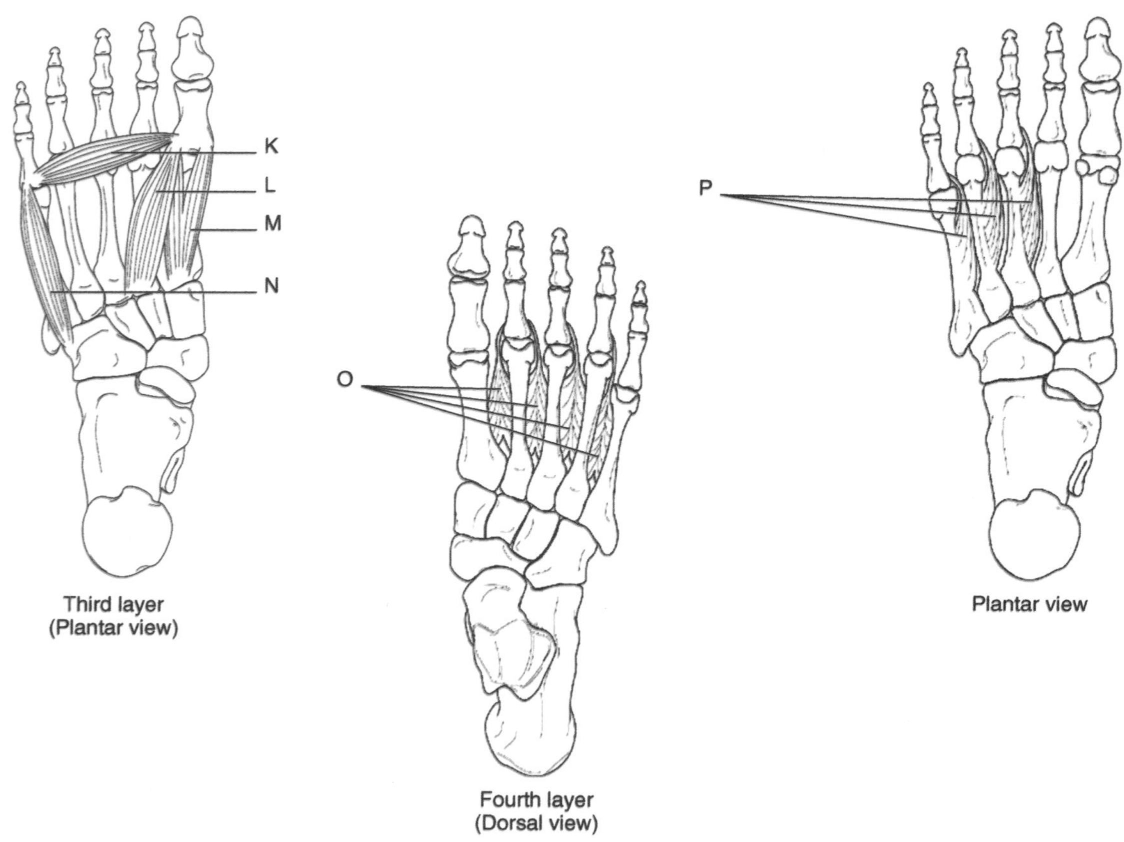

Figure 10-9 Continued.

Name ____________________ Course Number: __________ Lab Section ________

Lab 10: Muscles of the Limbs and Hip Worksheett

Write out the names of the muscles labeled in the selected figures.

Figure 10-1.

A. ____________________
B. ____________________
C. ____________________
D. ____________________
E. ____________________
F. ____________________
G. ____________________
H. ____________________
I. ____________________
J. ____________________
K. ____________________
L. ____________________
M. ____________________
N. ____________________
O. ____________________
P. ____________________
Q. ____________________
R. ____________________
S. ____________________
T. ____________________
U. ____________________
V. ____________________
W. ____________________
X. ____________________
Y. ____________________
Z. ____________________
AA. ____________________
BB. ____________________
CC. ____________________
DD. ____________________
EE. ____________________

Figure 10-3.

A. ____________________
B. ____________________
C. ____________________
D. ____________________
E. ____________________
F. ____________________
G. ____________________
H. ____________________
I. ____________________
J. ____________________
K. ____________________
L. ____________________
M. ____________________
N. ____________________
O. ____________________
P. ____________________
Q. ____________________
R. ____________________
S. ____________________
T. ____________________
U. ____________________
V. ____________________
W. ____________________
X. ____________________
Y. ____________________

Figure 10-4.

A. ____________________
B. ____________________
C. ____________________
D. ____________________
E. ____________________
F. ____________________
G. ____________________

H. ______________________

I. ______________________

J. ______________________

K. ______________________

Figure 10-5.

A. ______________________

B. ______________________

C. ______________________

D. ______________________

E. ______________________

F. ______________________

G. ______________________

H. ______________________

I. ______________________

J. ______________________

K. ______________________

Figure 10-6.

A. ______________________

B. ______________________

C. ______________________

D. ______________________

E. ______________________

F. ______________________

G. ______________________

H. ______________________

I. ______________________

J. ______________________

K. ______________________

L. ______________________

M. ______________________

N. ______________________

O. ______________________

P. ______________________

Q. ______________________

R. ______________________

S. ______________________

T. ______________________

U. ______________________

V. ______________________

W. ______________________

X. ______________________

Y. ______________________

Z. ______________________

AA. ______________________

BB. ______________________

CC. ______________________

DD. ______________________

EE. ______________________

FF. ______________________

GG. ______________________

HH. ______________________

II. ______________________

JJ. ______________________

KK. ______________________

LL. ______________________

MM. ______________________

NN. ______________________

OO. ______________________

PP. ______________________

QQ. ______________________

RR. ______________________

SS. ______________________

TT. ______________________

UU. ______________________

VV. ______________________

WW. ______________________

Figure 10-7.

A. ______________________

B. ______________________

C. ______________________

D. ______________________

E. ______________________

Figure 10-8.

A. ______________________

B. ______________________

C. ______________________

D. ______________________

E. ______________________

F. ______________________

G. ______________________

H. ______________________

I. ______________________

J. ______________________

K. ______________________

L. ______________________

M. ______________________

N. ______________________

Figure 10-9.

A. ______________________

B. ______________________

C. ______________________

D. ______________________

E. ______________________

F. ______________________

G. ______________________

H. ______________________

I. ______________________

J. ______________________

K. ______________________

L. ______________________

M. ______________________

N. ______________________

O. ______________________

P. ______________________

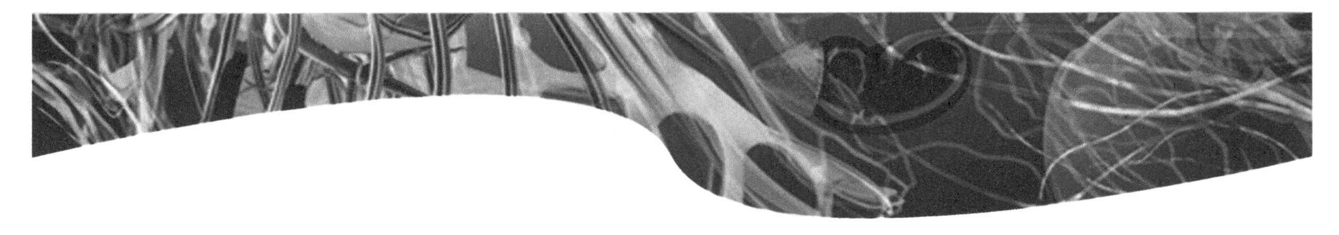

Lab 11 The Nervous System: Nervous Tissue and the Central Nervous System

Overview of the Nervous System

The **nervous system** includes organs that communicate with different parts of the body through electrical signals called **action potentials**, or **impulses**. The nervous system has two major subdivisions: the **central nervous system (CNS)**, which includes the *brain* and *spinal cord*, and the **peripheral nervous system (PNS)**, which includes *nerves, ganglia,* and *sensory receptors*. In this lab, you will be studying the cells of the nervous system and the major components of the central nervous system. An overview of the nervous system is shown in **Figure 11-1**.

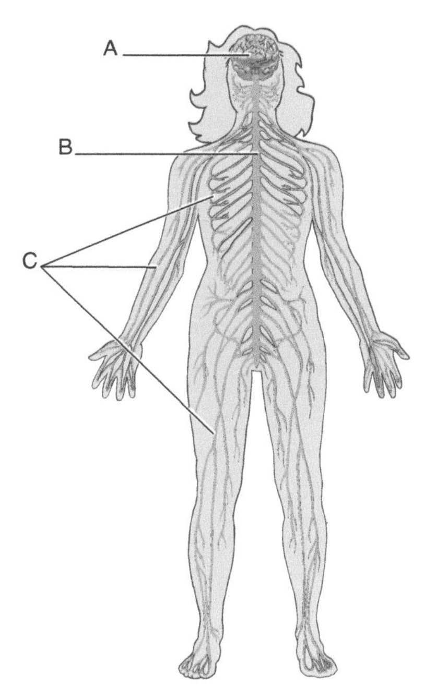

Figure 11-1 Overview of the nervous system.

Cells of the Nervous System

The two major groups of cells in the nervous system are *neurons* and *glial cells*. Neurons conduct impulses along their plasma membranes, but glial cells do not. Instead, glial cells may closely associate with neurons and provide support or protection, and certain glial cells can affect the speed at which a neuron conducts an impulse.

Neurons are classified in two ways: according to shape and according to function. Based on shape, a neuron can be *unipolar, bipolar, multipolar*, or *anaxonic*. Based on function, a neuron can be a *sensory, motor,* or *association* neuron. The basic characteristics of neurons and glial cells are presented in **Table 11.1** and **Figure 11-2**.

Parts of The Brain

The brain is the major control center for the nervous system, and it is housed within the cranial cavity of the skull. In this lab, you will

Table 11.1 Cells of the Nervous System	
Astrocyte	Most common glial cell; has projections that cover portions of neurons and blood vessels; control movement of materials out of blood vessels into nervous tissue
Axolemma	Plasma membrane on an axon
Axon	Also called a *nerve fiber*; long extension that conducts action potentials (impulses)
Axon hillock	Cone-shaped site where axon connects to the cell body
Axon terminal	Also called *synaptic knob* or *bouton*: releases neurotransmitter into a synapse
Axoplasm	Cytoplasm within an axon
Bipolar neuron	Has one dendrite and axon
Cell body	*Soma* or *perikaryon*; site of the nucleus and most other organelles
Dendrite	Short extension that responds to a stimulus
Ependymal cell	Secretes cerebrospinal fluid at choroid plexuses in brain ventricles
Ganglion	Collection of neuron cell bodies in the PNS
Gray matter	Collection of neuron cell bodies in the CNS
Interneuron	(Association neuron, internuncial neuron, connector neuron); a multipolar neuron that conducts impulses within the CNS; its axon is a tract in the brain or spinal cord
Microglial cell	Phagocytic glial cells in the CNS
Motor neuron	Multipolar neuron that conducts impulses out of the CNS
Multipolar neuron	Has many dendrites and one axon
Myelin sheath	Tightly wrapped lipid layer around certain axons (myelinated axons)
Nerve	Collection of axons (neuron fibers) in the PNS
Neurofibrils	Intermediate filaments in the cell body that maintain the neuron's shape
Nissl bodies	Rough ER in the cell body that stains dark with certain dye
Node of Ranvier	Space between Schwann cells along a myelinated axon
Oligodendrocyte	Glial cell that forms myelin sheaths in the CNS
Satellite cell	Surrounds a neuron within a ganglion
Schwann cell	Glial cell that forms myelin sheaths in the PNS
Sensory neuron	Conduct impulses toward the CNS; most are unipolar, but some are bipolar
Telodendron	Branch extending from the end of an axon and contain axon terminals
Tract	Bundle of axons in the CNS
Unipolar neuron	Has a single projection extending from cell body; its dendrite and axon form one extension
White matter	Light-colored collection of axons in the CNS

learn to recognize different parts of the brain and you will review their functions. For some parts of the brain, it is important to note the difference between a region that *perceives* a stimulus and one that *interprets* a stimulus. A region with the name "primary...cortex" is one that determines the *kind* of stimulus being detected; e.g., light, sound, touch. After perception occurs in one region, information is then sent to an "association area," which lets you know more about the stimulus. For example, seeing an apple is the job of the primary visual cortex, but knowing that the object seen is an apple and not a dog is the job of the visual association area.

As you look at models of the brain, first name a region and then state its functions. Be able to recognize the major parts of a preserved sheep brain, which your instructor will provide. Be sure to wear gloves when probing the preserved brains. **Table 11.2** lists major anatomical features of the brain in alphabetical order and **Figures 11-3 and 11-4** illustrate them.

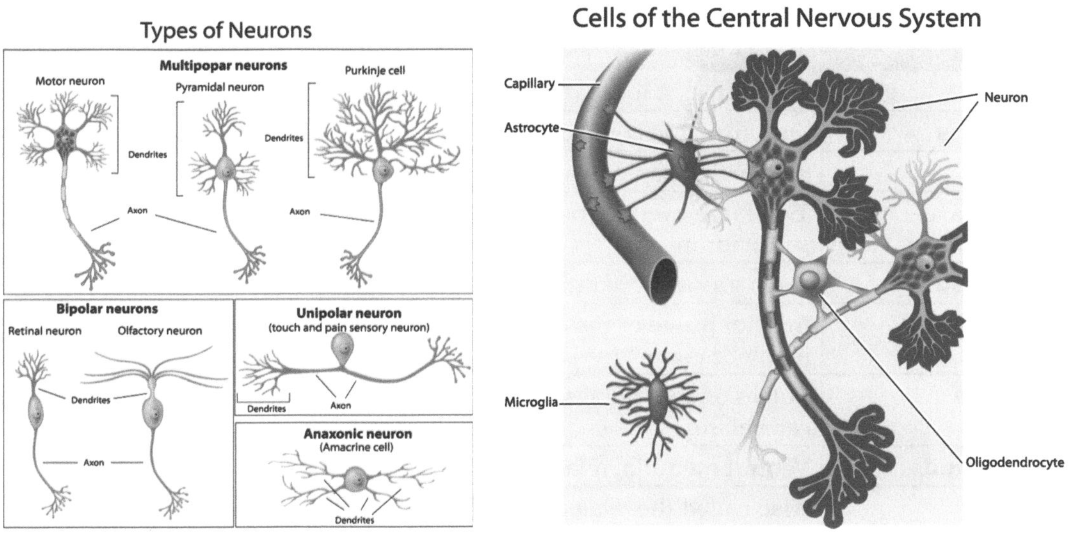

Figure 11-2 Neurons and selected glial cells.

Table 11.2 Gross Anatomical Features of the Brain

3rd ventricle	Slit-like cavity between the two halves of the hypothalamus
4th ventricle	Cavity between the pons and cerebellum
Arbor vitae	Branches of white matter in the cerebellum; seen in a medial view
Central sulcus	Sulcus that separates the frontal lobe from the parietal lobe
Cerebellum	Second largest part of the brain; located beneath the occipital lobes; consists of two cerebellar hemispheres and is important for maintaining balance and equilibrium
Cerebral aqueduct	Tube connecting 3rd and 4th ventricles; transports CSF
Cerebral peduncles	Anterior part of the mesencephalon with motor tracts from the cerebrum to the spinal cord
Cerebrum	Largest part of the brain; consists of a right and left cerebral hemisphere
Choroid plexus	Network of capillaries and ependymal cells within a ventricle; produces cerebrospinal fluid
Corpora quadrigemina	Four mound-like structures on the posterior portion of mesencephalon
Corpus callosum	Site of commissural fibers connecting cerebral hemispheres
Diencephalon	Between corpus callosum and midbrain; includes hypothalamus, thalamus, and epithalamus
Epithalamus	Superior portion of diencephalon; pineal body is attached to its posterior end
Frontal lobe	Anterior to the central sulcus; closest to the frontal bone; includes the precentral gyrus
Gyrus	Raised ridge on the surface of the cerebrum
Hypothalamus	Inferior portion of the diencephalon; center of autonomic nervous system function
Inferior colliculi	Two inferior mounds of the corpora quadrigemina that serve as reflex centers for sound

(Continued)

Table 11.2 Gross Anatomical Features of the Brain *(Cont'd)*	
Insula	Cerebral gyri located deep to the temporal lobe; site of gustatory cortex.
Intermediate mass	Structure that connects the right and left halves of the thalamus
Lateral sulcus	Sulcus that separates the temporal lobe from the parietal lobe
Lateral ventricles	Two large cavities, one in each cerebral hemisphere; separated by the septum pellucidum
Longitudinal fissure	Deep groove separating the right and left cerebral hemispheres
Medulla oblongata	Inferior bulge of the brainstem between the pons and spinal cord; contains many "centers" for involuntary control of muscles
Medullary olives	Bulges on the lateral side of the medulla; relays sensory information to the cerebellum
Medullary pyramids	Ventral medulla; relays motor information from cerebrum to spinal cord
Mesencephalon	Also called the *midbrain*: Located above the pons; contains cerebral peduncles anteriorly and corpora quadrigemina posteriorly
Occipital lobe	Posterior to the parieto-occipital sulcus and located closest to the occipital bone
Parietal lobe	Posterior to the central sulcus; closest to the parietal bone; includes the postcentral gyrus
Parieto-occipital sulcus	Separates the parietal lobe from the occipital lobe
Pineal gland (body)	Posterior end of epithalamus; secretes melatonin important in sleep-wake cycles
Postcentral gyrus	Gyrus immediately posterior to the central sulcus
Precentral gyrus	Gyrus immediately anterior to the central sulcus
Sulcus	Groove between adjacent gyri on the surface of the cerebrum
Superior colliculi	Two superior mounds of the corpora quadrigemina that serve as reflex centers for sight
Temporal lobe	Inferior to the lateral sulcus; closest to the temporal bone
Thalamus	Middle portion of diencephalon surrounding the intermediate mass
Ventricle	Cavity within the brain and it is filled with circulating cerebrospinal fluid

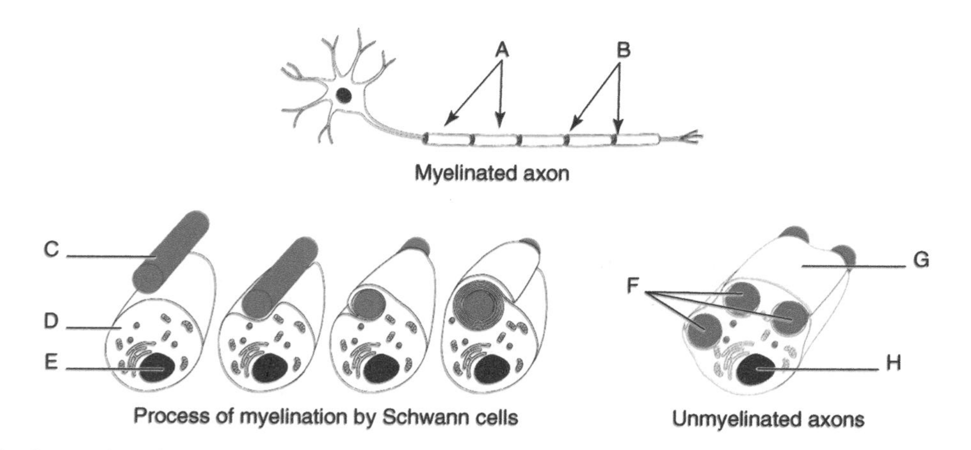

Figure 11-3 Myelinated and unmyelinated neurons in the PNS.

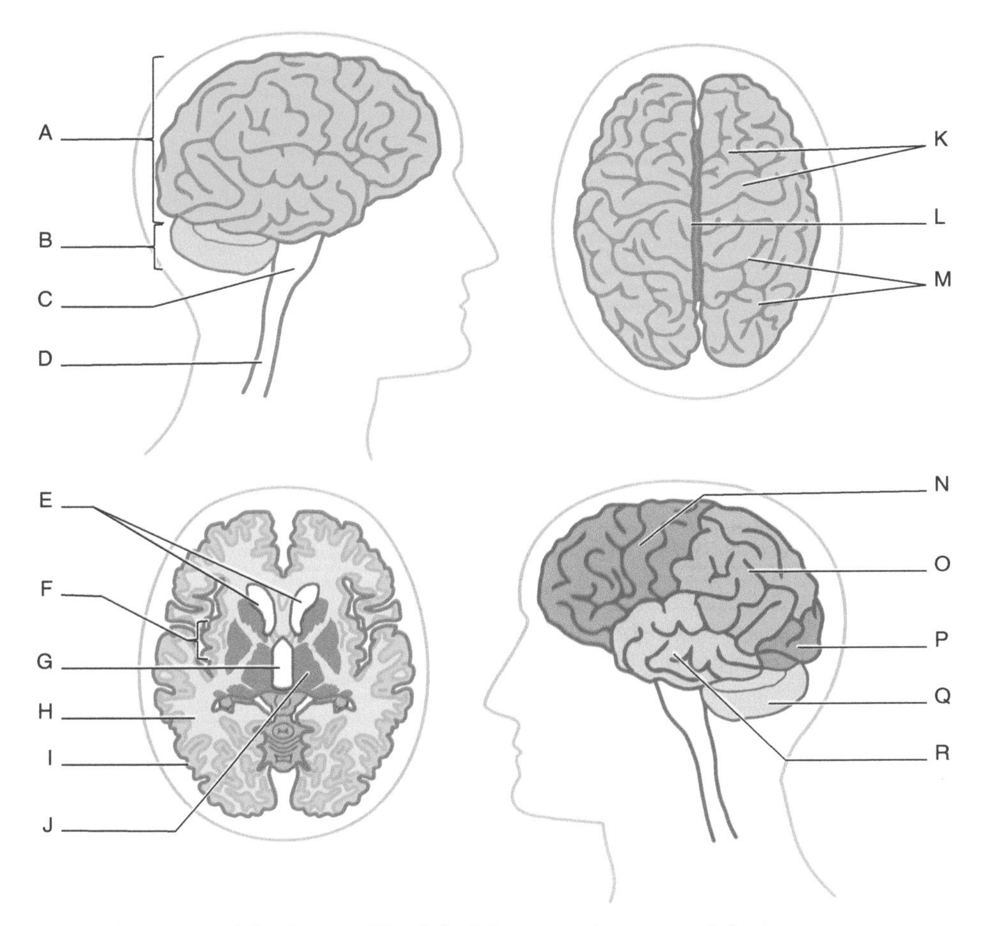

Figure 11-4 Selected features of the brain. The label lines are for general features.

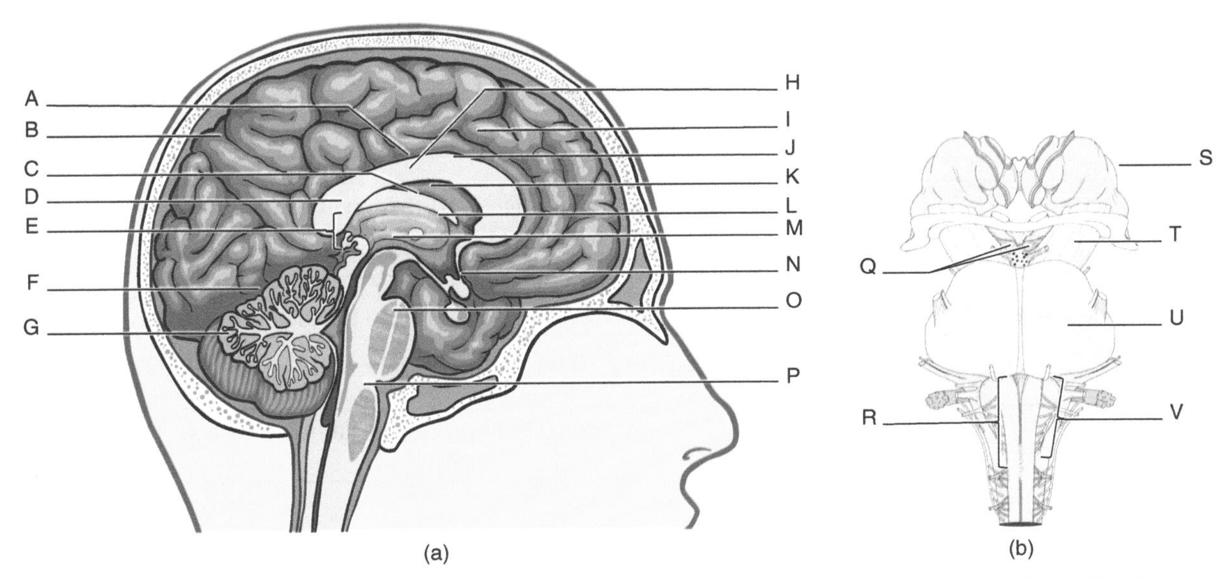

Figure 11-5 Selected anatomical features of the brain and spinal cord. (a) Midsagittal section of brain; (b) Anterior view of diencephalon and brainstem.

Table 11.3 Functional Regions of the Brain	
Amygdala	Part of the limbic system; in the temporal lobe; processes emotions for fear, anger, and pleasure; controls storage of emotional memories
Association fibers	Tracts that conduct impulses within a cerebral hemisphere
Auditory association area	In the temporal lobe; interpretation of sound
Auditory cortex	Gray matter for sound perception; in temporal lobes
Basal nuclei	Gray matter in the cerebrum and diencephalon; function in coordination and equilibrium; includes *caudate nucleus; putamen, globus pallidus, lentiform nucleus,* and *corpus striatum*
Broca's area	Speech center; controls motor movements for speech; usually in left frontal lobe
Caudate nucleus	A basal nucleus; not well understood
Commissural fibers	Tracts in the corpus callosum and connects right and left cerebral hemispheres
Corpus striatum	A basal nucleus; not well understood
Fornix	Tracts below the corpus callosum; connects upper and lower limbic system structures
Frontal association area	(Prefrontal area) Gray matter for thought processing; in frontal lobes
Frontal eye field	Gray matter that controls voluntary eye movement; in frontal lobes
Globus pallidus	A basal nucleus; not well understood
Gustatory cortex	Gray matter for taste perception; in the insula
Hippocampus	Part of the limbic system; controls conversion of short-term to long-term memory, emotional responses to memories, and spatial memory (remembering locations in space)
Insula	An "island" of cerebral cortex deep to each temporal lobe
Interventricular foramen	(Foramen of Monro); passage for cerebrospinal fluid moving from the lateral ventricles into the third ventricle
Lentiform nucleus	A basal nucleus; not well understood
Limbic system	Internal brain structures associated with emotions
Longitudinal fissure	Groove in the sagittal plane that separates the cerebral hemispheres
Mammillary bodies	Part of the limbic system; associated with the formation of memories
Motor cortex	Gray matter for voluntary muscle movement; in precentral gyrus of frontal lobe
Motor speech area	(Broca's area); gray matter for speech control; usually in lateral side of left frontal lobe
Nuclei	Gray matter in brain regions deep to the cortex
Olfactory cortex	Gray matter for smell perception; in temporal lobes
Optic chiasma	Beneath frontal lobes and anterior to pituitary gland; where the two optic nerves unite
Pons	Part of brain stem; relay center between medulla oblongata and higher brain centers

(Continued)

Table 11.3 Functional Regions of the Brain *(Cont'd)*

Postcentral gyrus	In parietal lobe immediately posterior to central sulcus; contains somatosensory cortex
Precentral gyrus	In frontal lobe immediately anterior to central sulcus; contains primary motor cortex
Prefrontal area	Gray matter for problem-solving, imagination, artistic skills; in frontal lobes
Premotor area	Gray matter for memory of reflexive motor skills; in frontal lobes
Projection fibers	Tracts that connect the cerebrum to lower brain centers
Putamen	A basal nucleus; not well understood
Septum pellucidum	Membrane in sagittal plane and separates the lateral ventricles
Somatosensory association area	Gray matter for interpretation of somatosensory input; in parietal lobes
Somatosensory cortex	Gray matter for perception of touch, temperature, pressure; in postcentral gyrus
Tracts	Bundles of white matter (neuron axons) in the CNS
Visual association area	Gray matter for interpretation of visual input; in occipital lobes
Visual cortex	Gray matter for light perception; in occipital lobes
Wernicke's area	(Auditory association area); gray matter for interpreting speech; in left temporal lobe

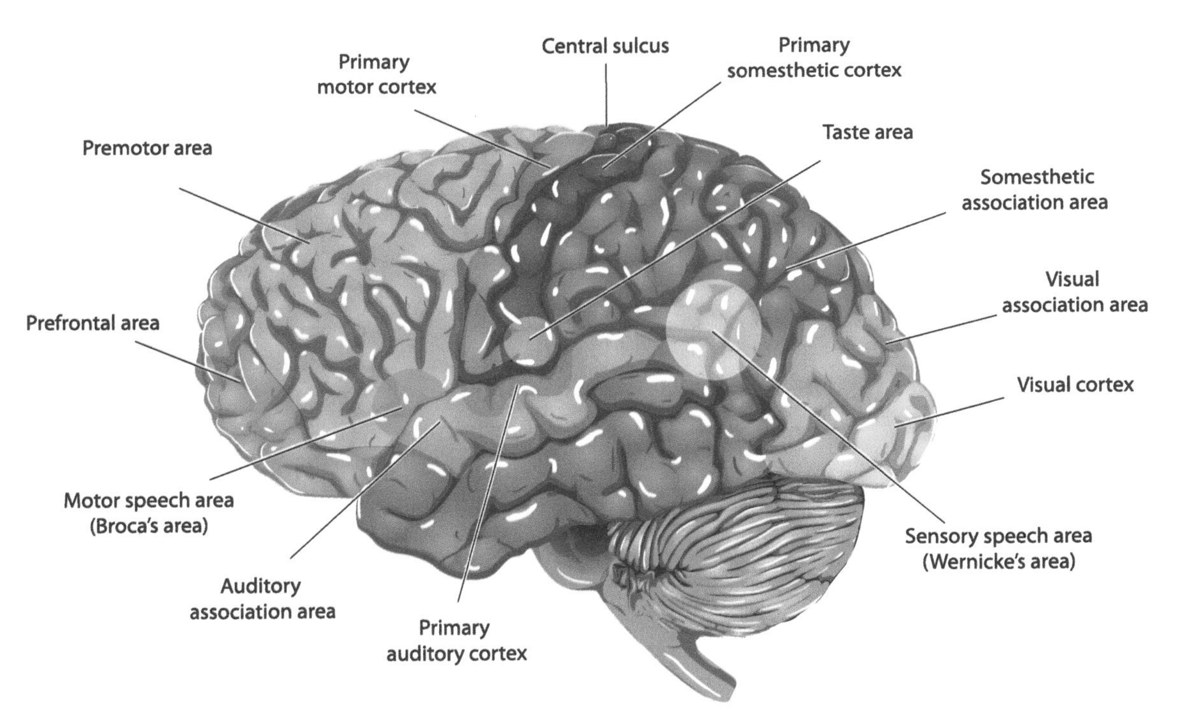

Figure 11-6 Functional regions of the cerebrum.

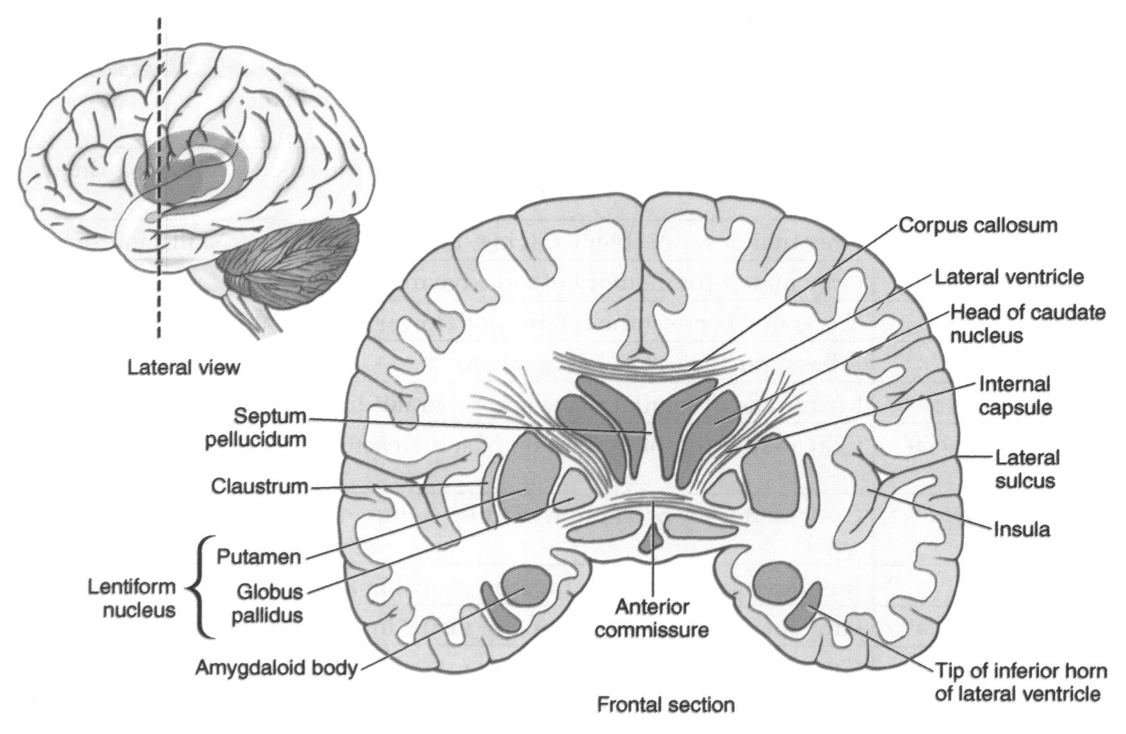

Figure 11-7 Nuclei deep within the cerebrum.

Table 11.4 Components Associated with the Spinal Cord	
Anterior column	White matter between anterior horns
Anterior horns	Gray matter projections on anterior side of cord
Cauda equina	Nerve roots at inferior end of cord; resembles horse's tail
Central canal	Small canal along entire length of cord
Conus medullaris	Posterior end of cord; at L2 in adult and L3 in infant
Denticulate ligament	Extensions of pia mater that anchors spinal cord to vertebrae
Filum terminale	Extensions of pia mater that anchors spinal cord to coccyx
Lateral columns	White matter around lateral horns
Lateral horns	Lateral projections of gray matter in cord
Posterior columns	White matter between posterior horns
Posterior horns	Gray matter projections on posterior side of cord

Table 11.5 Meninges and Associated Components	
Arachnoid mater	Immediately superficial to pia mater; resembles spider web
Arachnoid villi	Projections of the arachnoid into the superior sagittal sinus; drains CF into the blood
Dura mater	Most superficial meninx; tough and fibrous
Falx cerebri	Dura mater lying between the cerebral hemispheres; attaches to the crista galli in the skull
Pia mater	Deepest meninx; thin membrane attached to surface of brain and spinal cord
Subarachnoid space	Between arachnoid and pia mater
Subdural space	Between dura mater and arachnoid

Anatomy Of A Sheep Brain

Your lab instructor may provide for you a preserved sheep brain on which you can identify structures that are similar to those in the human brain. Be sure to wear gloves when you handle the sheep brains. **Figure 11-10** identifies the structures you need to know on the preserved brains.

Electroencephalography

Your lab may be set up with an instrument to monitor brain waves on a chart called an **electroencephalogram** (**EEG**; ē-LEK-trō-en-SEF-uh-lō-gram). The monitoring of the brain waves is called electroencephalography. If you have the opportunity to observe a demonstration for an electroencephalogram (EEG), the lab instructor will provide handouts and directions for this exercise.

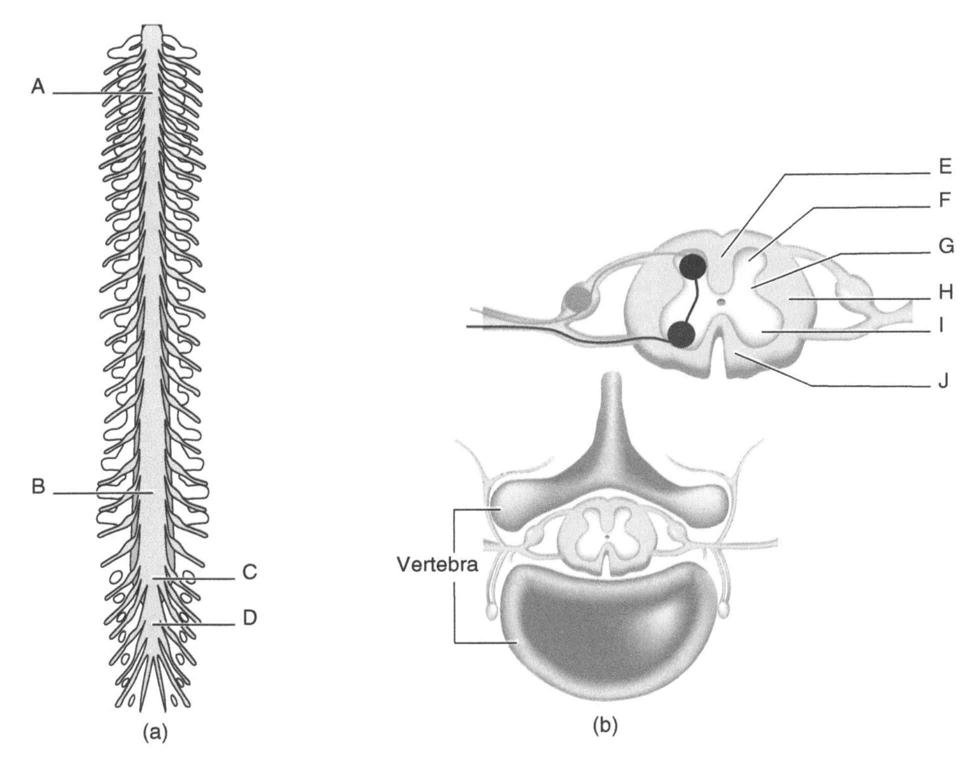

Figure 11-8 Anatomy of the spinal cord: (a) Dorsal view; (b) Cross section.

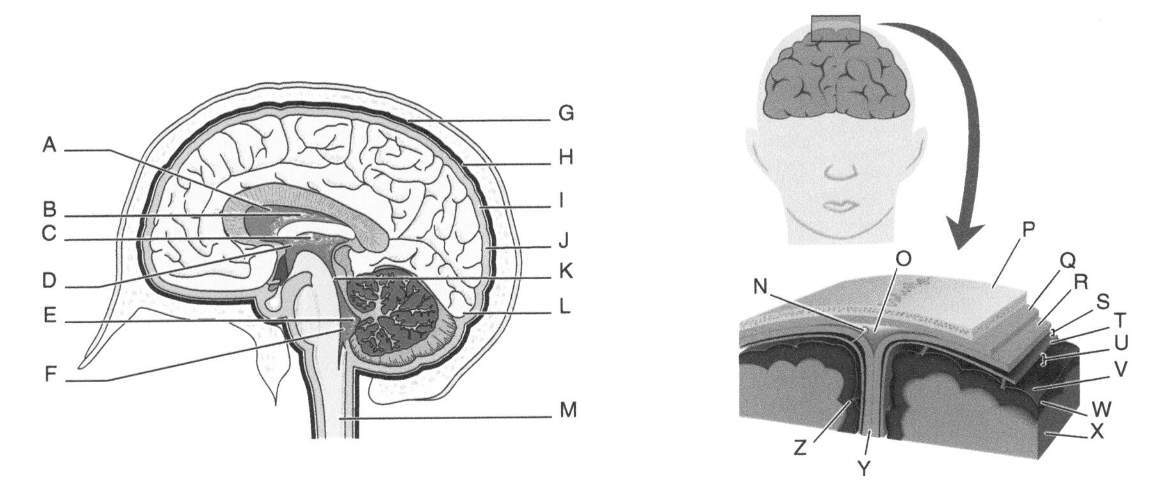

Figure 11-9 Meninges and cavities of the brain and spinal cord.

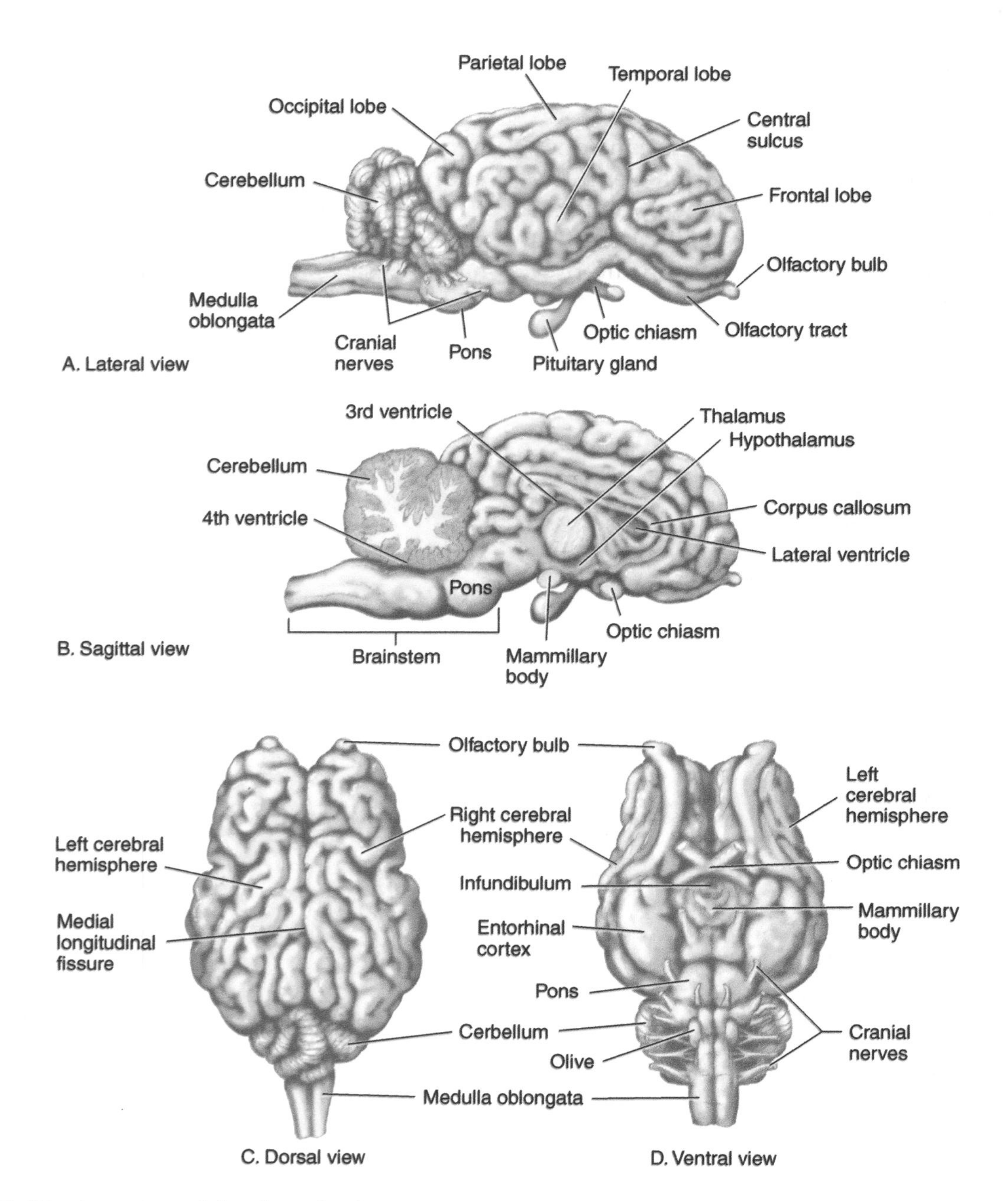

Figure 11-10 Anatomy of the sheep brain.

Name ____________________________ Course Number: _________ Lab Section _______

Lab 11: Nervous Tissue and Central Nervous System Worksheet

Write out the names of the labeled items in the selected figures.

Figure 11-1.

A. ______________________________
B. ______________________________
C. ______________________________

Figure 11-3.

A. ______________________________
B. ______________________________
C. ______________________________
D. ______________________________
E. ______________________________
F. ______________________________
G. ______________________________
H. ______________________________

Figure 11-4.

A. ______________________________
B. ______________________________
C. ______________________________
D. ______________________________
E. ______________________________
F. ______________________________
G. ______________________________
H. ______________________________
I. ______________________________
J. ______________________________
K. ______________________________
L. ______________________________
M. ______________________________
N. ______________________________
O. ______________________________
P. ______________________________
Q. ______________________________
R. ______________________________

Figure 11-5.

A. ______________________________
B. ______________________________
C. ______________________________
D. ______________________________
E. ______________________________
F. ______________________________
G. ______________________________
H. ______________________________
I. ______________________________
J. ______________________________
K. ______________________________
L. ______________________________
M. ______________________________
N. ______________________________
O. ______________________________
P. ______________________________
Q. ______________________________
R. ______________________________
S. ______________________________
T. ______________________________
U. ______________________________
V. ______________________________

Figure 11-8.

A. ______________________________
B. ______________________________
C. ______________________________
D. ______________________________
E. ______________________________
F. ______________________________
G. ______________________________
H. ______________________________
I. ______________________________
J. ______________________________

Figure 11-9.

A. ______________________________
B. ______________________________
C. ______________________________
D. ______________________________
E. ______________________________
F. ______________________________
G. ______________________________
H. ______________________________
I. ______________________________
J. ______________________________
K. ______________________________
L. ______________________________
M. ______________________________
N. ______________________________
O. ______________________________
P. ______________________________
Q. ______________________________
R. ______________________________
S. ______________________________
T. ______________________________
U. ______________________________
V. ______________________________
W. ______________________________
X. ______________________________

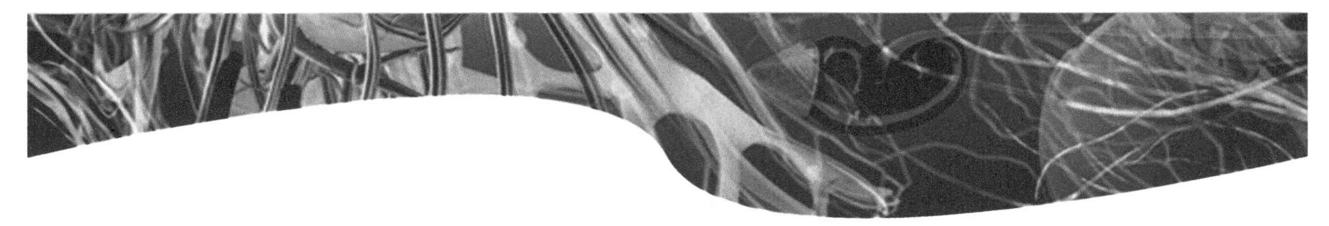

LAB 12 THE NERVOUS SYSTEM: PERIPHERAL NERVOUS SYSTEM

The **peripheral nervous system (PNS)** includes all nervous system components located outside of the brain and spinal cord. These components include *cranial nerves, spinal roots, rami, plexuses, spinal nerves, ganglia,* and *sensory receptors*. This lab covers all of the above except sensory receptors, which are covered in the next lab.

The PNS is subdivided into a *sensory* and a *motor* division. The motor division is further divided into a *somatic* and an *autonomic* division. Last, the autonomic division is subdivided into the *sympathetic* and *parasympathetic* divisions. The following tables (**Tables 12.1 through 12.3**) and figures (**Figures 12-1 through 12-3**) present information related to the PNS. You should become familiar with all of the information presented and be able to recognize PNS components on figures, models, and any preserved specimens that might be available in the lab.

Table 12.1 Major Components of the Peripheral Nervous System

Anterior (ventral) root	Ventral attachment of a spinal nerve to the spinal cord; contains motor axons
Dorsal ramus	Dorsal branch of a spinal nerve; contains axons connected to structures in the back
Dorsal root ganglion	Bulge on the dorsal root; contains cell bodies of sensory neurons
Endoneurium	Innermost layer of connective tissue in a nerve; insulates axons from one another
Epineurium	Outer connective tissue covering around a nerve
Fascicle	Bundle of axons within a nerve
Ganglion	Nodular structure containing cell bodies of neurons in the PNS
Motor neuron	Conducts impulses to an effector (muscle or gland)
Nerve	Collection of axons enclosed in connective tissue in the PNS
Paravertebral ganglia	(Chain ganglia); Sympathetic ganglia adjacent to the vertebral column; appear like a chain of beads and contain cell bodies of postganglionic neurons; also contain axons of preganglionic neurons that will synapse with postganglionic neurons in prevertebral ganglia located farther away from the spinal cord
Perineurium	Connective tissue covering surrounding a fascicle within a nerve
Plexus	Interconnected ventral rami in certain regions of the spinal column
Posterior (dorsal) root	Dorsal attachment of a spinal nerve to the spinal cord; contains sensory axons

(Continued)

Table 12.1 Major Components of the Peripheral Nervous System *(Cont'd)*

Postganglionic neuron	Conducts impulses from an autonomic ganglion to an effector; Ach from the preganglionic neuron binds to nicotinic, cholinergic receptors on the postganglionic neuron's dendrites
Preganglionic neuron	Conducts impulses from the CNS to postganglionic neurons inside an autonomic ganglion; releases acetylcholine (Ach) onto the postganglionic neurons
Prevertebral ganglia	**(Collateral ganglia);** Contain cell bodies of postganglionic neurons in the sympathetic division; located anterior to the spinal cord
Ramus (Dorsal/Ventral)	Connects a spinal nerve to a peripheral nerve
Ramus communicans	Small connections between a ramus and a parvertebral ganglion; *white ramus communicans* contains axons of preganglionic neurons; *gray ramus communicans* contains axons of postganglionic neurons
Sensory neurons	Conduct impulses from sensory receptors into the brain or spinal cord
Spinal root	Connects a spinal nerve to the spinal cord
Splanchnic nerve	Nerve that innervates visceral organs
Sympathetic chain	Vertically-oriented collection of sympathetic ganglia extending from the neck through the ventral cavity along the vertebral column
Ventral ramus	Ventral branch of a spinal nerve; contains axons connected to structures located anterior to the spinal column

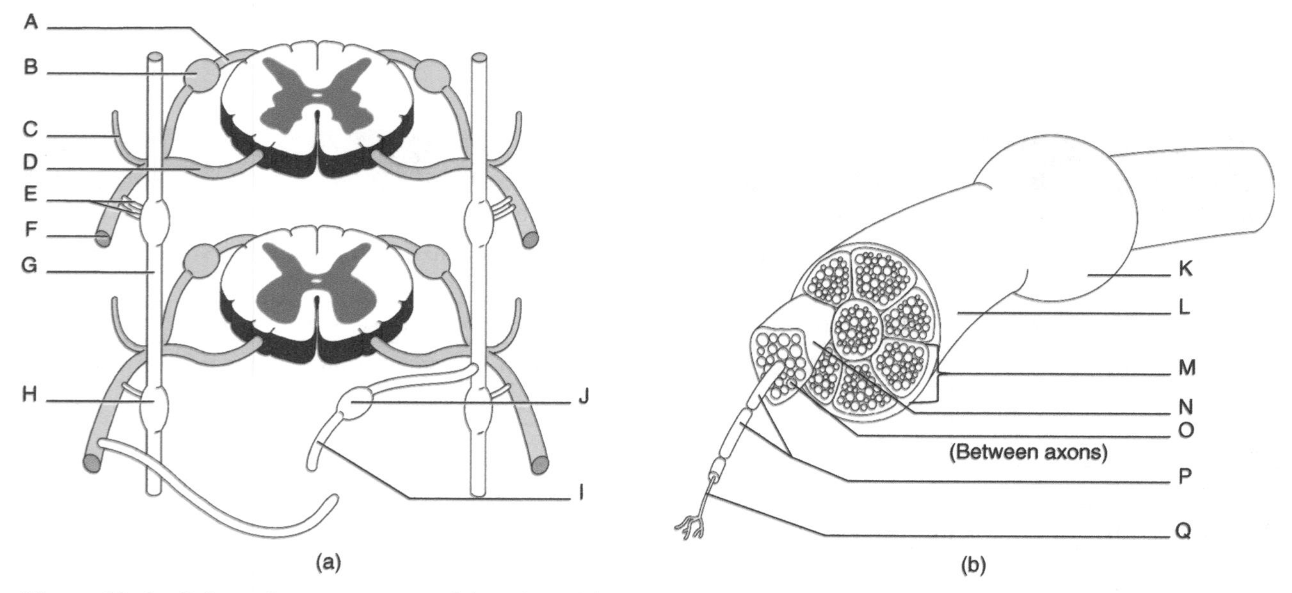

Figure 12-1 Selected components of the PNS: (a) Connections between the spinal cord and the PNS; (b) Components of a nerve.

Table 12.2 Cranial Nerves and Their Functions

Name of Nerve Pair	Type*	Its Axons Conduct Signals For
I: Olfactory	Sensory	Smell
II: Optic	Sensory	Vision
III: Oculomotor	Motor	Eye movements, pupil and lens changes for focusing; raising eyelids
IV: Trochlear	Motor	Looking down and to side
V: Trigeminal	Mixed	
Ophthalmic branch	Sensory	Touch, pain, temperature
Maxillary branch	Sensory	Touch, pain, temperature
Mandibular branch	Mixed	Touch, pain, temperature; chewing
VI: Abducens	Motor	Looking laterally
VII: Facial	Mixed	Taste (except bitter); facial expression, tear/saliva production
VIII: Vestibulocochlear	Sensory	Hearing and equilibrium
IX: Glossopharyngeal	Mixed	Bitter taste; touch, pain, temperature; blood pressure/chemistry, swallowing; saliva secretion
X: Vagus	Mixed	Touch, pain, temperature; blood pressure, chemistry, visceral organs; Stimulates muscle contractions in pharynx, neck, and viscera
XI: Accessory	Motor	Turning head and raising shoulders
XII: Hypoglossal	Motor	Moves tongue

*Sensory: contains only sensory axons; Mixed: has many sensory and motor axons; Motor: has *mostly* motor axons, but also a few sensory axons that conduct impulses from proprioceptors (stretch receptors in tendons)

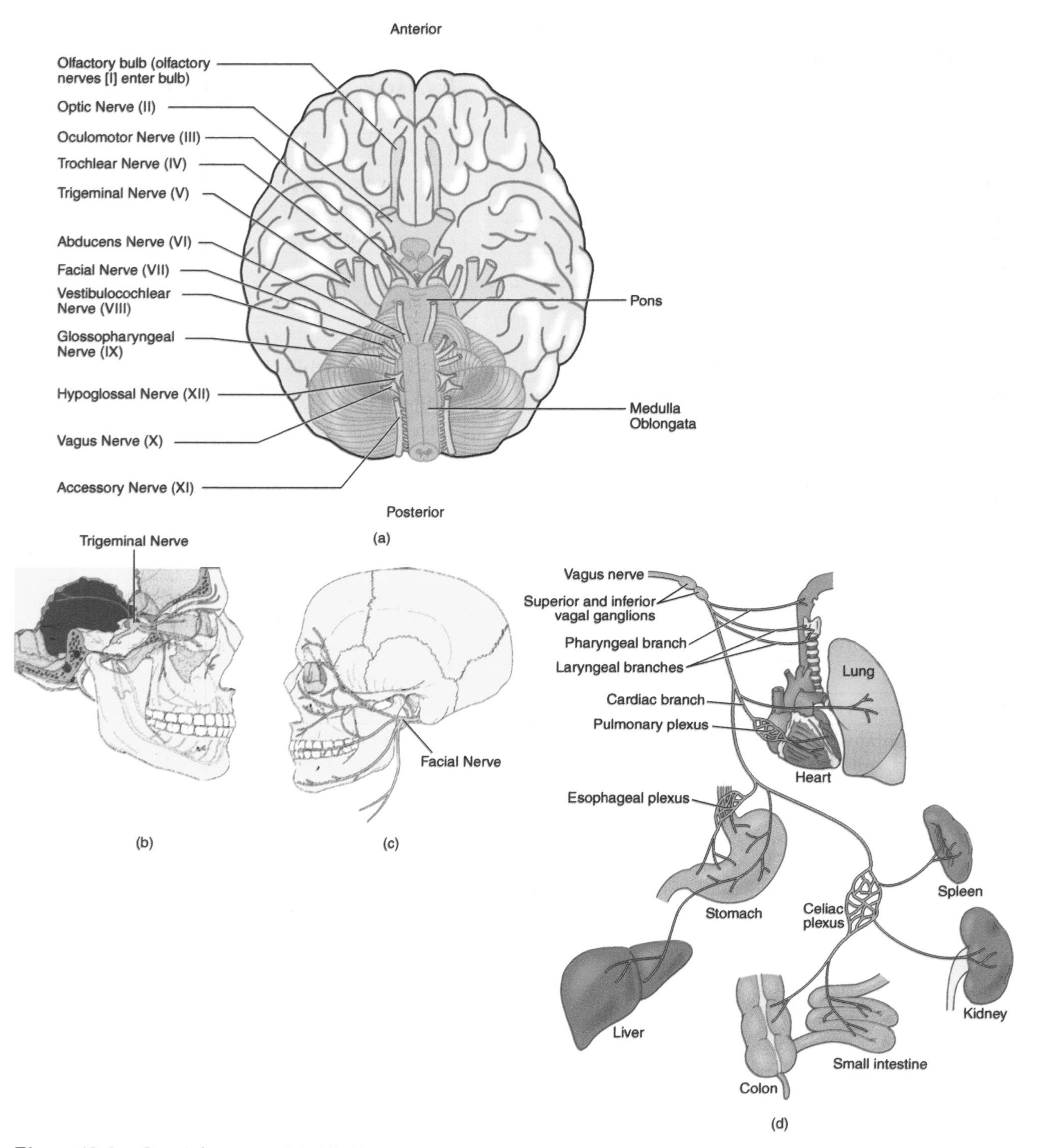

Figure 12-2 Cranial nerves: (s) All 12 pairs seen on the inferior surface of the brain; (b) trigeminal nerve's three branches; (c) facial nerve's five branches; (d) vagus nerve's branches.

Table 12.3 Ventral Rami, Nerves Plexuses, Peripheral Nerves and the Structures Innervated

Ventral Rami	Plexus Formed	Peripheral Nerves	Structures Innervated
C1-C5	**CERVICAL**	**Ansa cervicalis**	Deep neck muscles
		Auricular	Skin of ear
		Cervical	Muscles and skin of neck
		Occipital	Skin on back of head
		Phrenic	Diaphragm
C4-C8; T1	**BRACHIAL**	**Axillary**	Shoulder muscles
		Long thoracic	Serratus anterior muscle
		Median	Flexors in forearm; palm skin
		Musculocutaneous	Flexors in arm; skin of forearm
		Pectoral	Pectoralis muscles
		Radial	Extensors of arm and forearm; arm/forearm skin
		Ulnar	"Funny bone"; skin of forearm
		Suprascapular	Supraspinatus and infraspinatus muscles
		Subscapular	Teres major and subscapularis muscles
T1-T12	None	**Intercostal**	Intercostal muscles; skin of thorax, abdomen
L1-L4	**LUMBAR**	**Femoral**	Quadriceps muscles; skin on medial leg
		Obturator	Adductors; skin on medial thigh
L4-S4	**SACRAL**	**Pudendal**	Genitalia, external anal sphincter
		Sciatic	Hamstring muscles, fibular and tibial nerves
		Fibular (Peroneal)	Lateral leg and foot
		Tibial	Posterior leg and foot
S4-S5; C_O	**COCCYGEAL**	**Coccygeal**	Pelvic floor

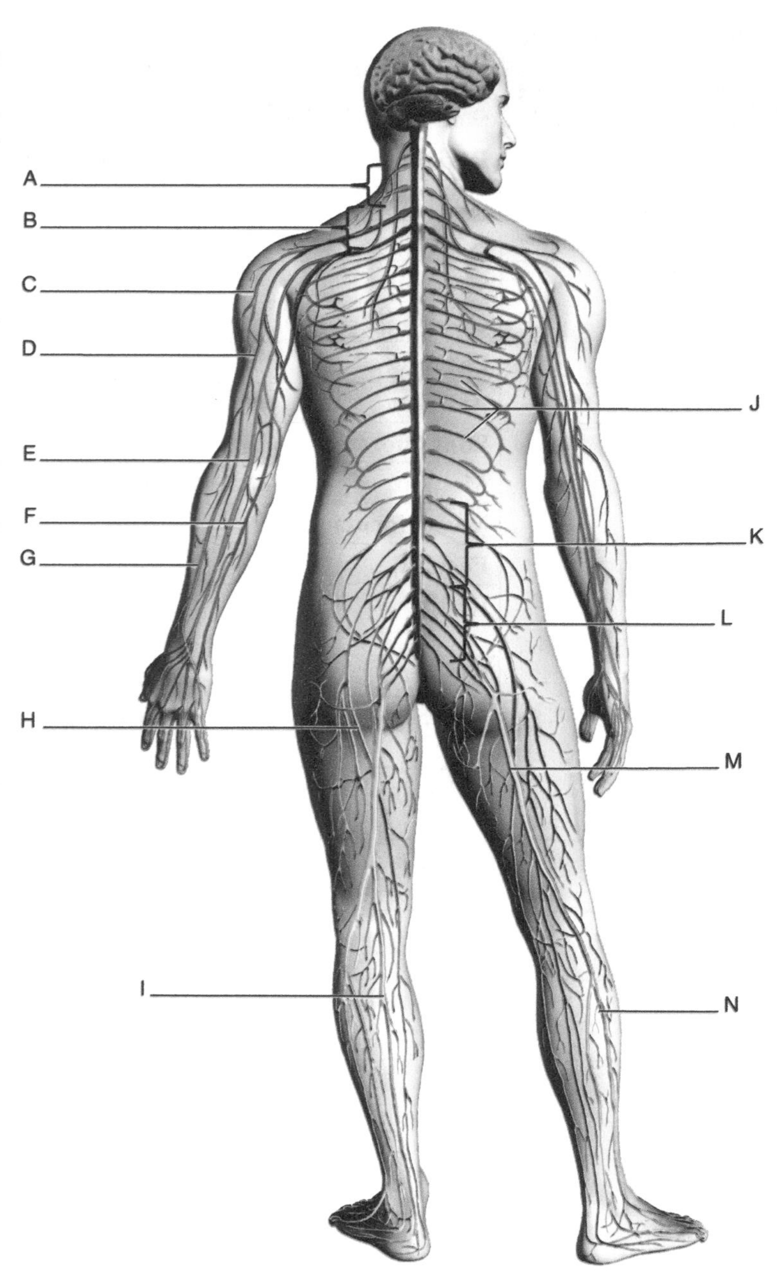

Figure 12-3 Plexus and major peripheral nerves.

Name ____________________ Course Number: ________ Lab Section ______

Lab 12: The Peripheral Nervous System Worksheet

Write out the names of the labeled items in the selected figures.

Figure 12-1.

A. ____________________
B. ____________________
C. ____________________
D. ____________________
E. ____________________
F. ____________________
G. ____________________
H. ____________________
I. ____________________
J. ____________________
K. ____________________
L. ____________________
M. ____________________
N. ____________________
O. ____________________
P. ____________________
Q. ____________________

Figure 12-3.

A. ____________________
B. ____________________
C. ____________________
D. ____________________
E. ____________________
F. ____________________
G. ____________________
H. ____________________
I. ____________________
J. ____________________
K. ____________________
L. ____________________
M. ____________________
N. ____________________

LAB 13 THE SENSES

A *sense*, or *sensation*, is a conscious awareness of a stimulus and can be classified as either a *general* or a *special* sense. **General senses** are in response to stimuli over large regions of the body and include feelings of touch, pressure, pain, stretch, and vibration. Be sure to review the tactile (touch) receptors in the skin and be able to identify them on images and models provided in the lab. **Special senses** are in response to stimulation in selected areas, such as the eye and ear, and include smell, taste, vision, hearing, and equilibrium. Senses are perceived in the brain, but only after impulses are sent to it from specialized structures called **sensory receptors**.

Classification of Sensory Receptors

Sensory receptors are classified based on their *structure* and on the *location of the stimulus* to which they respond. Related to structure, the simplest receptors are *free nerve endings*, which are bare dendrites of unipolar, sensory neurons. Slightly more complex are *encapsulated* receptors that have a surrounding *capsule* made of connective tissue. An example is a *lamellated corpuscle*, which contains multiple layers of connective tissue and looks sort of like an onion. The most complex receptors are those involved in the special senses and include components of the eye, nose, tongue, and inner ear. **Table 13.1** provides a general summary of receptors classified by structure and the types of stimuli to which receptors respond.

Receptors of the Skin

Sensory receptors in the skin may respond to pressure, temperature, and/or a variety of chemicals. In turn, the sensation brought about

Table 13.1 Classification of Sensory Receptors

CLASSIFICATION BY STRUCTURE			
Name of Receptor	**Structure**	**Location(s)**	**Sensitive to**
Free nerve endings	Free ending	Epidermis, papillary dermis	Touch, temperature, chemicals from damaged tissue
Hair follicle receptor	Free ending	Around hair follicles	Hair follicle movement
Krause corpuscle	Encapsulated	Mucous membranes, perineum	Light pressure, low-frequency vibration
Meissner's corpuscle	Encapsulated	Papillary dermis	Light pressure, low-frequency vibration
Merkel's disc	Tactile disc	Stratum basale	Light pressure
Pacinian (Lamellated) corpuscle	Encapsulated	Dermis and hypodermis	Deep pressure, high-frequency vibration

(Continued)

Table 13.1 Classification of Sensory Receptors *(Cont'd)*

CLASSIFICATION BY STIMULI	
Type of Receptor	**Responds to**
Chemoreceptor	Chemicals (e.g., CO_2 and H^+) in the surrounding fluid
Exteroceptor	Changes in the external environment
Mechanoreceptor	Stretch, touch, and/or motion
Nociceptor	Tissue damage and/or various chemicals to cause the brain to perceive pain
Osmoreceptor	Changes in cerebrospinal fluid's osmotic pressure in the hypothalamus
Photoreceptor	Light energy (electromagnetic radiation)
Proprioceptor	Stretch or motion; kinesthetic receptors, tendon organs, muscle spindles, receptors in the inner ear that respond to motion
Tactile receptor	Touch and pressure on the skin
Thermoreceptor	Changes in temperature
Visceroceptor	Changes within the body's blood vessels, visceral organs, or brain

by these responses include feelings of touch, pressure (light or deep), hot and cold, and/or pain. **Table 13.2** shows the major receptors in the skin. Be able to identify these receptors on images and models in the lab.

Somatic Reflex and Touch

A **reflex is a** quick, involuntary response to a stimulus and it requires a *sensory receptor*, an *afferent (sensory) pathway*, an *integration center* (such as the spinal cord), an *efferent (motor) pathway*, and an *effector*. A **visceral reflex involves** an effector other than a skeletal muscle, whereas, a **somatic reflex involves** skeletal muscles. A common somatic reflex that is tested in the clinical setting is the **knee-jerk reflex**. Tapping the patellar ligament (located between the patella and the tibial tuberosity) causes the patella to be drawn inferiorly, which stretches *muscle spindles* (receptors) located within the quadriceps muscles. As a result of impulses passing from the muscle spindles to the spinal cord and back to the quadriceps muscles, the quadriceps muscles contract and cause the knee to jerk. While sitting on the edge of table with your foot off the floor and your thigh relaxed, have a lab partner gently tap your patellar ligament to elicit a knee-jerk reflex (see **Figure 13-1**).

Two-Point Discrimination Test

This test will determine how close together tactile receptors are located in the skin in various locations. If tactile receptors are one inch apart in a particular region of the body, if two separate points less than one inch apart on the skin are touched at the same time, the person will perceive only one point. If the two points touched are farther apart than one inch, the person will perceive two separate points. Use a compass to determine the closest distance that two different, simultaneous points of touch can be distinguished as two separate points. The person being tested should not observe the compass at any time. The tester should begin

Tip of index finger = ______ mm   Palm = ______ mm   Forehead = _____ mm

Anterior forearm = ______ mm   Back of neck = ______ mm

Table 13.2 Classification of Receptors in the Skin

Ending Type	Free Nerve Ending	Root Hair Plexus	Merkel Disc
	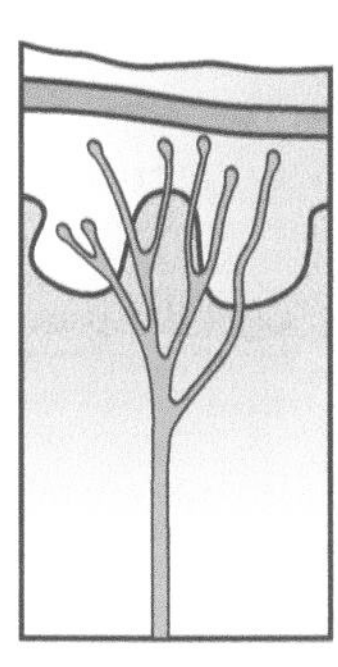		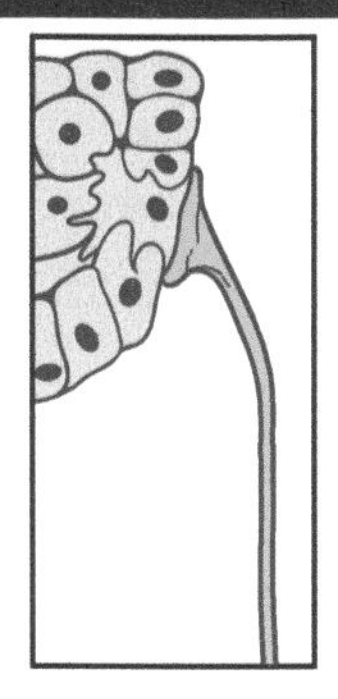
Location	Widespread in deep epidermis and papillary layer of the dermis	Surrounds hair follicles in the reticular layer of the dermis	Stratum basale of epidermis
Function	Detects pressure, change in temperature, pain, touch	Detects movement of the hair	Detects light touch, textures, and shapes
Modality of stimulus	Some are nociceptors; others are thermoreceptors or mechanoreceptors	Mechanoreceptors	Mechanoreceptors

Ending Type	Krause Bulb	Pacinian Corpuscle	Ruffini Corpuscle	Meissner's Corpuscle
	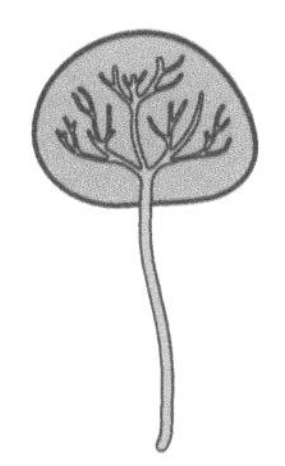	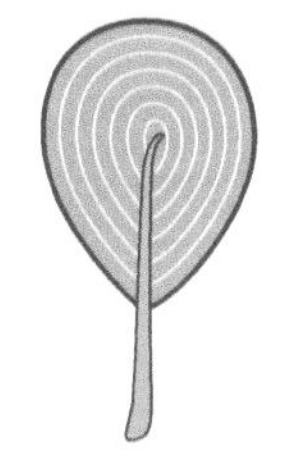	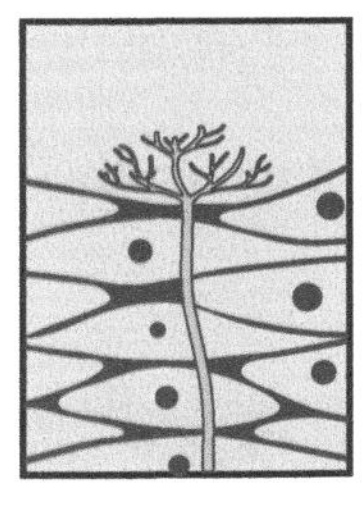	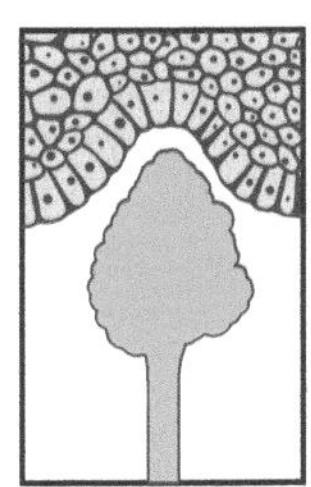
Location	Mucous membranes of oral cavity, nasal cavity, vagina, and anal canal	Dermis, subcutaneous tissue, synovial membranes, and some viscera	Dermis, and subcutaneous layer	Dermal papillae, especially in lips, palms, eyelids, nipples, genitals
Function	Detects light pressure and low-frequency vibration	Detects deep pressure and high-frequency vibration	Detects continuous deep pressure and skin distortion	Detects fine, light touch and texture
Modality of stimulus	Mechanoreceptors	Mechanoreceptors	Mechanoreceptors	Mechanoreceptors

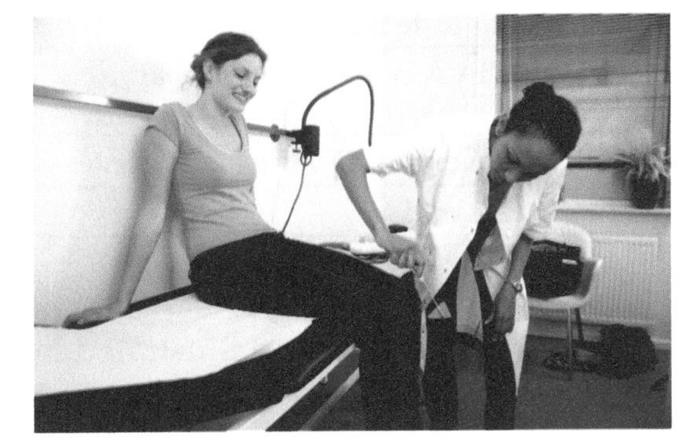

Figure 13-1 Testing a somatic reflex.

with the compass points being as close together as possible and gradually spread them apart until the person being tested can distinguish two points.

Smell and Taste

The senses of smell and taste are closely considered special senses, and they are related to one another. Try tasting a mint while holding your breath. You will notice that without the sense of smell, the sense of taste is greatly impaired. In the lab, you may have experiments set up to test your ability to recognize different odorants and test your sense of smell. The lab instructor will provide directions and handouts for these tests. Information about the anatomical features related to taste and smell is in **Tables 13.3** and **13.4**. After you become familiar with these terms, label **Figures 13-2** and **13-3**. Also, be able to identify these parts on other figures and models (if available).

Table 13.3 Terms Associated with the Sense of Smell	
Basal cells	In the olfactory epithelium; give rise to olfactory neurons
Cribriform plate	Flat region on either side of the crista galli located on the superior portion of the ethmoid bone; Holds the olfactory bulbs and contains olfactory foramina for passage of olfactory neuron axons into the olfactory bulbs
Granule cells	Inhibit the mitral cells leading to olfactory adaptation (getting accustomed to an odor)
Lamina propria	Areolar tissue beneath the olfactory epithelium; contains mast cells
Mitral cell	Neurons in an olfactory bulb that receives impulses from olfactory neurons and conducts impulses to olfactory cortex located partly in the temporal and frontal lobes
Odorant	Airborne chemical that stimulates olfactory neurons; chemical that we can "smell"
Olfactory bulb	Anterior swelling of an olfactory tract beneath the frontal lobe; Contains mitral cells and receive impulses from olfactory neurons
Olfactory cilia	Non-motile cilia projecting into nasal mucus from dendrites of olfactory neurons; contains receptors for odorants (chemicals that we smell)
Olfactory gland	Multicellular gland in lamina propria beneath olfactory epithelium; secretes mucus
Olfactory neuron	Part of the nasal mucosa; depolarize in response to odorants in the nasal cavity
Olfactory epithelium	Contains olfactory neurons, supporting cells, and basal cells; in the superior nasal cavity
Supporting cells	Simple columnar epithelial cells that surround the olfactory neurons

Table 13.4 Terms Associated with the Sense of Taste	
Basal cells	In the gustatory epithelium; give rise to gustatory cells
Gustatory cells	Taste cells inside a taste bud; stimulate sensory neurons in response to certain substances dissolved in the saliva
Taste bud	Structure embedded in a tongue papilla; contains gustatory, supporting, and basal cells
Tongue papillae	
Circumvallate	Large and circular; on posterior part of tongue; innervated by the glossopharyngeal nerve
Filiform	Most numerous papillae; long and thin and do not contain taste buds
Foliate	Ridges on posterior, lateral margin of tongue; innervated by facial and glossopharyngeal nerves
Fungiform	Mushroom-shaped mainly near the sides of the tongue; innervated by the facial nerve
Supporting cell	Columnar cells that surround gustatory cells within a taste bud

The Sense of Sight

Sight, hearing, and equilibrium are the most complex senses and, thus, have the most complex sensory organs. Anatomical features of the eye are listed in **Table 13.5** and illustrated in **Figure 13-4**. Accessory structures of the eye are located outside of the eyeball and are important to the normal functioning of the eye. They are listed in **Table 13.6.** and illustrated in **Figure 13-5.**

Dominant Eye Test

Most individuals do not make equal use of both eyes. They depend more heavily on one eye, the dominant eye. The dominant eye can be identified as follows: Make a tube about 1.5 inches in diameter from a sheet of notebook paper. Look

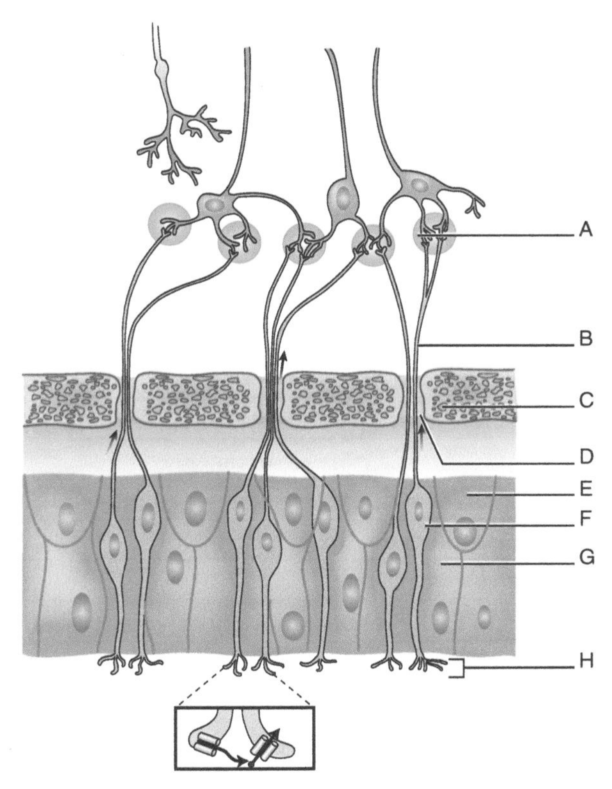

Figure 13-2 Receptors and sensory pathways for the sense of smell.

through it at some object across the room with both eyes. Hold the tube steady, close one eye at a time. The eye with which you see through the tube is the dominant eye.

Which eye is dominant? _________.

Blind Spot Demonstration

Hold the dot about 20 inches directly in front of your right eye. Keeping your left eye closed, slowly move the dot closer to your right eye while looking directly at it with your right eye. As the dot gets closer to the eye, the image of the dot remains fixed on the fovea, but the image of the cross moves along the right eye's retina as the angle at which its image enter the right eye changes. At a certain distance, the cross will disappear from the right eye's field of vision because its image is now focused on the blind spot (optic

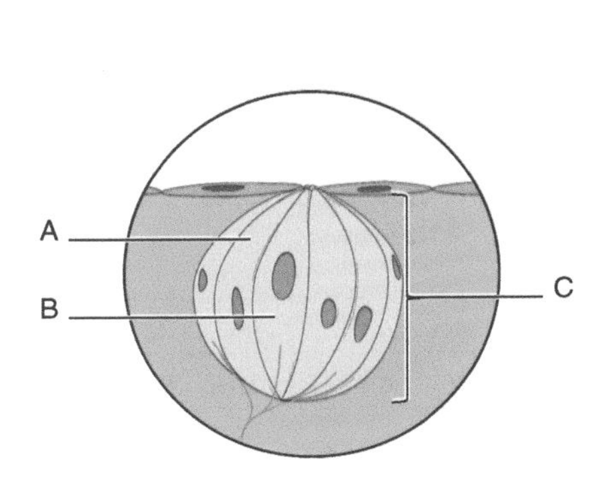

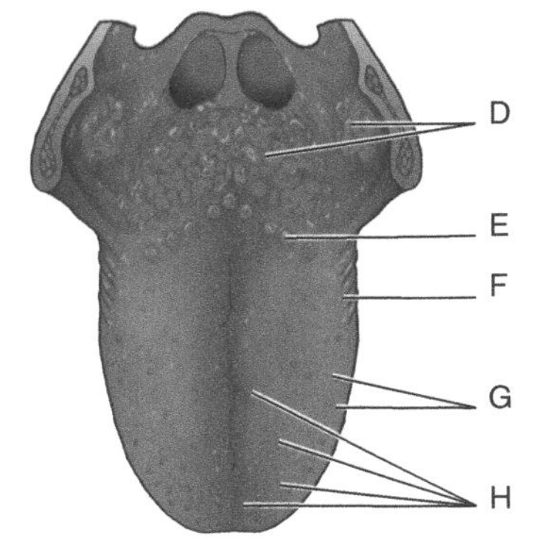

Figure 13-3 Receptors for sense of taste.

Table 13.5 Components of the Eyeball	
Amacrine cells	Retinal cells that modify the way in which ganglion cells and bipolar cells interact
Anterior aqueous chamber	Between the iris and cornea; contains aqueous humor
Aqueous humor	Watery fluid secreted from the ciliary body; passes through the retina and drains into the scleral venous sinus; an excess of this fluid causes glaucoma
Bipolar cell	Retinal cells that are inhibited by photoreceptors; stimulates ganglion cell
Bulbar conjunctiva	Stratified columnar epithelium covering the anterior surface of sclera

(Continued)

Table 13.5 Components of the Eyeball *(Cont'd)*	
Choroid coat	Vascularized tissue layer between the sclera and retina
Ciliary body	Muscle ring attached to the lens by suspensory ligaments; flattens the lens when taut
Cone	Photoreceptor used in color vision; only receptor in the fovea; contains iodopsin
Cornea	Clear, anterior surface of the eyeball
Fibrous tunic	Outermost tissue layer; includes the sclera, cornea, and bulbar conjunctiva
Fovea centralis	Small pit in the center of the macula lutea; site of visual acuity; contains cones
Ganglion cells	Retinal cells that are stimulated by bipolar cells; have axons in optic nerve
Horizontal cells	Retinal cells that modify the way in which photoreceptors and bipolar cells interact
Iris	Colored part of the eye around the pupil; regulates the pupil's diameter
Lens	Biconvex disc made of crystallin proteins; changes shape to focus image on fovea
Macula lutea	Yellow spot located lateral to the optic disc; contains the fovea centralis
Nervous tunic	Innermost tissue layer formed by the retina
Optic disc	Blind spot; site where the optic nerve connects to the retina; lacks photoreceptors
Optic nerve	Contains axons of ganglion cells; conducts impulses from the retina to the brain
Ora serrata	Jagged, anterior border of photoreceptor portion of the retina
Pigmented layer	Part of the retina adjacent to the choroid coat; contains melanocytes
Posterior aqueous chamber	Between the lens and iris; contains aqueous humor
Pupil	Opening in the iris through which light passes into the vitreous chamber
Retina	Innermost layer of cells adjacent to the vitreous humor; contains photoreceptors, bipolar cells, ganglion cells, amacrine cells, horizontal cells, and blood vessels
Rod	Photoreceptor used mostly in dim light; contains rhodopsin
Sclera	Outer white part of the eyeball
Scleral venous sinus	Canal of Schlemm at the edge of the iris; drains aqueous humor into the blood
Suspensory ligament	Connects the ciliary body to the edge of the lens
Vascular tunic	Middle tissue layer; includes the choroid coat, iris, and ciliary body
Vitreous chamber	Largest chamber in the eye between the lens and retina; contains vitreous humor
Vitreous humor	Jellylike fluid in the vitreous chamber

disk), which does not contain photoreceptors. When the "x" disappears, record the distance the page is from the eye. For the right eye, it is ______ cm. Continue to move your right eye closer to the paper and the cross will reappear because its image has moved off the blind spot and onto the photosensitive part of the retina.

Next, repeat the process, but holding the cross directly in front of the left eye and keeping the right eye closed. When performing this test, the dot will disappear then reappear as the page is brought closer to the left eye. : When the dot disappears, record the distance the page is from the eye. For the left eye, it is ______ cm.

Visual Accommodation

The individual whose eyes are normal can clearly see distant objects when the ciliary muscles are relaxed, which means the lens is more rounded. However, ciliary muscles must contract to flatten the lens so that near objects can be brought into focus on the retina. This is called lens accommodation and it is automatic but limited. The distance from the eye to

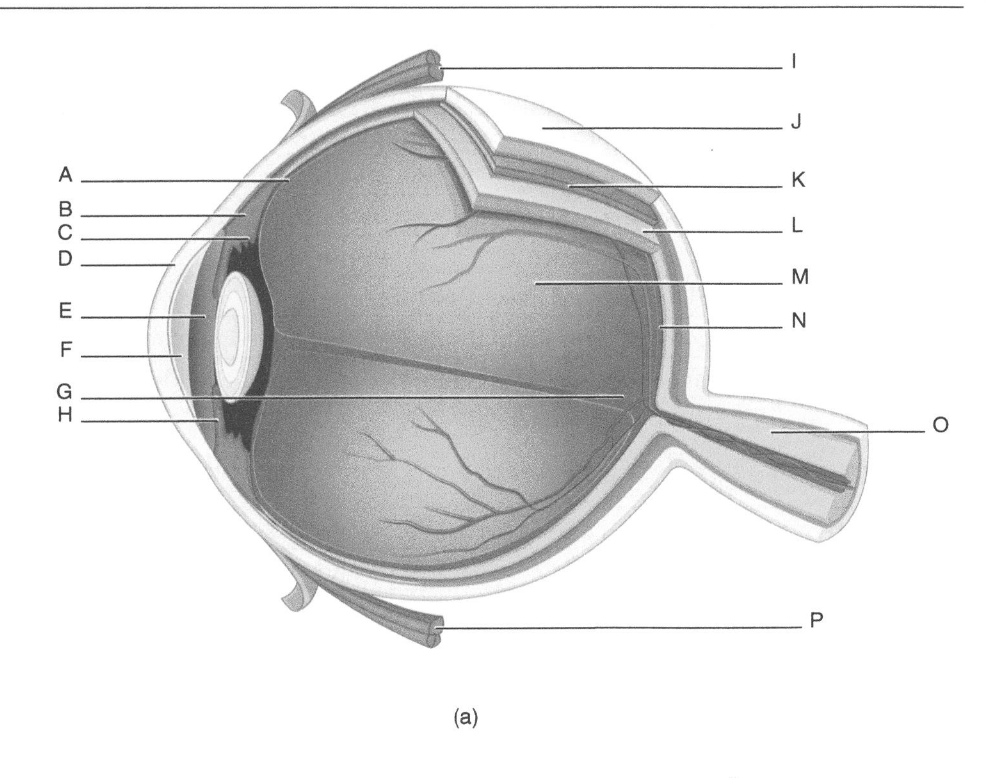

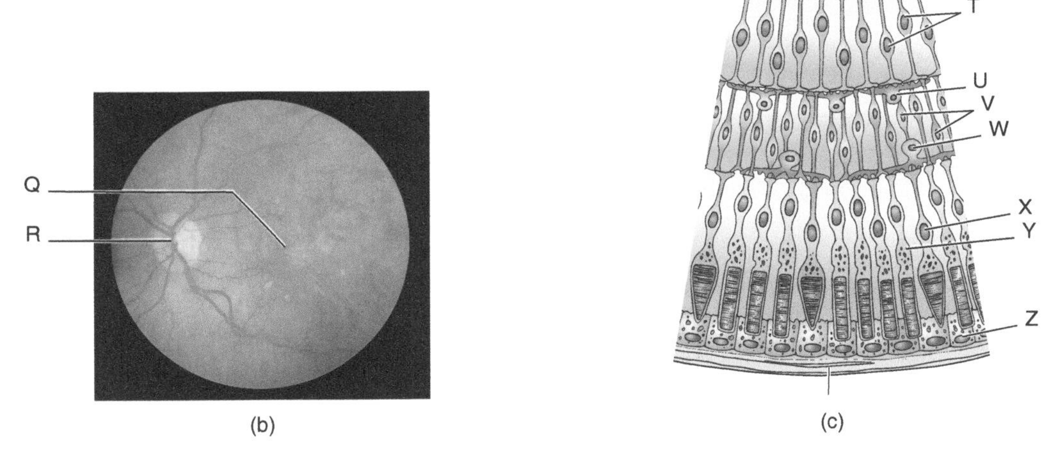

Figure 13-4 Anatomy of the eye: (a) Internal view; (b) retina viewed through an ophthalmoscope; (c) cells of the retina.

Table 13.6 Accessory Structures of the Eye	
Eyebrow	Short terminal hairs on supraorbital ridge
Eyelash	Short terminal hairs on eyelid margin
Extrinsic eye muscles	
Inferior oblique	Attached to inferior surface of eyeball; elevates eyeball
Inferior rectus	Attached to inferior surface of eyeball; inferior rotation
Lateral rectus	On lateral side of eyeball; lateral rotation
Medial rectus	On medial side of eyeball; medial rotation

(Continued)

Table 13.6 Accessory Structures of the Eye (Cont'd)	
Superior oblique	On superior part of eyeball; depresses eyeball
Superior rectus	On superior part of eyeball; elevation of eye
Lacrimal canals	Tiny tubes that drain tears into nasolacrimal duct
Lacrimal caruncle	Fleshy protuberance at medial canthus; secretes oily fluid
Lacrimal gland	Tear gland located in superior-lateral part of orbit
Lacrimal puncta	Tiny openings to lacrimal canals on medial edge of eyelids
Lateral canthus	Lateral connection between upper and lower eyelids
Levator palpebrae muscle	Elevates upper eyelid
Palpebrae	Eyelids
Palpebral conjunctiva	Thin membrane on inner surface of eyelids
Medial canthus	Medial connection between upper and lower eyelids
Nasolacrimal duct	Tube between lacrimal punctum and nasal cavity; drains tears
Retinal artery and vein	Enter the vitreous chamber through the optic nerve
Tarsal (Meibomian) gland	In eyelid; secretes lubricating fluid to keep eyelids from sticking together
Tarsal plate	Gives rigidity to eyelid and its margin

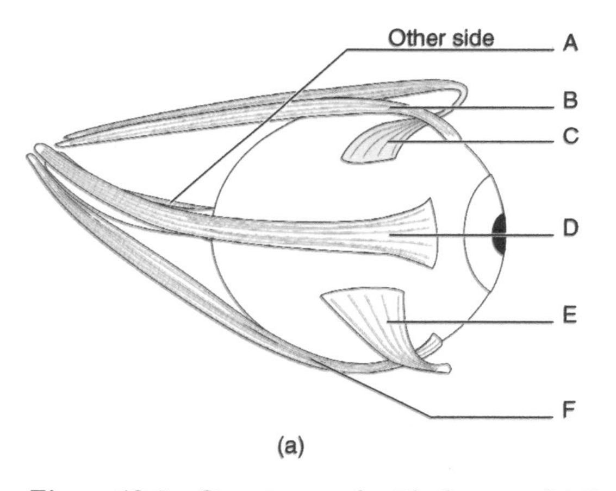

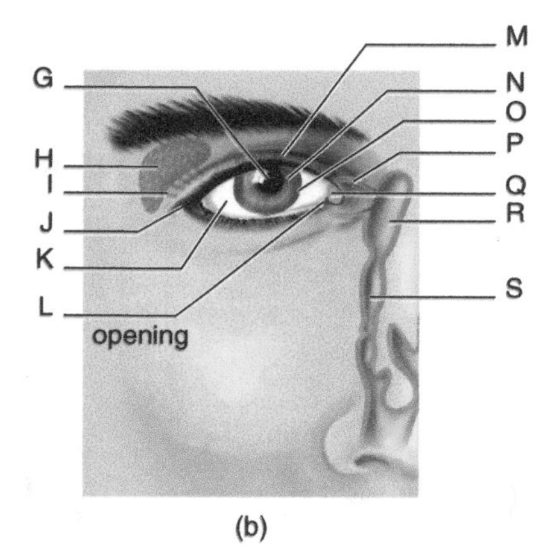

Figure 13-5 Structures of with the eye: (a) Extrinsic eye muscles; (b) accessory structures and parts of eye.

the nearest object that can be clearly focused is the **near point of vision.** Place one hand over an eye and focus the other on a straight pin held at arm's length. Gradually bring the pin closer to your eye, focusing continually until the pin is no longer in sharp focus. Measure the distance from the eye to the pin. This is the near point. Repeat the process with the other eye.

Near point of vision for right eye = _______ mm

Near point of vision for left eye = _______ mm

Vision rating

Normal vision is 20/20; i.e., the person is able to see an object clearly from 20 feet away that the "average" person can see from the same distance. The first number indicates the vision of the person being tested; the second number is for an average person. If a person has 20/40 vision, an "average" person can see clearly see an image 40 feet away, but the person being tested must be closer (20 feet away) in order to see the same object clearly. In other words, he or she is near-sighted.

Stand at the tape on the floor marked "20 feet." Cover one eye and try to read one line on the eye chart. If you cannot read it from 20 feet, slowly move toward the chart until you can read the line. Repeat with other eye.

Vision in right eye = _____/_____

Vision in left eye = _____/_____

20/200 vision is (NEAR/FAR)-sighted.

20/10 vision has (BETTER/POORER) vision compared to the average person.

A person who is (NEAR/FAR)-sighted may see distant objects as well as the average person but has problems reading a book without corrective lenses.

Depth Perception

Determine the accuracy with which you can correctly align two arrows on a depth perception device placed on a table. Take the cord and sit approximately 8 feet from the device. Have your partner place the stationary arrow at some point along the track (do not watch while the arrow is being positioned). With your left eye closed, look toward the device and move the other arrow with the cord until you think both arrows are at the same location. Repeat with the other eye, then using both eyes.

Difference using right eye only = _____ mm;

Difference using left eye only = ______ mm

Difference when using both eyes = ________ mm.

The Ear

Now we turn our attention to the ear, which consists of parts devoted to the sense of hearing and also the sense of balance and equilibrium. The ear has three major regions: external, middle, and inner ear. Review the parts of the external and middle ear in **Table 13.7** and then identify the parts in **Figure 13-6**.

Table 13.7 Components of the External and Middle Ear

EXTERNAL EAR	
Antihelix	Inner rim of pinna
Antitragus	Projection just above earlobe
Ear concha	Curved depressions in the pinna
Ear lobule	Fleshy inferior portion of pinna
External auditory meatus	**(Auditory canal)**; canal between the pinna and eardrum
Helix	Outer curved portion of pinna
Pinna	**(Auricle)**; fleshy part of outer ear on side of head)
Tragus	Projection on anterior side of the opening to external auditory meatus
Tympanum	**(Eardrum)**; vibrates in response to sound waves and moves the malleus (hammer) ossicle
MIDDLE EAR	
Auditory tube	**(Eustachian tube)**; connects middle ear with throat allowing for pressure equilibration
Hair cells	Sensory cells in the maculae, cristae, and organ of Corti
Helicotrema	Connection between scala vestibuli and scala tympani
Incus	**(Anvil)**; middle ear ossicle between malleus and stapes
Malleus	**(Hammer)**; middle ear ossicle between tympanum and incus
Stapedius muscle	Smallest skeletal muscle; attached to stapes and prevents excessive movement
Stapes	**(Stirrup)**; middle ear ossicle between incus and oval window
Tensor tympani muscle	Tiny skeletal muscle attached to tympanum; prevents excessive movement

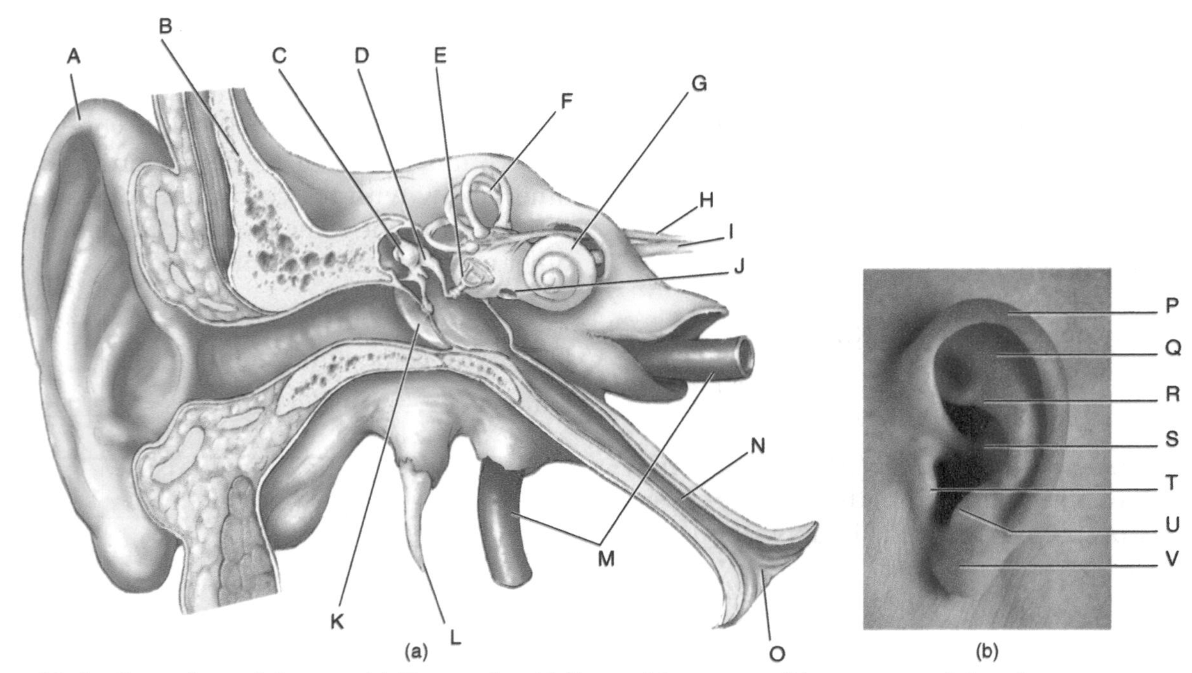

Figure 13-6 Overview of the ear: (a) External, middle, and inner ear; (b) anatomy of the pinna.

Table 13.8 Components of the Inner Ear

COCHLEAR PART	
Bony labyrinth	Fluid-filled series of cavities inside the petrous portion of the temporal bone; includes semicircular canals, vestibule, and cochlea
Cochlear nerve	Part of the vestibulocochlear nerve (CN 8); carries impulses for hearing
Helicotrema	Region of the cochlea where the scala vestibuli meets the scala tympani
Membranous labyrinth	Fluid-filled series of tubes inside the bony labyrinth; includes semicircular ducts, utricle, saccule, and cochlear duct
Oval window	Membrane-covered opening between stapes and scala vestibuli
Round window	Membrane-covered opening at end of scala tympani; relieves pressure in cochlea
Scala media	Also called the cochlear duct; part of membranous labyrinth that contains the spiral organ (of Corti)
Scala tympani	Inferior chamber in cochlea
Scala vestibuli	Superior chamber in cochlea
Spiral organ (of Corti)	Sensory organ of hearing; inside cochlear duct
Tectorial membrane	Gelatinous mass overlying the organ of Corti
Vestibular membrane	Separates scala vestibuli from scala media (cochlear duct)
SEMICIRCULAR CANAL PART	
Ampulla	Sac at base of semicircular duct; contains endolymph and a crista
Crista ampullaris	Structure inside an ampulla that responds to dynamic movement (e.g., spinning, tumbling, etc.); consists of hair cells with cilia embedded in cupula
Cupula	Gelatinous mass in which the cilia of a crista ampullaris is embedded
Kinocilium	Long cilium on a hair cell within the crista
Semicircular canal	Ring-like tube in bony labyrinth; contains perilymph and a semicircular duct
Semicircular duct	Part of the membranous labyrinth inside a semicircular canal; contains endolymph and a crista ampullaris

VESTIBULAR PART	
Kinocilium	Long cilium on a hair cell within the macula
Macula	Structure inside a utricle and saccule that responds to head position along with acceleration and deceleration of the body
Otolithic membrane	Gelatinous mass on top of maculae inside utricle and saccule
Otoliths	"Ear stones" embedded in otolithic membrane
Saccule	Vertical sac-like structure inside the vestibule; part of the membranous labyrinth and contains a macula
Stereocilia	Microvilli (extensions of a hair cell's membrane)
Utricle	Horizontal sac-like structure inside the vestibule; part of the membranous labyrinth and contains a macula
Vestibular nerve	Contains axons from the maculae and cristae
Vestibule	Part of bony labyrinth between semicircular canals and cochlea

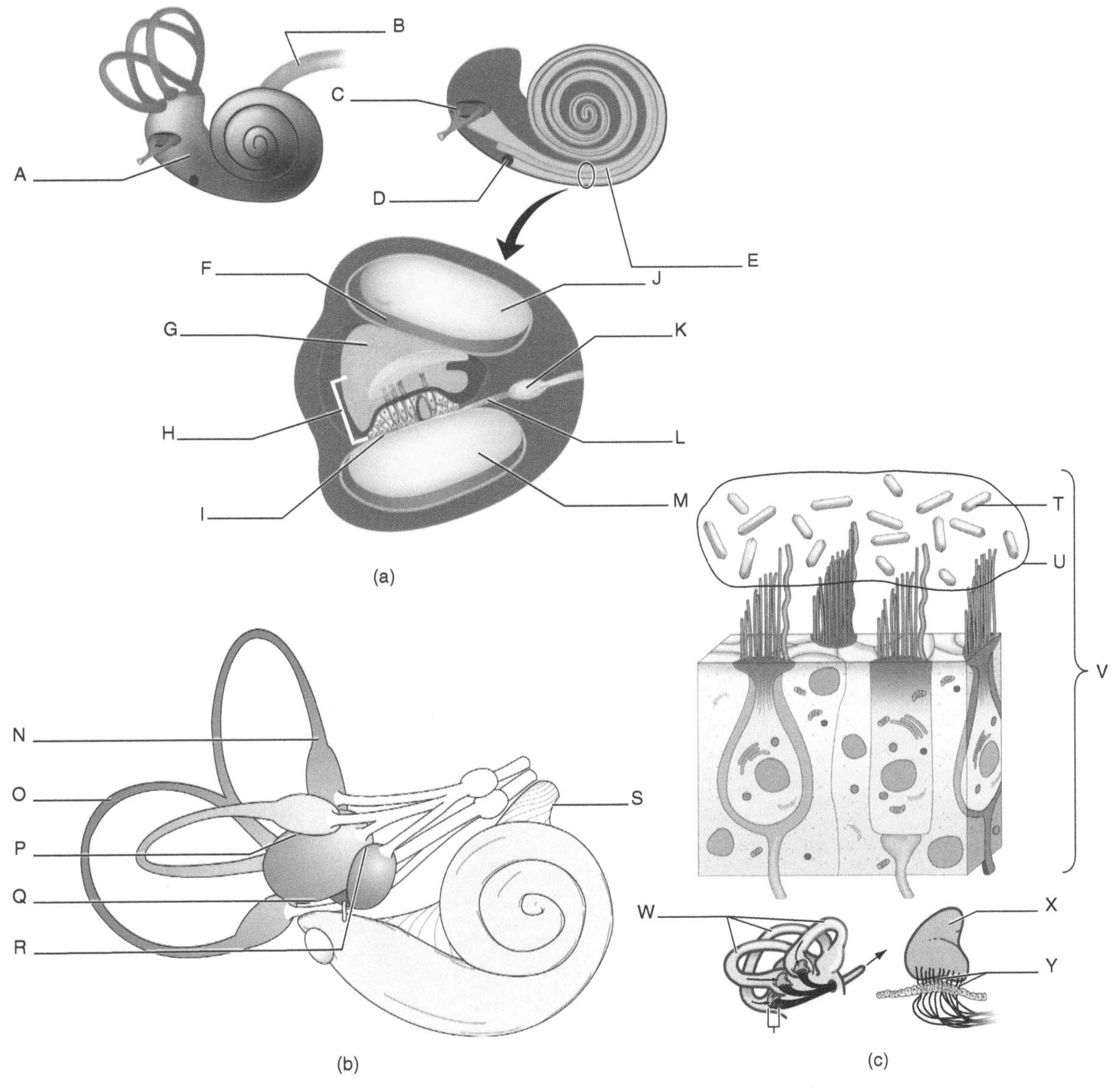

Figure 13-7 Structures of the inner ear: (a) Chambers within cochlea; (b) membranous labyrinth; (c) hair cells in maculae & cristae.

Reaction Time

The time it takes for a person to respond to a stimulus is called reaction time. In this test, you will measure your reaction time to a visual and an audible stimulus. A visual response will be measured electronically, using a reaction timer, and with a meter stick, while the audible response will be measured electronically. Have a partner activate the reaction timer for visual stimuli, as directed by your instructor, and then measure your response time. Next, have your partner activate the timer for an audible stimulus. Perform several runs and average your best reaction time in 1/100s of a second. Additionally, have your partner hold the meter stick at the top while you hold your thumb and index finger about 1-inch apart at the bottom end. While you watch the bottom end, your partner will release the stick. When you realize that the stick is dropping, squeeze your thumb and index finger together to stop the stick. The increments on the meter stick are in milliseconds (msec). Try this several times and record your best time. The goal of your partner is to try and not be predictable about dropping the stick. You do not want to begin squeezing your thumb and index finger together prematurely in anticipation of the stick dropping.

Visual reaction time (using timer) = ________ second

Visual reaction time (using stick) = ________ second

Audible reaction time = ________ second

Name ____________________ Course Number: ________ Lab Section ______

Lab 13: Senses Worksheet

Write out the names of the labeled items in the selected figures.

Figure 13-2.

A. ______
B. ______
C. ______
D. ______
E. ______
F. ______
G. ______
H. ______

Figure 13-3.

A. ______
B. ______
C. ______
D. ______
E. ______
F. ______
G. ______
H. ______

Figure 13-4.

A. ______
B. ______
C. ______
D. ______
E. ______
F. ______
G. ______
H. ______
I. ______
J. ______
K. ______
L. ______
M. ______
N. ______
O. ______
P. ______
Q. ______
R. ______
S. ______
T. ______
U. ______
V. ______
W. ______
X. ______
Y. ______
Z. ______

Figure 13-5.

A. ______
B. ______
C. ______
D. ______
E. ______
F. ______
G. ______
H. ______
I. ______
J. ______
K. ______
L. ______
M. ______
N. ______
O. ______
P. ______
Q. ______
R. ______
S. ______

Figure 13-6.

A. ______
B. ______
C. ______
D. ______
E. ______
F. ______
G. ______
H. ______
I. ______
J. ______
K. ______
L. ______
M. ______
N. ______
O. ______
P. ______
Q. ______
R. ______

S. ____________________
T. ____________________
U. ____________________
V. ____________________

Figure 13-7.

A. ____________________
B. ____________________
C. ____________________
D. ____________________
E. ____________________
F. ____________________
G. ____________________
H. ____________________
I. ____________________
J. ____________________
K. ____________________
L. ____________________
M. ____________________
N. ____________________
O. ____________________
P. ____________________
Q. ____________________
R. ____________________
S. ____________________
T. ____________________
U. ____________________
V. ____________________
W. ____________________
X. ____________________
Y. ____________________

Lab 14 Metabolism

Overview of Metabolism

Metabolism is the sum of all chemical reactions in the body, and it includes ***anabolism*** (synthesis reactions) and ***catabolism*** (decomposition reactions). Since human anatomy and physiology presents information related to various aspects of metabolism, it is important to have an appreciation for factors that affect the rates at which metabolic reactions occur. A general overview of where glycolysis and cellular respiration take place within a cell are shown in **Figure 14-1**, while a more detailed view of a mitochondrion is shown in **14-2**. Use these figures as you define and identify the location of each item in **Table 14.1**.

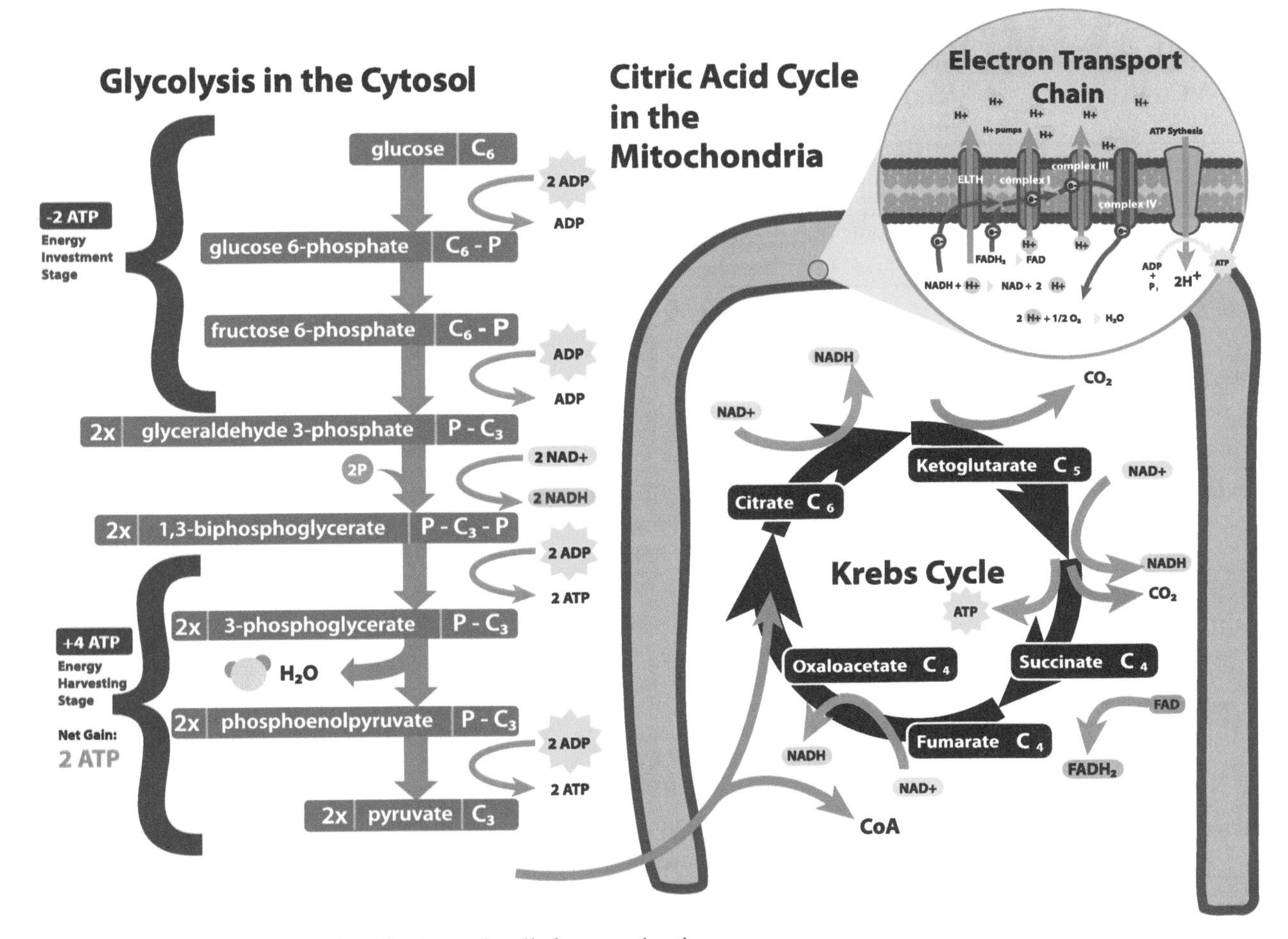

Figure 14-1 Reactions in glycolysis and cellular respiration.

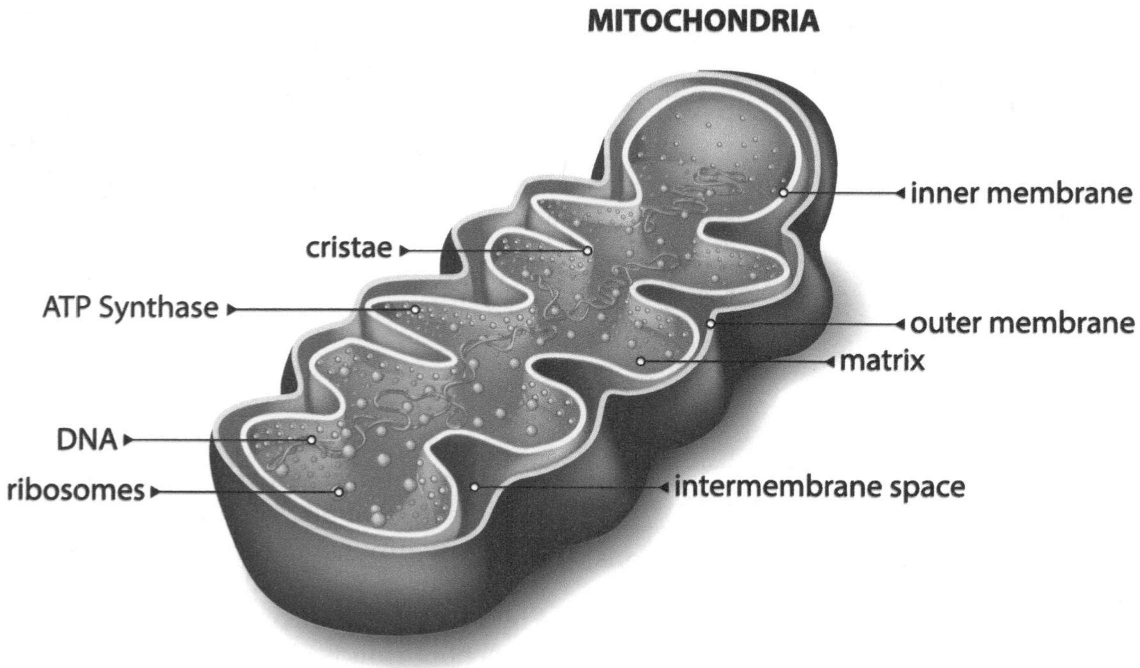

MITOCHONDRIA

inner membrane
cristae
ATP Synthase
outer membrane
matrix
DNA
ribosomes
intermembrane space

Figure 14-2 Components of a mitochondrion.

Table 14.1 Items to Know for Metabolism	
Acetyl CoA	Acetyl compound bound to coenzyme A (derived from pantothenic acid); binds to oxaloacetate in the mitochondrial matrix to form citrate
Aerobic respiration	Process in which molecular oxygen is used in the process of ATP production within the mitochondrion; involves ***oxidative phosphorylation*** of ADP
ATP synthases	Enzymes in the mitochondrion's inner membrane that use energy from diffusing protons (H^+ ions) to phosphorylate ADP
Citrate	Ionized form of ***citric acid***; first compound formed in the Kreb's cycle when oxaloacetate binds with acetyl CoA
Coenzyme A	Coenzyme derived from pantothenic acid (vitamin B_5) that binds to acetyl compounds in the mitochondrial matrix to form acetyl CoA
Cytochromes	Specific proteins within the ETC
Electron shuttles	Transport electrons from cytosolic NADH through the mitochondrion's intermembrane space and delivers them to the ETC; ***malate-aspartate shuttle*** is in most cells while brown fat cells use the ***glycerol-phosphate shuttle***
Electron transport chain	***ETC***; also called ***electron transport system (ETS)***; has four complexes (I, III, and IV function as proton pumps while II contains FAD); oxidizes NADH to NAD^+ and $FADH_2$ to FAD; pumps protons into intermembrane space; reduces molecular oxygen causing formation of water
FAD	***Flavin adenine dinucleotide***; a coenzyme derived from riboflavin (vitamin B_2) and found in complex II of the ETC; receives electrons from the Kreb's cycle and delivers them to complex III; oxidized form: FAD; reduced form: $FADH_2$
Fermentation	Anaerobic reaction in which pyruvate is reduced to form lactate; in yeast this reaction involves decarboxylation and generates alcohol and CO_2 instead of lactate
Glycolysis	Initial stages of glucose oxidation in the cytosol; requires 2 ATPs but forms 4 ATPs; also forms pyruvic acid and reduces NAD^+ to NADH

Table 14.1 Items to Know for Metabolism *(Cont'd)*

Kreb's cycle	Also called ***citric acid cycle*** and ***tricarboxylic acid (TCA) cycle***; occurs in the mitochondrion; reduces NAD^+ to NADH and FAD to $FADH_2$; forms GTP that converts ADP to ATP; begins when acetyl CoA binds to oxaloacetic acid to form citric acid
Lactate	3-carbon ionized form of ***lactic acid***; formed when pyruvate becomes reduced by NADH in the cytosol; allows NAD^+ to return to glycolysis reactions
Mitochondrion	Membranous organelle in which aerobic respiration occurs
NAD	***Nicotinamide adenine dinucleotide***; coenzyme derived from niacin (vitamin B_3) that transports electrons in the cytoplasm (including the mitochondria); oxidized form is NAD^+ and reduced form is NADH
Oxaloacetate	Ionized form of **oxaloacetic acid**; binds to coenzyme A to form citrate
PGAL	**Phosphoglyceraldehyde** or **glyceraldehyde phosphate**; intermediate compound in glycolysis
Pyruvate	3-carbon ionized form of **pyruvic acid**; formed in glycolysis and enters mitochondrion where it is decarboxylated and oxidized to form a 2-carbon acetyl compound

Factors Affecting Metabolic Rate

For this lab, teams will conduct experiments using unicellular, fungal organisms called **yeasts** to determine how certain variables, including temperature and the concentrations of enzymes and substrates, affect the rate at which metabolism produces particular products. For these experiments, the yeast will be the source of enzymes that break down a substrate, which will be common table sugar (sucrose). Since the complete breakdown of sugar produces carbon dioxide (CO_2), metabolic rate will be based on the amount of CO_2 the yeasts produce in a certain amount of time. In order to make logical conclusions about each variable's effect, it is important to investigate only one variable at a time.

Effect of Temperature

The effect of temperature on CO_2 production will be observed in mixtures having the *same amount of yeast* and the *same amount of sugar*. The different temperatures for this experiment will be *warm* (in an oven), *room temperature* (at your lab table), and *cold* (in ice).

Effect of Enzyme Concentration

The effect of enzyme concentration will be determined by varying the amount of yeast in different samples. The effect of yeast concentration on CO_2 production will be observed in mixtures having the *same amount of sugar* and the *same temperature*.

Effect of Substrate Concentration

The effect of substrate concentration on metabolic rate will be determined in mixtures having the *same amount of yeast* and the *same temperature*.

Team 1: Mix 10 mL of yeast suspension with 40 mL of sugar solution, and then place mixture in the **oven**.
Team 2: Mix 10 mL of yeast suspension with 40 mL of sugar solution, and then place mixture **at your table**.
Team 3: Mix 10 mL of yeast suspension with 40 mL of sugar solution, and then place mixture in the **ice**.

Team 4: Mix 25 mL of yeast suspension with 25 mL of sugar solution, and then keep mixture in the **oven**.
Team 5: Mix 25 mL of yeast suspension with 25 mL of sugar solution, and then keep mixture at **your table**.
Team 6: Mix 25 mL of yeast suspension with 25 mL of sugar solution, and then keep mixture in the **ice**.

Team 7: Mix 40 mL of yeast suspension with 10 mL of sugar solution, and then place mixture in the **oven**.
Team 8: Mix 40 mL of yeast suspension with 10 mL of sugar solution, and then place mixture **at your table**
Team 9: Mix 40 mL of yeast suspension with 10 mL of sugar solution, and then place mixture in the **ice**.

Your lab instructor will divide the class into 9 teams, and each team will prepare test tubes with various yeast-sugar combinations, as described below. The number of milliliters of yeast and sugar depends on the team to which you are assigned. Your instructor will show you how to position your team's test tubes in the oven, at your table, or in the ice container.

1. Procedure for preparing yeast-sugar mixtures
   a. In a plastic beaker, mix a specified amount of yeast suspension with a specified amount of sugar solution.
   b. Fill a small test tube to overflowing with the yeast-sugar mixture.
   c. Place a larger test tube down over the smaller tube then quickly invert the two tubes (as shown below).
   d. Use a ruler to measure the height of the displaced yeast-sugar mixture in the outer tube. Initial height = ______ mm.
   e. Place the two test tubes in the location designated for your team and leave for 25 minutes.
   f. After 25 minutes, remove the tubes and measure the height of the displaced yeast-sugar mixture in the outer tube.

      Final height = _____ mm.
   g. Subtract the initial height from the final height and record the difference. Total fluid displaced = _____ mm.
   h. Calculate the displaced per minute by dividing the total displacement by 25. Total displacement/minute = ______ mm
   i. Record your data on the board at the front of the room.

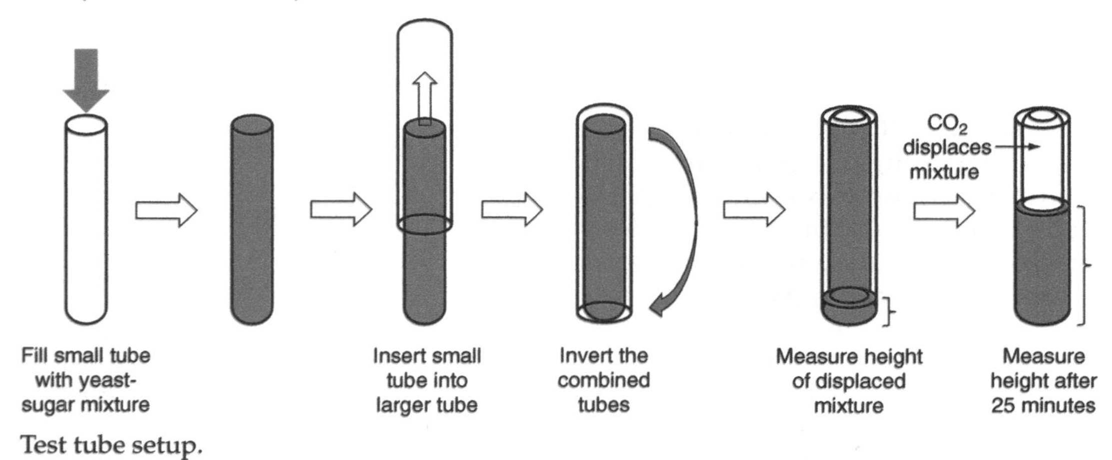

Test tube setup.

2. Graph the effect of temperature on yeast metabolic rate.
   a. Looking at the data recorded on the board, assign values to the Y-axis ranging from 0 to the maximum displacement.
   b. Plot three data points for each graph and draw a reasonable "line of best fit" through the data.

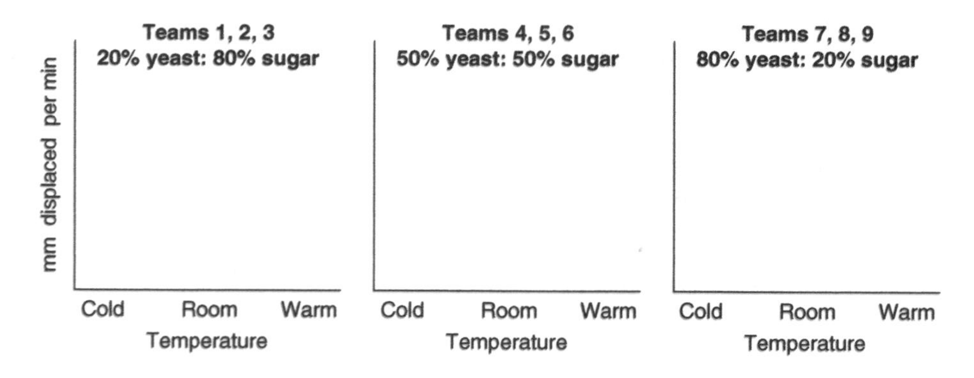

3. Graph the effect of yeast concentration on CO_2 production rate.

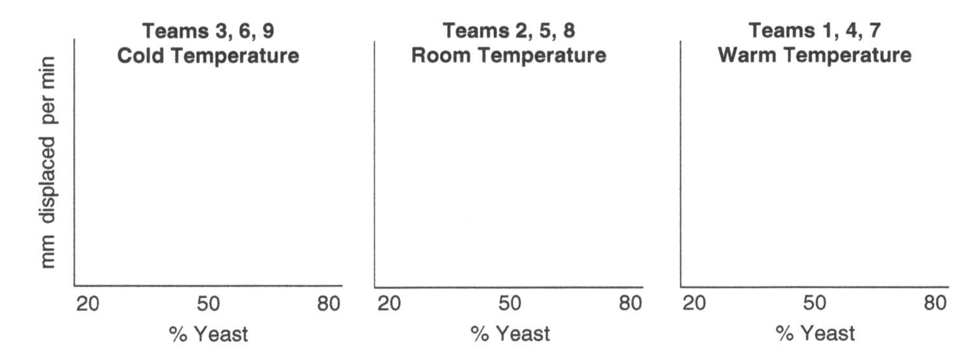

Did temperature affect the metabolic rate of yeasts? ________ If so, what can you conclude?

Did <yeast> affect the amount of CO_2 produced in 25 minutes? ________ If so, what can you conclude?

Did the <sugar> affect the amount of CO_2 produced in 25 minutes? _______ If so, what can you conclude?

Surface Area / Volume Ratio

When studying humans and other endotherms (all mammals and birds), one finds that small individuals usually have higher metabolic rates than larger individuals at the same temperature. You may wonder why a baby has a faster respiration rate than an adult or why mice have faster respiration rates than elephants. The reason is that small organisms have more surface area exposed to the external environment *relative* to each unit of volume in their bodies. (Note the term *relative*.) Larger organisms have more surface area and more volume, but when calculating the quotient SA/V, the value is larger for smaller organisms than for larger organisms having the same basic shape. Therefore, at normal room temperature, smaller endotherms lose heat more quickly than do larger individuals. Complete the Lab 14 Worksheet on the next page.

Name ____________________ Course Number: ________ Lab Section ______

Lab 14: Metabolism Worksheet

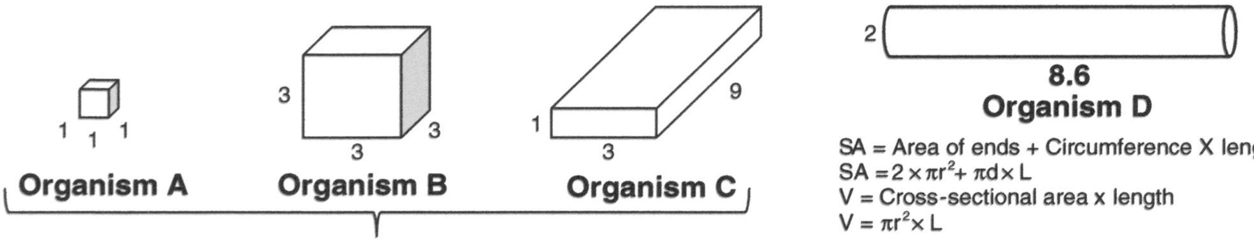

1 1 1 **Organism A** 3 3 3 **Organism B** 1 3 9 **Organism C**

SA = Area of each side × Number of sides
V = Length × Width × Height

2 **8.6** **Organism D**

SA = Area of ends + Circumference X length
SA = $2 \times \pi r^2 + \pi d \times L$
V = Cross-sectional area x length
V = $\pi r^2 \times L$

1. **Surface Area to Volume Ratio**: Calculate the SA/V for each organism above. Show your calculations on each line.

Organism A: SA = ______________ V = ______________; SA/V = ______________
Organism B: SA = ______________ V = ______________; SA/V = ______________
Organism C: SA = ______________ V = ______________; SA/V = ______________
Organism D: SA = ______________ V = ______________; SA/V = ______________

2. **SA/V and Heat Loss in a Cold Environment**: Each cubic unit of volume in the organisms can produce 1 kcal of heat per **MINUTE**. Assume that in a cold environment 600 Kcal of heat escapes from each unit of surface area in a **DAY**. Calculate whether each organism will live or die in this cool environment.

Organism A is generating ______________________________ kcal/hour
Organism A is losing ______________________________ kcal/hour
Organism A would likely (LIVE/DIE) in this environment.
Organism B is generating ______________________________ kcal/hour
Organism B is losing ______________________________ kcal/hour
Organism B would likely (LIVE/DIE) in this environment.
Organism C is generating ______________________________ kcal/hour
Organism C is losing ______________________________ kcal/hour
Organism C would likely (LIVE/DIE) in this environment.
Organism D is generating ______________________________ kcal/hour
Organism D is losing ______________________________ kcal/hour
Organism D would likely (LIVE/DIE) in this environment.

3. **SA/V and Heat Retention in a Warm Environment**: The hypothetical organisms above were placed into a very warm environment, one that requires removal of excess body from the body to prevent hyperthermia (overheating). This environment allows a maximum of 0.008 kcal of heat to radiate out through each unit of surface area each **SECOND**. As before, each cubic unit of volume is still generating 1 kcal of heat per **MINUTE**.

Organism A is generating ______________________________ kcal/hour
Organism A is losing ______________________________ kcal/hour
Organism A would likely (LIVE/DIE) in this environment.
Organism B is generating ______________________________ kcal/hour

Organism B is losing ________________ kcal/hour
Organism B would likely (LIVE/DIE) in this environment.
Organism C is generating ________________ kcal/hour
Organism C is losing ________________ kcal/hour
Organism C would likely (LIVE/DIE) in this environment.
Organism D is generating ________________ kcal/hour
Organism D is losing ________________ cal/hour
Organism D would likely (LIVE/DIE) in this environment.

4. **SA/V and Nutrient Acquisition**: Now, assume the hypothetical organisms are unicellular and are placed into a jar of water containing nutrients they need for survival. Assume each unit of volume requires 0.001 units of nutrients per **SECOND** and that each unit of surface area allows diffusion of 1.6 units of nutrients into the cell per **HOUR**.

Organism A needs ________________ nutrient units/hour
Organism A can absorb ________________ nutrient units/hour
Organism A would likely (LIVE/DIE) in this environment.
Organism B needs ________________ nutrient units/hour
Organism B can absorb ________________ nutrient units/hour
Organism B would likely (LIVE/DIE) in this environment.
Organism C needs ________________ nutrient units/hour
Organism C can absorb ________________ nutrient units/hour
Organism C would likely (LIVE/DIE) in this environment.
Organism D needs ________________ nutrient units/hour
Organism D can absorb ________________ nutrient units/hour
Organism D would likely (LIVE/DIE) in this environment.

5. **Surface Area/Volume Ratio and Toxin Exposure**: What if the hypothetical organisms above are unicellular and are placed into a jar of water containing a toxin compound that will readily diffuse into the cell? Assume that each unit of volume in these organisms has a tolerable limit of 3 units of toxin per **HOUR** (more than this is fatal) and that each unit of surface area allows for the diffusion of 0.025 units of toxin into the cell per **MINUTE**.

Organism A can tolerate ________________ toxin units/hour
Organism A will absorb ________________ toxin units/hour
Organism A would likely (LIVE/DIE) in this environment.
Organism B can tolerate ________________ toxin units/hour
Organism B will absorb ________________ toxin units/hour
Organism B would likely (LIVE/DIE) in this environment.
Organism C can tolerate ________________ toxin units/hour
Organism C will absorb ________________ toxin units/hour
Organism C would likely (LIVE/DIE) in this environment.
Organism D can tolerate ________________ toxin units/hour
Organism D will absorb ________________ toxin units/hour
Organism D would likely (LIVE/DIE) in this environment

LAB 15 ENDOCRINE SYSTEM

Hormone-Secreting Organs

The endocrine system is a control system, as is the nervous system, but instead of stimulating other systems with electrical signals, the endocrine system stimulates them with chemicals called *hormones*. Cells that secrete hormones are endocrine cells, and they secrete hormones directly into the blood. Cells that respond to a particular hormone are the *targets* for that hormone. In most cases, endocrine cells are found within *endocrine glands*, which have as their primary function the production of hormones. However, some hormone-secreting organs also secrete other substances. For example, the stomach produces substances needed for food digestion, but it also secretes several hormones, including *gastrin* and *ghrelin*. In this lab, you should be able to recognize the major hormone-producing organs listed in **Table 15.1** and illustrated in **Figures 15-1, 15-2, and 15-3**. Also, be able to name the hormones secreted from each structure and discuss their functions, which are shown in **Table 15.2**.

Table 15.1 Hormone-Secreting Organs

Organ	Hormones Secreted
Adenohypophysis	*Adrenocorticotropic hormone* (ACTH or corticotropin), *Follicle-stimulating hormone* (FSH), *Growth hormone* (GH or *somatotropin*), *Luteinizing hormone* (LH), *Prolactin* (PRL or *lactogenic hormone*), *Thyroid-stimulating hormone* (TSH or *thyrotropin*)
Adipose tissue	Leptin
Adrenal cortex	Mineralocorticoids (*Aldosterone*), glucocorticoids (*Cortisol* or *hydrocortisone*), gonadocorticoids (*androgens*)
Adrenal medulla	*Epinephrine* (E or *adrenaline*), *Norepinephrine* (NE or *noradrenaline*)
Duodenum	*Gastrin, Cholecystokinin* (CCK), *Glucose insulinotropic peptide* (GIP or *gastric inhibitory peptide*), *Secretin, Vasoactive intestinal peptide* (VIP)
Heart	*Atrial natriuretic peptide* (ANP)
Hypothalamus	*Corticotropin releasing hormone* (CRH), *Growth hormone releasing hormone* (GHRH), *Growth hormone inhibiting hormone* (GHIH or *Somatostatin*), *Gonadotropin-releasing hormone* (GnRH), *Prolactin-releasing hormone* (PRH), *Thyrotropin-releasing hormone* (TRH)
Kidneys	*Erythropoietin* (EPO) and *calcitriol* (only activated in the kidney)
Neurohypophysis	*Antidiuretic hormone* (ADH or *Vasopressin*), *Oxytocin* (OT)
Ovaries	*Estrogen, Progesterone, Relaxin*
Pancreas	*Glucagon, Insulin, Growth hormone-inhibiting hormone* (**GHIH** or *somatostatin)*
Parathyroids	*Parathormone* (PTH)
Pineal	*Melatonin*

(Continued)

Table 15.1	Hormone-Secreting Organs (Cont'd)
Placenta	*Human Chorionic Gonadotropin (HCG)*
Stomach	*Gastrin* and *Ghrelin*
Testes	*Testosterone, Inhibin*
Thymus	*Thymosin*
Thyroid	*Calcitonin* (CT), *Thyroxine*

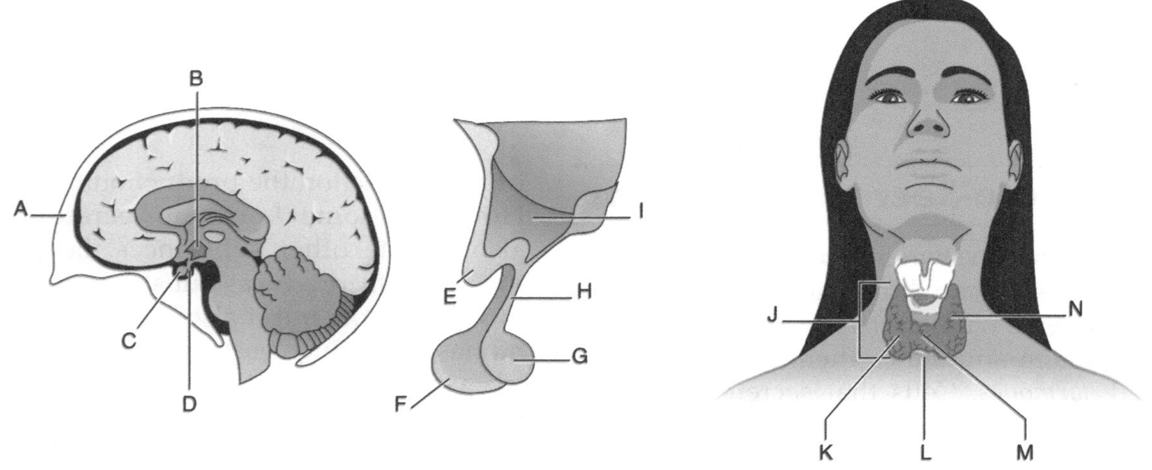

Figure 15-1 Hormone-secreting structure in head and neck.

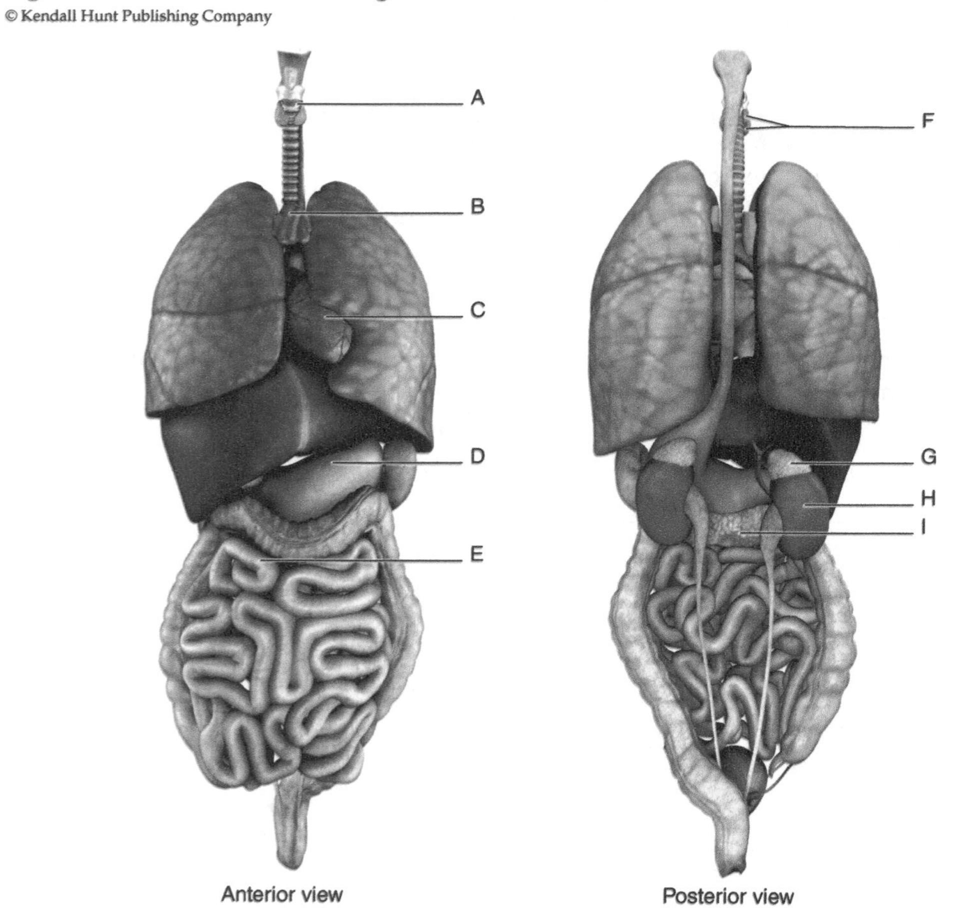

Figure 15-2 Hormone-secreting organ of the thorax and abdomen (reproductive organs not shown).

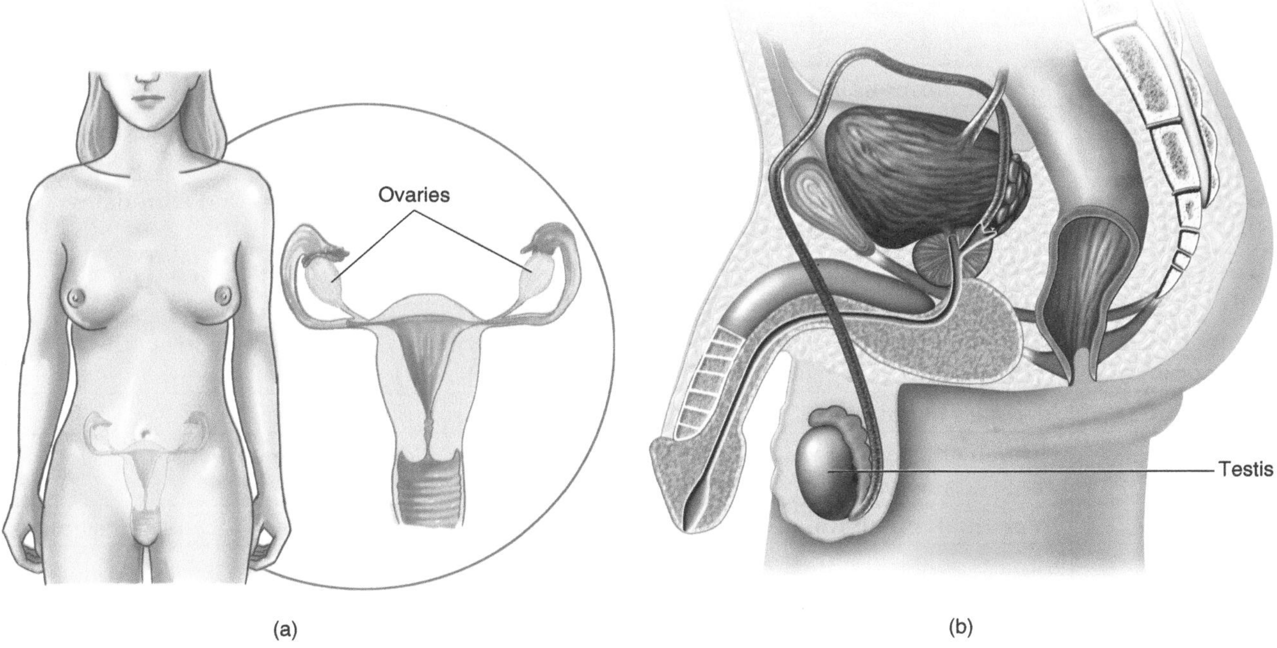

Figure 15-3 Hormone-secreting reproductive organs: (a) Female; (b) Male.

Table 15.2 Functions of selected hormones. See Table 1 for complete names of hormones

ACTH	**Adrenocorticotropic hormone** or **corticotropin**; stimulates the adrenal gland to release corticosteroids, mainly the glucocorticoids
ADH	**Antidiuretic hormone** or **vasopressin**; promotes water conservation in the kidneys and vasoconstriction throughout the body
Aldosterone	Promotes conservation of Na^+ and excretion of K^+ in the kidneys
ANP	**Atrial natriuretic peptide**; inhibits aldosterone secretion and causes vasodilation
Angiotensin II	Activated from angiotensin I in lung capillaries; vasoconstrictor and stimulates release of aldosterone and vasopressin
Calcitriol	Active vitamin D; derived from dehydrocholesterol in the skin
CCK	**Cholecystokinin;** stimulates gall bladder contraction and enzyme secretion from pancreas
Cortisol	**Hydrocortisone;** Promotes lipolysis and proteolysis, and also gluconeogenesis (in the liver)
CRH	**Corticotropin-releasing hormone;** Stimulates the pituitary gland to release its hormones
CT	**Calcitonin**; decreases blood calcium by stimulating osteoblasts and inhibiting osteoclasts
Epinephrine	Fight-or-flight hormone; vasoconstrictor (except in heart); promotes glycogenolysis
Erythropoietin	**EPO;** Stimulates blood cell production in bone marrow
Estrogen	Promotes development of female characteristics; stimulates uterine wall to thicken
FSH	**Follicle-stimulating hormone**; stimulates follicle development in the ovaries and sperm production in the testes
Gastrin	Stimulates stomach activity and relaxes the ileocolic valve
GHIH	**Growth hormone-inhibiting hormone** or **somatostatin;** inhibits release of GH, glucagon, and insulin.
Ghrelin	Released before a meal and stimulates hunger sensation

(Continued)

Table 15.2 Functions of selected hormones. See Table 1 for complete names of hormones *(Cont'd)*

GHRH	**Growth hormone-releasing hormone**; stimulates pituitary to release hGH
GIP	**Gastric inhibitory peptide** or **glucose insulinotropic peptide**; Stimulates pancreas to release insulin and inhibits stomach activity
GnRH	**Gonadotropin-releasing hormone**; Released from the hypothalamus and stimulates anterior pituitary to release gonadotropins
Gonadocorti-coids	Androgenic steroids released from the adrenal cortex; promote development of male characteristics and responsible for libido (sex drive) in females
Growth hormone	**GH or hGH** or **somatotropin**; Stimulates cell growth and promotes protein synthesis, lipolysis, gluconeogenesis
Glucagon	Increases blood glucose by promoting glycogenolysis
HCG	**Human chorionic gonadotropin**; Released from the placenta and maintains progesterone production in the ovaries
ICSH	**Interstitial cell-stimulating hormone** (same as LH); stimulates testes to secrete testosterone
Inhibin	Inhibits sperm production in the testes
Insulin	Lowers blood glucose by promoting glucose uptake by most cells
Leptin	Released shortly after a meal; binds to receptors in hypothalamus causing feeling of satiation
LH	**Luteinizing hormone**; Promotes ovulation in females; in males, it is called ICSH (see above)
Melatonin	Helps regulate sleep-wake cycles
Norepinephrine	Virtually the same effects as epinephrine
Oxytocin	Promotes uterine and lactiferous duct contractions
PRH	**Prolactin-releasing hormone**; Stimulates pituitary to release prolactin
Progesterone	Stimulates uterine wall to thicken
Prolactin	**PRL**; Stimulates mammary glands to produce milk
PTH	**Parathormone**; increases blood calcium by stimulating osteoclasts and inhibiting osteoblasts; needed for activation of vitamin D in kidneys; promotes conservation of calcium in kidneys
Relaxin	Causes softening of the pubic symphysis during childbirth
Secretin	Stimulates bile secretion in liver and bicarbonate secretion from pancreas
Testosterone	Promotes development of male characteristics; thus, it is an *androgenic* hormone
TH	**Thyroid hormone** or **thyroxine**; The major calorigenic (heat-promoting) hormone and regulator of basal metabolic rate (BMR)
Thymosin	Stimulates lymphocytes called T cells
TRH	**Thyrotropin-releasing hormone**; Stimulates pituitary to release thyrotropin (TSH)
TSH	**Thyroid-stimulating hormone** or **thyrotropin**; Stimulates thyroid gland to release thyroxine
VIP	**Vasoactive intestinal peptide**; Stimulates the small intestine to secrete alkaline fluid

Basal Metabolic Rate

Basal metabolic rate (BMR) is a measure of inherent cellular chemistry under basal conditions, which means the person (1) is relaxed, (2) has not eaten in the last 12 hours, and (3) is comfortable with a room temperature ~68–80°F. The BMR is the minimum amount of energy required by the body to maintain homeostasis. It is traditionally recorded as kilocalories per square meter of body surface area per hour **(Kcal/m2/hr)**. Reporting the metabolic rate per unit of size (surface area or weight) allows individuals of different size to be logically compared. When comparing individuals of similar size, one might record the metabolic rate simply as Kcal/hr (excluding the size variable). Normal BMR for a human is considered ±10% of the average value for people in the same age group. Factors affecting BMR include gender, with males typically have a higher rate due to testosterone, exercise, climate or temperature outside the body, emotions, hormones (primarily thyroxine and epinephrine), food intake, height and weight, and body temperature. In the following worksheet, you will be calculating BMR using a known variable, including weight, oxygen consumption, and surface area.

Name ____________________ Course Number: ________ Lab Section ______

Lab 15: Endocrine Basal Metabolic Rate Worksheet

You will be calculating basal metabolic rate (BMR) using a known variable, including weight, oxygen consumption, and body surface area. A mathematical equation, known as the DuBois formula*, is commonly used to calculate body surface area (BSA) using weight and height data. Use this equation (or a nomogram table provided by your instructor) to determine BSA.

$$\mathbf{BSA = 0.007184 \times W^{0.425} \times H^{0.725}}$$

where W is the weight in kg, and H is the height in cm
(1 kg = 2.2 pounds, and 1 cm = 0.394 inches)

Calculation of BMR using Weight: The average BMR for females is approximately **0.7 Kcal/kg/hr**. The average for males is approximately **1.0 Kcal/kg/hr**. Calculate your expected BMR in Kcal/day using your weight.

Calculation of BMR Using Oxygen Consumption: Approximately **4.825** Kcal of heat are liberated for each liter of oxygen consumed. Using standardized charts for surface area and average BMR values for different age groups, and by determining how much oxygen is consumed in a given amount of time under basal conditions (using an oxygen-filled spirometer), one can determine if a person's BMR is high, low, or normal.

Problem 1: Betty is 23 years old, 5′5″ tall; weighs 120 pounds; and consumes 500 milliliters of oxygen in two minutes under basal conditions. Is her BMR high, low, or normal?

a. Betty's surface area, based on the nomogram, is ________ m^2

b. According to the chart provided, the "normal" BMR (Kcal/m^2/hr) for Betty's age group is __________ Kcal/m^2/hr

c. Calculate Betty's BMR, in Kcal/m^2/hr.

d. Is Betty's BMR normal, high, or low? Calculate as follows $\frac{\text{(Actual BMR} - \text{Normal BMR)}}{\text{Normal BMR}} \times 100$

e. The normal BMR (calculated for surface area) for your age group and your size = ________ Kcal/day

f. The normal BMR (calculated for weight) for your age group and your size = ________ Kcal/day

g. Using the data from part "f," your expected BMR is ______________ Kcal/kg/hr.

h. Using data from the nomogram, your expected BMR is ______________ Kcal/m^2/hr.

---

*Du Bois, D, and Du Bois EF. 1916. "A formula to estimate the approximate surface area if height and weight be known." *Archives of Internal Medicine* 17 no. 6: 863–71.

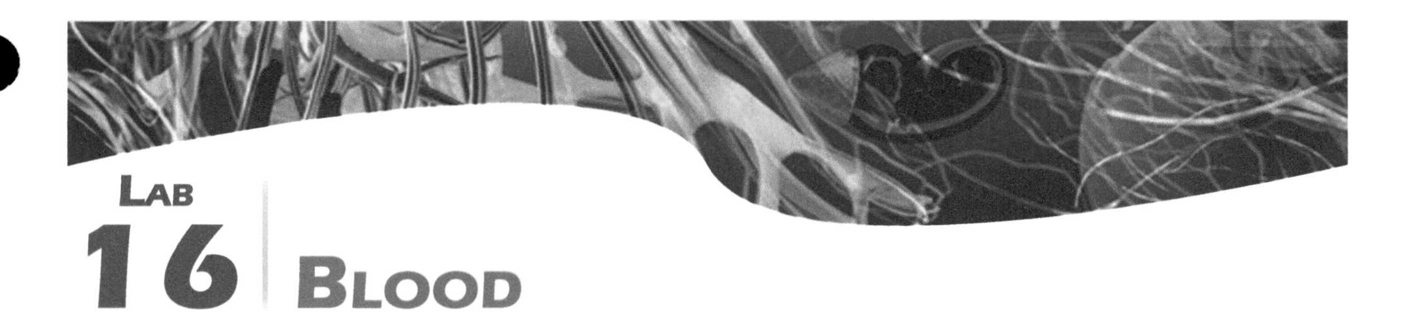

LAB 16 BLOOD

Blood is a liquid tissue that circulates through the body within vessels of the cardiovascular system. Its functions include delivery of oxygen, nutrients, hormones, and chemicals of defense to the body's cells, and removal of cellular wastes. The blood consists of cellular components, known collectively as **formed elements**, and an extracellular liquid component called **plasma**. The formed elements include **erythrocytes** (or **red blood cells**), **leukocytes** (or **white blood cells**), and **thrombocytes** (or **platelets**). The plasma is mostly water that contains a wide variety of solutes, including ions and plasma proteins. **Figure 16-1** shows how plasma and the formed elements are separated using a centrifuge.

In this lab, you will learn to recognize the different formed elements on microscope slides and figures. In addition, you will learn how to ascertain a person's blood type and how possible blood transfusions are determined. Last, you will learn how genetics gives rise to a person's blood type. **Table 16.1** provides some introductory information related to blood. Become familiar with this information and learn to recognize the elements on microscope slides and images provided in the lab (see **Figure 16-2**). **Table 16.2** provides information on blood types and blood typing.

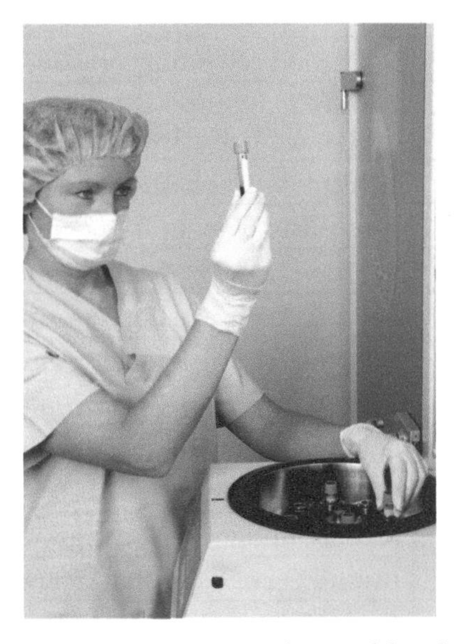

Figure 16-1 Centrifuging blood.

General Rule For Blood Tranfusions

When giving transfusions, it is crucial that the agglutinins (antibodies) in the recipient's plasma must not cause the agglutinogens (antigens) on the donor's red blood cells to agglutinate. Generally, clinicians are not so concerned about the possibility of antibodies in the donor's blood causing significant clumping of the recipient's antigens since the donor's antibodies become very diluted in the recipient's plasma. Moreover, in many cases, whole blood is not transfused into the patient. Instead, only the formed elements suspended in an isotonic solution are present in the transfused blood. Based on the information provided thus far in the lab, complete **Table 16.3**.

Determining Blood Type

In this lab, you will be using simulated blood to determine the blood type of a hypothetical patient. The procedure is similar to that used to determine the blood type of real blood. This

Table 16.1 Items Related to the Blood's Formed Elements	
0.9% NaCl	Isotonic saline (physiological saline; NaCl solution that is isotonic to blood cells; it contains approximately ~280 mOsm of solute per liter of solution, recorded as 280 mOsm/L
280 mOsm/L	Normal osmolarity of human tissue fluid; equivalent to 0.9% NaCl or 5% glucose solutions
5% glucose	Glucose (dextrose) solution that is isotonic to blood cells; approximately 280 mOsm/L
Agranulocytes	White blood cells with a non-grainy cytoplasm; includes lymphocytes and monocytes
Basophil	Granulocyte; least abundant leukocyte; releases histamine
Crenation	Shriveling of a cell due to loss of water when surrounded by a hypertonic solution
Eosinophil	Type of granulocyte; next to last in abundance of leukocytes
Erythrocyte	Red blood cell; transports oxygen to systemic tissues and carbon dioxide to the lungs
Granular leukocyte	Type of leukocyte with a "grainy" cytoplasm; also called PMNs (polymorphonuclear cells) includes neutrophils, eosinophils, and basophils
Hematocrit	Percent of the blood that is cellular (determined primarily by erythrocytes)
Hemolysis	Rupturing of an erythrocyte due to being surrounded by a hypotonic solution
Hyperosmotic	Solution having a higher osmotic pressure (high solute particle concentration)
Hypertonic	Solution having a higher osmotic pressure than a cell; causes a cell to crenate
Hypoosmotic	Solution having a lower osmotic pressure (low solute particle concentration)
Hypotonic	Solution having a lower osmotic pressure than a cell; causes a cell to swell
Isoosmotic	Solution having the same osmotic pressure as a cell; does not cause swelling or crenation
Leukocytes	White blood cells
Lymphocyte	Most abundant type of agranulocyte and second most abundant leukocyte
Lysis	Rupturing of a cell that may occur when it is surrounded by a hypotonic solution
Monocyte	Type of agranulocyte that becomes a macrophage; third most abundant leukocyte
Neutrophil	Type of granulocyte; most abundant leukocyte
Osmotic pressure	Positively correlated with the amount of solute present: the higher the solute particle concentration, the higher the osmotic pressure; hypertonic solutions have a higher osmotic pressure than hypotonic solutions; water diffuses from lower osmotic pressure solutions into higher osmotic pressure solutions.
Thrombocytes	Platelets; important in coagulation

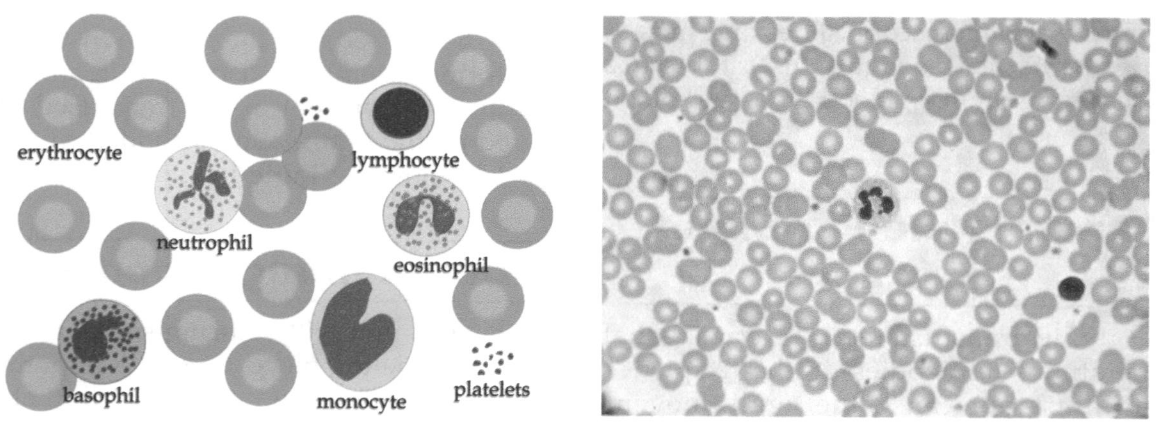

Figure 16-2 Formed elements in blood; the photograph at right is a blood smear at 400X.

Table 16.2 Terms Related to Blood Typing and Transfusions

ABO group	Group of polysaccharides found on the surface of certain erythrocytes; "O" is not a polysaccharide but represents not having "A" or "B" present
Agglutinin	Blood type **antibody** that can cause particular blood type antigens to agglutinate (clump together); anti-A agglutinins cause blood cells displaying type A agglutinogens to clump; anti-B agglutinins cause blood cells displaying type "B" agglutinogens to clump; anti-D (Rh) agglutinins cause blood cells displaying type "D" (major Rh) agglutinogens to clump.
Agglutinogen	Polysaccharide found on the surface of certain erythrocytes; also called blood type **antigens**; the major agglutinogens are A, B, and D (Rh); "O" and "-" are not polysaccharides but represent not having A, B, or D present
Antiserum A	An IgM antibody (agglutinin) that causes blood cells with A agglutinogens to clump
Antiserum B	An IgM antibody (agglutinin) that causes blood cells with B agglutinogens to clump
Antiserum D	An IgG antibody (agglutinin) that causes blood cells with Rh agglutinogens to clump
Blood type	Refers to the type of agglutinogens (antigens) present on the surface of red blood cells
Rh factor	Group of polysaccharides found on the surface of certain erythrocytes; the "-" is not a polysaccharide but represents a lack of the major Rh (D) factor
Transfusion	Clinical introduction of blood into the venous system of a patient
Type A^-	Blood that has only the A agglutinogens on the surface of the red blood cells
Type A^+	Blood that has the A and Rh agglutinogens on the surface of the red blood cells
Type AB^-	Blood that has the A and B agglutinogens on the surface of the red blood cells
Type AB^+	Blood that has the A, B, and Rh agglutinogens on the surface of the red blood cells
Type B^-	Blood that has only the B agglutinogens on the surface of the red blood cells
Type B^+	Blood that has the B and Rh agglutinogens on the surface of the red blood cells
Type O^-	Blood that has no major agglutinogens on the surface of the red blood cells
Type O^+	Blood that has only the Rh agglutinogens on the surface of the red blood cells
Universal donor	Person with type O- blood, which contains no major agglutinogens (A, B, or D); in most cases, this person's blood can be donated to anyone, provided there are no significant interactions among this blood's antibodies and the patient's antigens or significant interactions among some of the minor agglutinogens and agglutinins.
Universal recipient	Person with type AB^+ blood, which contains no major agglutinins (anti-A, anti-B, or anti-D); in most cases, this person can receive any blood type, provided there are no significant interactions among this blood's antigens and the donor's antibodies or significant interactions with some of the minor agglutinogens and agglutinins.

Table 16.3 Possible Blood Transfusions

Type	Agglutinogens	Agglutinins	Can Accept from These Types	Can Donate to These Types
A+				
A-				
B+				
B-				
AB+				
AB-				
O+				
O-				

Figure 16-3 Agglutination in a blood.

is done by mixing a small sample of simulated blood with the three major agglutinins: *anti-A*, *anti-B*, and *anti-D*. At your table, you should have a dropper bottle containing simulated blood for a hypothetical patient, one tray with three depressions, and three dropper bottles containing anti-sera (antibodies). Use the following procedure to determine each patient's blood type.

1. Place one drop of blood in each depression of the plastic tray.
2. Mix one drop of anti-A with the first drop blood; one drop of anti-B with the second drop of blood; and one drop of anti-Rh (D) with the third drop of blood.
3. Mix each blood-antisera mixture with separate tooth picks. After several minutes, look at each depression for signs of agglutination (clumping). Agglutination indicates that a particular antigen is present on the red blood cells. For example, if the red blood cells with which anti-A antibodies are mixed agglutinate, then the "A" antigen must be the present on the cells. Agglutination in a real blood sample is shown in **Figure 16-3**. Determine the blood type of stimulated blood at your table and record your data on the board. Then record the data from other teams and complete **Table 16.4**.

Genetics of Blood Type

Blood type is genetically controlled in that specific genes code for specific enzymes, which in turn synthesize the polysaccharides responsible for blood type. The gene that codes for either type A or B is on chromosome #9, while the gene that codes for the major Rh factor (D) is on chromosome #1. If a #9 chromosome lacks the gene for A or B, then it codes for type O; there is no polysaccharide called type O. The type O blood cell simply does not have A or B present on its surface. Likewise, if a #1 chromosome lacks the gene for type D, then that chromosome codes for Rh-negative (Rh^-) blood. If the #1 chromosome has the gene for type D, then it codes for Rh-positive (Rh^+) blood.

Since the ABO group and Rh factor genes are located on different chromosomes, they are inherited separately and are non-linked. In other words, inheriting a gene for either type A or type B does not influence whether the person is Rh^+ or Rh^-. However, since we inherit two copies of each chromosome (a maternal and a paternal copy), we all have two genes for both the ABO group and the Rh factor traits.

Genotype and Phenotype

Understanding the genetics of blood type requires an understanding of the difference between inheriting a gene and expressing that gene. In this case, **genotype** refers to all the genes we inherit for blood type, while **phenotype** refers to how our inherited genes bring about the actual blood type. The genes for type A and B are **codominant**, which means inheriting both an A and B, resulting in the AB genotype, causes the person to have type AB blood (which is the phenotype). In contrast, type O is **recessive,** which means inheriting A and O genes gives rise to the type A phenotype. In this case, the genotype is either AO or OA, depending on whether the maternal or paternal chromosome has the A gene. In this case, when the inherited genes are different, the person is said to be **heterozygous** (het-er-ō-ZĪ-gus) type A. If the person's genotype is AA, he or she is **homozygous** (hō-mō-ZĪ-gus) type A. Likewise, a type B person could be either heterozygous (BO or OB) or homozygous (BB). The only way a person can have type O blood is if both of his or her #9 chromosomes lack genes for type A or B. The type O person's genotype will always be OO.

For the Rh factor, having a gene for Rh+ blood is dominant, while lacking this gene is recessive. Thus, the Rh^+ person's genotype

Table 16.4 Results of Blood Type Determination

Team	Antisera A Reaction	Antisera B Reaction	Antisera D Reaction	Blood Type	Can Accept from	Can Donate to
1						
2						
3						
4						
5						
6						

Table 16.5 Blood Types and Possible Genotypes

Type	Phenotype	Possible Genotypes
A+		
A-		
B+		
B-		
AB+		
AB-		
O+		
O-		

is either ++, +–, or –+, while the Rh^- person's genotype must be ––. Based on this information, a person whose blood type is O^- has the genotype OO––.

Complete **Table 16.5** by listing all possible genotypes for any given phenotype. You do not need to list whether a particular gene is maternal or paternal. For example, for this lab, we will consider AO, BO, and +– to be the same as OA, OB, and –+, respectively, and you do not need to list the alternatives.

Inheritance Problems

When trying to determine the possible blood types for children when the parents' blood types are known requires knowledge of the parents' genotypes. For example, assume that Billy has type A+, his father has type AB+ and his mother has O–. Since his mother could donate only O and –, Billy's genotype must be **AO+–**. Billy marries Betty, who has type B– (her father is A+ and her mother is B+). Since Betty must receive two ABO genes and two Rh genes from each parent, her genotype must be **BO––**. What is the chance that Billy and Betty will have a child who is type O+?

To solve this problem, we can use a simple Punnett Square that shows all the different kinds of gametes (sex cells), with reference to the blood type genes, produced by Billy (at the top) and Betty (at the left). The phenotypes (blood types) resulting from the union of the different gametes is shown within the box. As you can see, the chance that Billy and Betty would have a child with type O+ blood is 1 out of 8, or 12.5%.

	A+	A–	O+	O–
B–	AB+	AB–	B+	B–
O–	A+	A–	O+	O–

Name ____________________ Course Number: ________ Lab Section ______

Lab 16: Blood Worksheet

Now, work through the following problems that will check your understanding of blood type genetics and transfusion possibilities.

Problem 1: Jack has type AB^- blood (His mother is type A^+ and his father is type B^+). Jack marries Sally, who has type O^+ (her mother is A^+ and her father is B^-). What is the chance that Jack and Sally will have a child with type A^-? ____________________

Problem 2. What is the chance that Jack and Sally will have a child that cannot give blood to either parent? ____________________

Problem 3. What is the chance that they will have a child who cannot accept blood from either parent? ____________________

Problem 4. What is the chance of having a child who can accept blood from Jack, but not Sally?

____

Problem 5. Billy has type B^- blood and Betty has type B^+ blood. Assume that Billy and Betty have all the agglutinins that they could ever make. If a drop of Billy's blood mixes with a drop of Betty's blood on a glass slide, then (Billy's / Betty's) blood cells will agglutinate. Circle the correct answer and explain why you chose that answer. ____________________

Problem 6. Buddy has type AB+ (his mother is AB+ and his father is B–). He marries Susan, who has type A+ (her mother is B+ and her father is A–). What is the chance that Buddy and Susan will have a child who cannot safely accept a blood transfusion from Buddy or Susan?

____

Problem 7. What is the chance Buddy and Susan will have a child who can safely donate blood to Buddy but not to Susan? ____________________

LAB 17 THE HEART

Overview of the Heart

The heart provides the driving force for moving blood to tissues throughout the body, and it is the only organ that contains cardiac muscle tissue. After this lab, you should be able to (1) identify and describe the anatomical features of the heart, (2) explain the heart's sounds, (3) explain the parts of an electrocardiogram, and (4) correlate the heart's mechanical events with its electrical activity. Let's first consider the heart's gross anatomical features. Review the external and internal features of the heart listed in **Tables 17.1** and **17.2** and shown in **Figures 17-1** and **17-2**, respectively. Be able to identify these parts, when possible, on models and preserved specimens in the lab.

Table 17.1 External Anatomy of the Heart

Apex of heart	Inferior, pointed end of the heart
Base of heart	Broad, superior portion of the heart; includes the atria and large blood vessels
Epicardium	(Visceral pericardium) Simple squamous epithelium on the outer surface of the heart
Fibrous pericardium	Dense connective tissue that forms the outer portion of the pericardial sac
Left auricle	Flap-like structure resembling a dog's ear on top of the left atrium; allows expansion of the atrium when it is filling with blood
Left atrium wall	Encloses the heart chamber that receives oxygenated blood from the pulmonary veins
Left ventricle wall	Encloses the heart chamber that receives oxygenated blood from the left atrium; pumps blood into the systemic circulation; its wall is thicker than that of the right ventricle
Ligamentum arteriosus	Small fibrous cord connecting the pulmonary trunk to the aorta; it is a remnant of the *ductus arteriosus* in the fetus that allows oxygenated blood to bypass the lungs
Parietal pericardium	Serous membrane lining the inner surface of the fibrous pericardium
Pericardial cavity	Virtual cavity between the parietal and visceral pericardia; filled with pericardial fluid
Pericardial sac	Encloses the heart; consists of the fibrous pericardium and parietal pericardium
Right atrium wall	Encloses the heart chamber that receives deoxygenated blood from rest of body
Right auricle	Flap-like structure resembling a dog's ear on top of the right atrium; allows expansion of the atrium when it is filling with blood
Right ventricle wall	Encloses the heart chamber that receives deoxygenated blood from right atrium

Table 17.2 Internal Anatomy of the Heart	
Aortic semilunar valve	(Aortic valve) Three-cusp valve located between the left ventricle and aorta
Bicuspid valve	(Mitral valve; left AV valve) Two cusp valve between the left atrium and left ventricle
Chordae tendineae	Long, fibrous cords of connective tissue that connect the papillary muscles to the AV valves; they prevent the AV valves from prolapsing during ventricular systole
Endocardium	Simple squamous epithelium that forms the inner lining of the heart chambers
Interatrial septum	Muscular partition that separates the right and left atria
Interventricular septum	Muscular partition that separates the right and left ventricles
Left atrium	Heart chamber that receives oxygenated blood from the pulmonary veins
Left ventricle	Heart chamber that receives oxygenated blood from the left atrium; pumps blood into the systemic circulation; its wall is thicker than that of the right ventricle
Myocardium	The muscular wall around the heart chambers; consists of cardiac muscle tissue
Papillary muscles	Projections of cardiac muscle within the ventricles; attached to chordae tendineae
Pulmonary semilunar valve	(Pulmonary valve) Three-cusp valve between the right ventricle and pulmonary trunk
Right atrium	Heart chamber that receives deoxygenated blood from rest of body; **pectinate muscles** form ridges on inner wall
Right ventricle	Heart chamber that receives deoxygenated blood from right atrium
Trabeculae carneae	Ridges on the inner walls of the ventricles
Tricuspid valve	(Right AV valve) Three-cusp valve between the right atrium and right ventricle

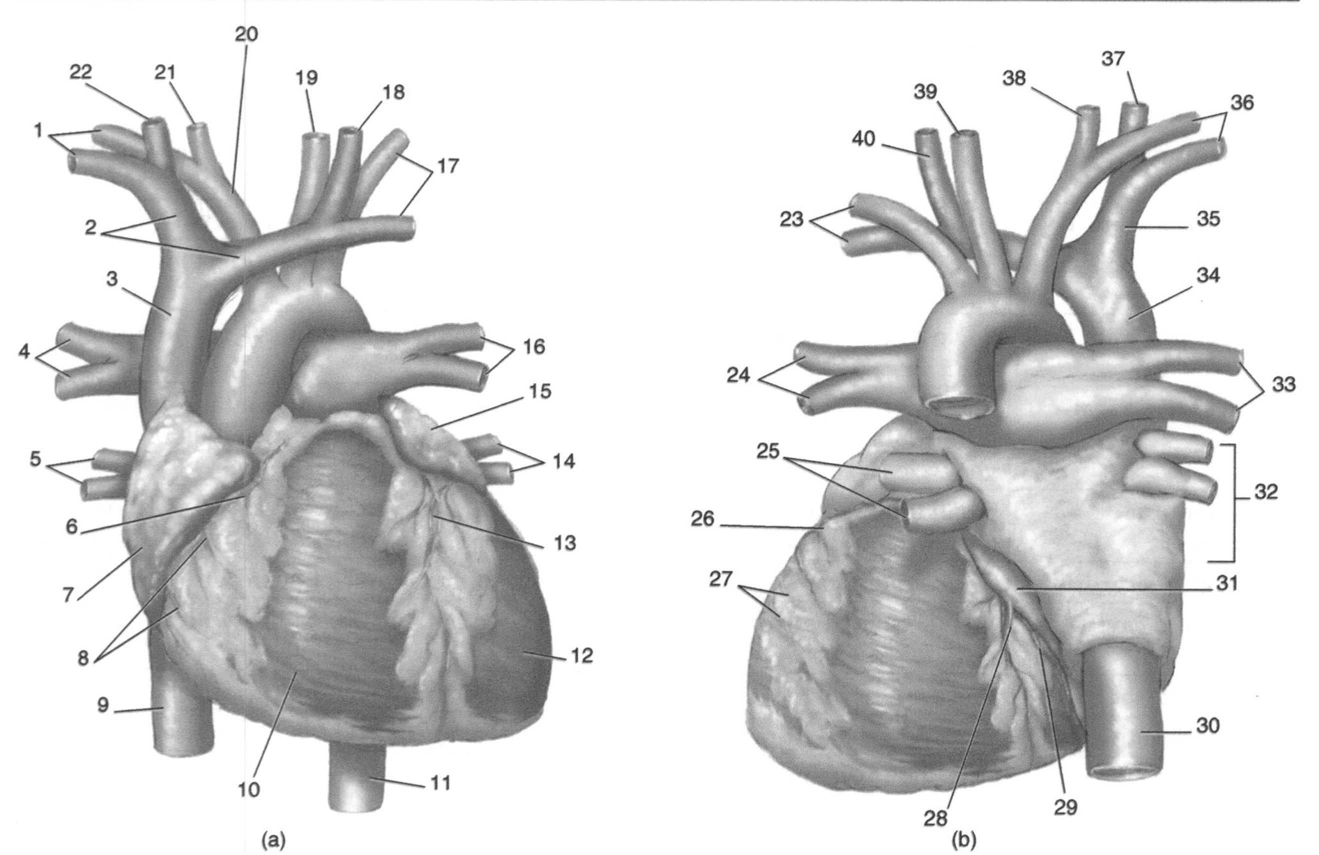

Figure 17-1 External anatomy of the heart: (a) Anterior view, (b) posterior view.

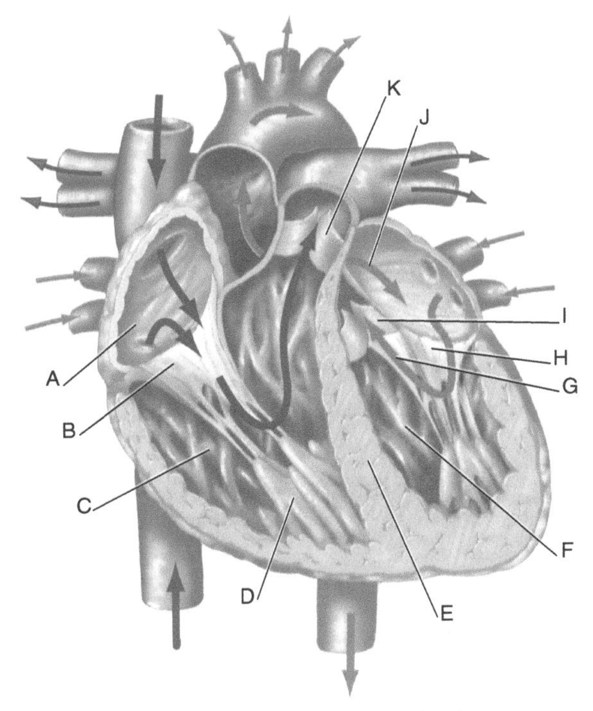

Figure 17-2 Internal anatomy of the heart.

Great Vessels of the Heart

The heart's *great vessels* are attached to the base of the heart, and they include several large veins, which carry blood to the heart, and arteries, which carry blood away from the heart. Vessels depicted in red carry blood that is *oxygenated* blood (high in oxygen), whereas vessels depicted in blue carry blood that is *deoxygenated* (low in oxygen). In the adult human, the only arteries depicted in blue are the pulmonary arteries because they are carrying deoxygenated blood to the lungs for oxygenation. In like fashion, the only veins depicted in red are the pulmonary veins because they are carrying oxygenated blood from the lungs back to the heart. The great vessels are listed in **Table 17.3** and illustrated in **Figures 17-1** and **17-2**.

Heart-Lung Blood Trace

In this section, you will trace the path of blood through the heart chambers and to and from the lungs. Blood enters the heart's right side and eventually reaches the lungs where it picks up oxygen and releases carbon dioxide. Blood returns from the lungs to the heart's left side before flowing out into the systemic circulation. Blood exchanges gases (oxygen and carbon dioxide) with surrounding tissues through small blood vessels called *capillaries*. Using **Table 17.3** and **Figure 17-2** as references, fill in the following blanks for tracing blood through the heart and also to and from the lungs.

1. Vena cava
2. ______________
3. ______________
4. Right ventricle
5. ______________
6. ______________
7. Pulmonary artery
8. Capillaries of the lungs
9. ______________
10. ______________
11. ______________
12. Left ventricle
13. ______________
14. Aorta

Table 17.3 Great Vessels of the Heart

Aortic arch	Receives blood from the left ventricle
Inferior vena cava	Delivers systemic blood from below the heart to the right atrium
Pulmonary arteries	Branch off the pulmonary trunk and carries blood to the lungs
Pulmonary trunk	Receives blood from the right ventricle and carries blood to the pulmonary arteries
Pulmonary veins	Bring oxygenated blood to heart from the lungs; empties blood into the left atrium
Superior vena cava	Delivers systemic blood from above the heart to the right atrium

Coronary Circulation

Due to the thickness of the myocardium, blood, while it is within the heart's chambers cannot provide adequate nourishment or remove waste products efficiently enough to maintain homeostasis in the myocardium. Instead, the myocardium must pick up nutrients and remove cellular wastes through capillaries that are part of a special set of vessels on the heart's outer surface. These vessels comprise the *coronary circulation* and its components are listed in **Table 17.4** and **Figure 17-3**.

Blood Tracing

Trace 1: Trace a drop of blood from the inferior vena cava to the capillaries of the left ventricle's lateral wall and back to the right atrium.

1. Inferior vena cava
2. Right side of heart ➔ Lungs ➔ Left side of heart
3. Aorta

Table 17.4 Vessels of the Coronary Circulation

CORONARY ARTERIES	
Anterior interventricular artery	On the anterior left side of the heart, between the ventricles; receives blood from the left coronary artery; supplies blood to the anterior sides of both ventricles; forms anastomoses (connections) with the posterior interventricular artery at the heart's apex
Circumflex artery	Major branch of the left coronary artery; extends laterally around the left side of the heart to supply blood to the left atrium and posterior side of the left ventricle
Left coronary artery	Branches off the aorta and extends downward toward the left ventricle; branches to form the circumflex artery and anterior interventricular artery
Left marginal artery	Branches off the circumflex artery; supplies blood to left lateral side of the heart
Posterior interventricular artery	Branches off the right coronary artery on the posterior side of heart between the ventricles; supplies blood to most of the posterior, middle and right side myocardium; unites with the anterior interventricular artery at the heart's apex
Right coronary artery	Branches off the aorta and extends laterally around to the posterior side of the heart; delivers blood to the right marginal artery; branches off on the lateral side of the heart
Right marginal artery	Branches off the right coronary artery; supplies blood to the right lateral side of the heart
CARDIAC VEINS	
Anterior cardiac veins	Small veins that drain anterior surface of right atrium; empty blood into right atrium
Coronary sinus	Large horizontal vessel on the posterior side of the heart immediately beneath the atria; collects blood from cardiac veins and empties it into the right atrium
Great cardiac vein	Parallel to the anterior interventricular artery on the anterior side of the heart between the ventricles; collects blood from anterior side of the heart; superiorly it runs laterally around the left side of heart to empty blood into the coronary sinus
Middle cardiac vein	Drains the posterior side of the heart; empties blood into the coronary sinus
Posterior cardiac vein	Drains the left lateral side of the heart; empties blood into the coronary sinus
Small cardiac vein	On the heart's right side; courses around the right side; empties blood into the coronary sinus

4. ____________________

5. ____________________

6. Capillaries of posterior left ventricle

7. ____________________

8. ____________________

9. Right atrium

Trace 2: Trace a drop of blood from the pulmonary trunk to the right atrium via the posterior interventricular artery.

1. Pulmonary trunk
2. Pulmonary artery ➔ Lungs ➔ Left side of heart
3. Aorta
4. ____________________
5. ____________________
6. Capillaries of posterior right ventricle
7. ____________________
8. ____________________
9. Right atrium

Trace 3: Trace a drop of blood from the superior vena cava to the right ventricle via the anterior interventricular artery.

1. Superior vena cava
2. Right side of heart ➔ Lungs ➔ Left side of heart
3. Aorta
4. ____________________
5. ____________________
6. Capillaries on anterior side of ventricles
7. ____________________
8. ____________________
9. ____________________
10. ____________________
11. Right ventricle

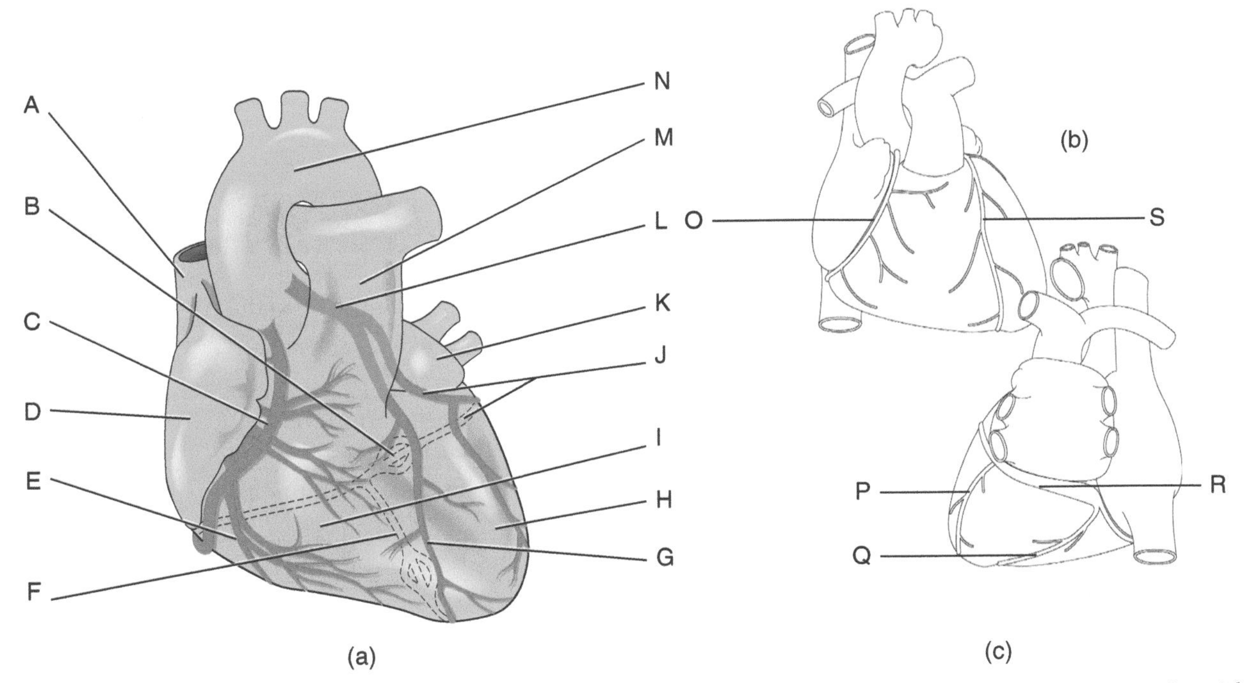

Figure 17-3 (a) Coronary arteries, (b) Right and left coronary arteries (c) Cardiac veins on posterior side.

Electrocardiography

Before cardiac myofibers contract, they must experience an electrical change (i.e., depolarize and then repolarize), and the measurement of this electrical activity is **electrocardiography**. The heart's electrical activity is the result of impulses (action potentials) spreading through the myocardium, and it is recorded as an **electrocardiogram**, or **ECG**. Impulses move from one cardiac muscle cell to another as ions pass to adjacent cells through *gap junctions* within intercalated discs. The path taken by impulses through the heart is the **cardiac conduction system** and its parts are shown in **Table 17.5** and **Figure 17-4**.

Mechanical vs. Electrical Events

It is important to differentiate between the heart's electrical activity, shown on an ECG, and its mechanical activity, shown on a **myogram**. The heart's mechanical activity includes the following: contraction and relaxation of the atria and ventricles, opening and closing of the heart's valves, and pressure changes within the chambers of the heart and within the great vessels. Contraction of a heart chamber's walls is known as **systole** (SIS-tō-ē) and it follows depolarization of the muscle tissue. Relaxation is known as **diastole** (dī-AS-tō-lē) and it follows repolarization of the muscle tissue. Remember that mechanical events are not recorded on an ECG, but it is possible to show the approximate times when they occur relative to the electrical events. Using your class notes, place the numbers of the following 33 heart-related events at appropriate locations on the generic ECG chart. In addition, fill in **Table 17.6** showing the 33 events in chronological order. Include in the table the original numbers in the alphabetized list.

Alphabetized List of Heart Events

1. 1/4 of the way through the quiescent period
2. 3/4 of the way through the period when all atrial contractile cells are in systole
3. 3/4 of the way through the time when all ventricular contractile cells are in systole
4. Arterial pressure first exceeds ventricular pressure
5. Atrial pressure first exceeds ventricular pressure but AV valve not yet open
6. AV valves begin to open
7. AV valves close
8. Best time to measure EDV
9. Best time to measure ESV
10. Bundle branches begin depolarizing

Table 17.5 Components of the Heart's Conduction System

Atrioventricular (AV) node	Mass of pacemaker cells at the inferior portion of the right atrium; receives impulses from the SA node via the internodal pathways; sends impulses to the AV bundle
Atrioventricular (AV) bundle	(Bundle of His) Receives impulses from the AV node and sends them to the bundle branches
Bundle branches	Two pathways that arise from the AV bundle and conduct impulses through the interventricular septum to the Purkinje fibers at the heart's apex
Internodal pathways	Myoconduction fibers spanning between the SA node and AV node
Purkinje fibers	Myoconduction fibers; transmit impulses through ventricle walls
Sinoatrial (SA) node	"Pacemaker" of heart in right atrium

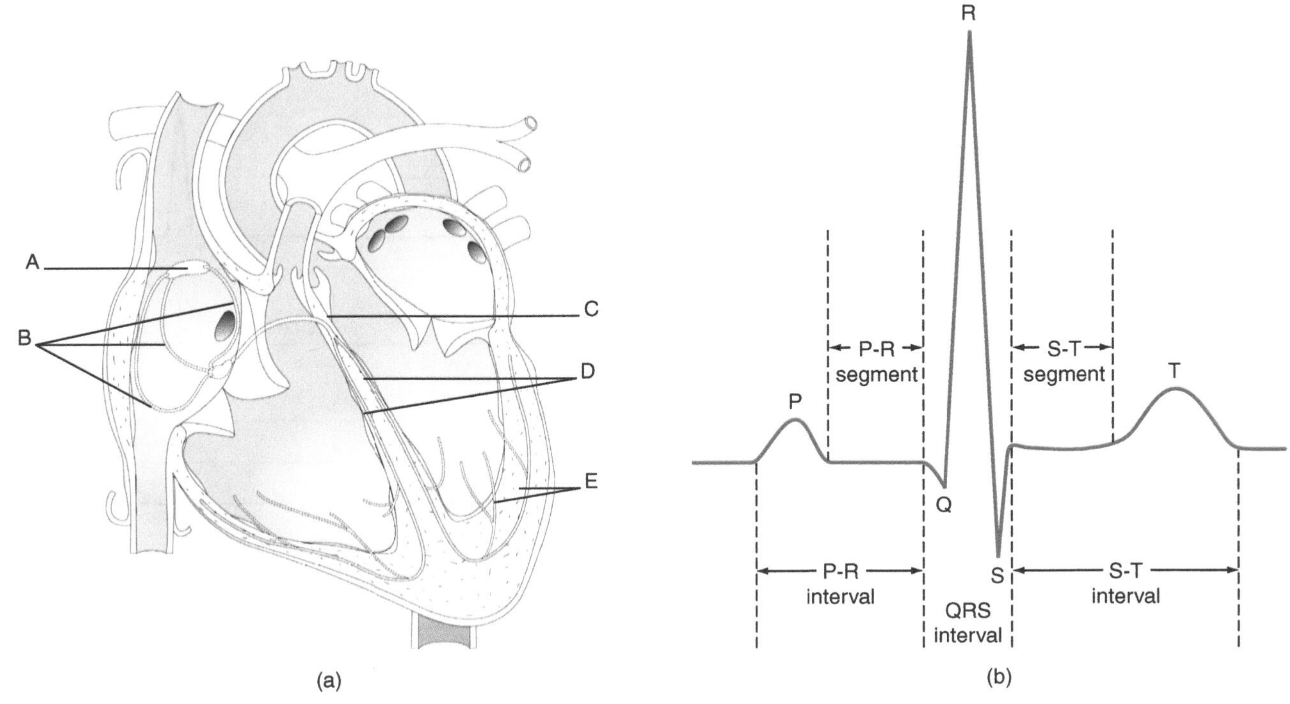

Figure 17-4 Electrical activity of the heart: (a) Cardiac conduction system, (b) an electrocardiogram.

11. Fast calcium channels begin to open
12. Fast sodium channels first open in atria
13. First heart sound heard
14. First moment when all atrial contractile cells have reversed polarity
15. First moment when all ventricular contractile cells are depolarized
16. First moment when all ventricular contractile cells are in diastole
17. First moment when all ventricular contractile cells are repolarizing
18. Halfway through atrial kick
19. Halfway through period of isovolumetric contraction
20. Halfway through period of isovolumetric relaxation
21. Halfway through period of ventricular ejection
22. Halfway through period of ventricular filling
23. Heart's apex begins systole
24. Immediately before autorhythmic cells in SA node reach threshold
25. P wave completed on ECG
26. Pacemaker potential reaches threshold
27. Purkinje fibers begin depolarizing
28. Second heart sound heard
29. Semilunar valves close
30. SL valve begins to open and blood begins flowing through
31. T wave completed
32. Ventricular pressure first exceeds arterial pressure but SL valve not yet open
33. Ventricular pressure first exceeds atrial pressure

Table 17.6 Chronological Events in the Heart

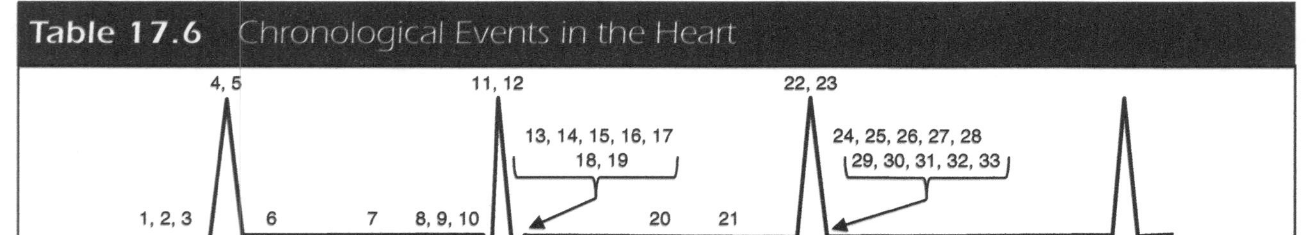

Event on ECG	Original #	Event	Electrical	Mechanical
1	24	Immediately before autorhythmic cells in SA node reach threshold	X	
2				
3				
4				
5				
6				
7				
8				
9				
10				
11				
12				
13				
14				
15				
16				
17				
18				
19				
20				
21				
22				
23				
24				
25				
26				
27				
28				
29				
30				
31				
32				
33	22	Halfway through period of ventricular filling		X

Name ___________________________ Course Number: _________ Lab Section _______

Lab 17: Heart Worksheet

A.	Aorta	______	prevents blood backflow from left ventricle to left atrium
B.	Aortic Semilunar Valve	______	transports blood to left lungs
C.	Bicuspid (mitral) Valve	______	returns venous blood from coronary system to right atrium
D.	Chordae Tendineae	______	major artery leading from left ventricle
E	Inferior Vena Cava	______	major vein leading into lower right atrium
F.	Left Atrium	______	major vein leading into upper right atrium
G.	Left Pulmonary Artery	______	prevents blood backflow from aorta to left ventricle
H.	Left Ventricle	______	pumps blood to pulmonary artery
I.	Myocardium	______	major artery leading from right ventricle
J.	Coronary Arteries	______	pumps blood to left ventricle
K.	Coronary Sinus	______	major veins leading into left atrium
L.	Papillary Muscles	______	prevents blood backflow from right ventricle to right atrium
M.	Pulmonary Artery (trunk)	______	attach cuspid valve cusps to papillary muscles
N.	Pulmonic Semilunar Valve	______	pumps blood to right ventricle
O.	Pulmonary Veins	______	transport blood aorta to cardiac muscles
P.	Right Atrium	______	anchor sites for the chordae tendineae
Q.	Right Ventricle	______	pumps blood to aorta
R.	Superior Vena Cava	______	myocardium that separates the two ventricles
S.	Tricuspid Valve	______	prevents blood backflow from pulmonary artery to right ventricle
T.	Ventricular Septum	______	muscle wall of the heart

The two receiving chambers for blood are ______________________________________

The two discharging chambers for the blood are the ______________________________

The ____ separates the heart chambers.

The LEFT side of the heart RECEIVES blood FROM the ____________________________

The RIGHT side of the heart RECEIVES blood FROM the ___________________________

The LEFT side of the heart PUMPS blood TO the _________________________________

The RIGHT side of the heart PUMPS blood TO the ________________________________

Heart Chamber or Blood Vessel	Oxygenated or Deoxygenated Blood
Left Ventricle	
Right Ventricle	
Left Atrium	
Right Atrium	
Pulmonary Artery	
Pulmonary Vein	
Superior Vena Cava	
Inferior Vena Cava	
Aorta	

4) Where does blood go AFTER it leaves the …

Right Atrium ____________________

Aorta ____________________

Left Atrium ____________________

Superior Vena Cava ____________________

Right Ventricle ____________________

Inferior Vena Cava ____________________

Left Ventricle ____________________

Lung ____________________

Pulmonary Veins ____________________

Pulmonary Arteries ____________________

5) Where did blood come from BEFORE it entered the …

Right Atrium ____________________

Aorta ____________________

Left Atrium ____________________

Superior Vena Cava ____________________

Right Ventricle ____________________

Inferior Vena Cava ____________________

Left Ventricle ____________________

Lungs ____________________

Pulmonary Veins ____________________

Pulmonary Arteries ____________________

Lab

18 Systemic Blood Vessels

Overview of System Blood Vessels

Systemic blood vessels deliver blood to and from tissues other than those supplied by the *pulmonary circulation*. Recall that the pulmonary circulation moves blood to and from capillaries in the lungs where gas exchange occurs with the atmosphere. The coronary blood vessels that supply blood to the myocardium are part of the systemic circulation, whereas the heart-lung blood trace comprises the pulmonary circulation. While the heart's thinner-walled, right ventricle pumps blood into the pulmonary circulation, the stronger, thicker-walled left ventricle pumps blood into the systemic circulation.

As you study the body's blood vessels, it is important to keep several points in minds. First, **arteries** carry blood away from the heart and **veins** carry blood to the heart. Under normal conditions in the adult, systemic arteries carry oxygenated blood (depicted in red), while systemic veins carry deoxygenated blood (depicted in blue). Exchange of nutrients, gases, and wastes between the blood and surrounding tissues takes place in the smallest vessels called **capillaries**. In this lab, you will be studying various pathways in the systemic circulation. Before learning the names of specific systemic vessels and their pathways, you should become familiar with the basic organization of blood vessels shown in **Table 18.1** and **Figure 18-1**.

Table 18.1 Terms Related to Blood Vessels	
Adventitia	Outer connective tissue covering of a blood vessel
Anastomosis	Joining together of two or more vessels
Arteriole	Receives blood from an artery and delivers it to a metarteriole
Arteriovenous anastomosis	Direct connection between an artery and vein
Artery	Transports blood away from the heart
Capillary bed	Network of capillaries in a tissue
Conducting artery	Large elastic artery (e.g., aorta)
Continuous capillary	Have tight junctions between endothelial cells
Distributing artery	Muscular artery farther away from heart
Elastic artery	Conducting artery that can propel blood into distributing arteries
End artery	Only source of blood for a certain organ; no collateral circulation
Endothelium	Simple squamous epithelium making up the tunica interna of a vessel
Fenestrated capillary	Contains openings in the endothelial cells; important in filtration

(Continued)

Table 18.1	Terms Related to Blood Vessels *(Cont'd)*
Lumen	Cavity within a blood vessel
Metarteriole	Receives blood from an arteriole and delivers it to a capillary
Muscular artery	Distributing artery farther away from the heart
Portal system	Capillary bed → vein → capillary bed → another vein
Precapillary sphincter	Ring of smooth muscle that regulates how much blood enters capillary
Pulmonary circulation	Circulation between heart and lungs; oxygenates blood in lungs
Sinusoid	Large-lumen capillary
Smooth myofiber	Smooth muscle cell
Thoroughfare channel	Receives blood from a capillary and delivers it to a venule
True capillary	Site of material exchange between the blood and another tissue
Tunica adventitia	Outer connective tissue covering of a vessel
Tunica externa	Same as tunica adventitia
Tunica intima	Endothelium inner lining of a vessel
Tunica media	Smooth muscle layer within a blood vessel
Vasa vasorum	Tiny vessels that nourish large vessels
Vein	Transports blood toward the heart
Venous valve	Prevents backflow of blood within a vein
Venule	Receives blood from a thoroughfare channel and delivers it to a vein

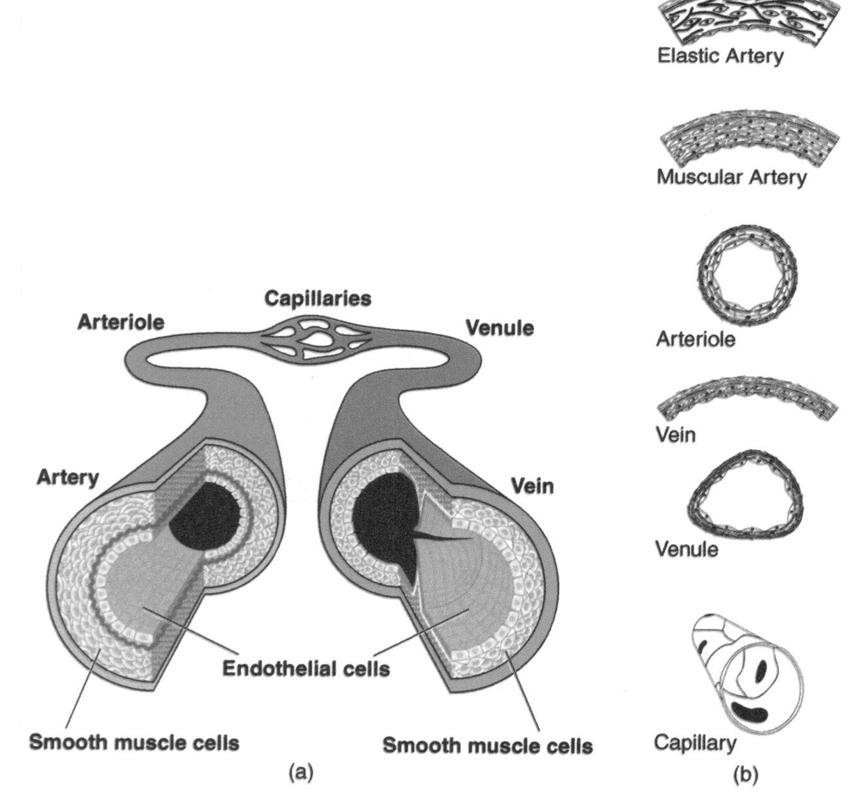

Figure 18-1 Blood vessel organization: (a) Basic vessel pathway, (b) anatomy of vessel walls.

Systemic Vessels

As you study the body's systemic vessels, you will notice that many of the vessels have names that you have heard before. For example, since you already know terms for major regions of the body, it should not be difficult to locate vessels such as the *axillary artery*, *brachial vein*, *radial* and *ulnar arteries*, or the *femoral* and *anterior tibial veins*. In addition to locating and identifying major systemic vessels, you will also need to know how to trace blood from one part of the body to another using some of the major pathways. The major systemic arteries (arteries of interest) that you need to learn to recognize on images and models are listed in **boldface** in the middle column of **Table 18.2**. The other arteries listed in italics in the middle column may be used for future reference. The artery in the left-hand column delivers blood to the artery of interest, which then delivers blood to the arteries or other organs in the right-hand column.

After learning the major systemic arteries, you can turn your attention to the major systemic veins, which are shown in **Table 18.3**. Many of the major arteries and veins are illustrated in **Figures 18-2** and **18-3**.

Table 18.2 Major Systemic Arteries

This **ARTERY** supplies blood to ➔	this **ARTERY**, which delivers blood to ➔	these **ARTERIES** (or other organs listed)
Thoracic Aorta	**Aorta (abdominal)**	phrenic, celiac, suprarenal, superior mesenteric, renal, gonadal, inferior mesenteric, lumbar, common iliac, median sacral
Aortic arch	**Aorta (thoracic)**	pericardial, bronchial, esophageal, mediastinal, posterior intercostal, phrenic
Heart	**Aortic arch**	brachiocephalic, left common carotid & subclavian
Dorsalis pedis	**Arch (dorsal)**	metatarsal a, ankle and top of foot
Posterior tibial	**Arch (plantar)**	digital arteries in toes
Subclavian	**Axillary**	thoracoacromial, lateral thoracic, subscapular, humeral circumflex, brachial
Vertebral	**Basilar**	cerebellum, pons, inner ear, posterior cerebral.
Axillary	**Brachial**	deep brachial, radial and ulnar
Aortic arch	**Brachiocephalic**	right common carotid, right subclavian
Aortic arch	**Carotid (left common)**	left internal and external carotid
Brachiocephalic	**Carotid (right common)**	right internal and external carotid
Common carotid	**Carotid (external)**	superficial temporal, facial, superior thyroid, lingual, pharyngeal, auricular, occipital, maxillary, dura mater
Common carotid	**Carotid (internal)**	ophthalmic a, anterior and middle cerebral
Abdominal aorta	**Celiac**	left gastric, splenic, hepatic
Internal carotid	**Cerebral (anterior)**	medial cerebral hemispheres (except occipital)
Internal carotid	**Cerebral (middle)**	lateral part of frontal, temporal, parietal lobes
Basilar, circle of Willis	**Cerebral (posterior)**	occipital lobe
Internal carotid, basilar	**Circle of Willis**	unites major anterior and posterior brain arteries

(Continued)

Table 18.2 Major Systemic Arteries *(Cont'd)*		
Inferior mesenteric	**Colic (left)**	transverse colon, descending colon
Superior mesenteric	**Colic (middle)**	ascending colon, transverse colon
Superior mesenteric	**Colic (right)**	cecum, ascending colon
Plantar arch (foot)	**Digital (foot)**	toes
Palmar arch (hand)	**Digital (hand)**	fingers
Anterior tibial	**Dorsalis pedis**	ankle, top of foot, arcuate
External carotid	**Facial**	muscles and skin of face
External iliac	**Femoral**	deep femoral, thigh muscles, femoral circumflex
Posterior tibial	**Fibular (Peroneal)**	peroneal muscles, heel
Celiac	**Gastric** (left)	lesser curvature of stomach, stomach fundus
Internal iliac	**Gluteals**	gluteal muscles
Abdominal aorta	**Gonadal**	gonads
Celiac	**Hepatic**	liver, gastroduodenal, right gastric
Superior mesenteric	**Ileocolic**	ileum, appendix, cecum, ascending colon
Abdominal aorta	**Iliac (common)**	internal and external iliac
Common iliac	**Iliac (external)**	anterior abdominal wall, femoral
Common iliac	**Iliac (internal)**	pelvic wall and viscera, pudendal
Internal thoracic	**Intercostal (anterior)**	intercostal region of anterior thorax
Thoracic aorta	**Intercostal (posterior)**	posterior thorax muscles, vertebrae, spinal cord
Abdominal aorta	**Lumbar**	posterior abdominal wall
Abdominal aorta	**Median sacral**	sacrum, coccyx
Abdominal aorta	**Mesenteric (inferior)**	left colic, sigmoidal, Rectal
Abdominal aorta	**Mesenteric (superior)**	intestinal, ileocolic, right colic, middle colic
External carotid	**Occipital**	posterior scalp
Internal carotid	**Ophthalmic**	eyeball, orbit, forehead, nose
Radial and ulnar	**Palmar arch**	hand, digital
Pudendal	**Penile**	penis
Thoracic aorta	**Phrenic (superior)**	superior surface of diaphragm
Posterior tibial	**Plantar arch**	bottom of foot, digital
Femoral	**Popliteal**	knee region, anterior and posterior tibial
Internal iliac	**Pudendal**	external genitalia in males and females
Brachial	**Radial**	lateral forearm, thumb, index finger
Inferior mesenteric	**Rectal**	rectum
Abdominal aorta	**Renal**	kidney

Table 18.2 Major Systemic Arteries *(Cont'd)*

Inferior mesenteric	**Sigmoidal**	descending colon and sigmoid colon
Celiac	**Splenic**	spleen
Aortic arch	**Subclavian (left)**	vertebral, thyrocervical, costocervical, axillary
Brachiocephalic	**Subclavian (right)**	vertebral, thyrocervical, costocervical, axillary
External carotid	**Temporal (superficial)**	most of scalp
Subclavian	**Thoracic (internal)**	anterior intercostals, mammary glands
Popliteal	**Tibial** (anterior)	tibialis anterior m., anterior leg, dorsalis pedis
Popliteal	**Tibial** (posterior)	calf muscles, posterior leg, fibular, plantar
Brachial	**Ulnar**	medial forearm, fingers 3,4,5.
Internal iliac	**Uterine**	uterus
Internal iliac	**Vaginal**	vagina
Subclavian	**Vertebral**	cervical spinal cord, cerebellum, basilar
Internal iliac	**Vesicular**	urinary bladder

Table 18.3 Major Systemic Veins

This **VEIN** receives blood from ➔	this **VEIN**, which receives blood from ➔	these **VEINS** (or other organ listed)
Subclavian	**Axillary**	basilic, brachial, cephalic veins
Superior vena cava	**Azygos**	right ascending lumbar, right posterior intercostals, hemiazygos, and accessory hemiazygos veins; thoracic viscera, chest
Axillary	**Basilic**	dorsal venous arch, arm & forearm, median cubital vein
Axillary	**Brachial**	radial, ulnar, forearm
Superior vena cava	**Brachiocephalic**	subclavian, internal jugular, vertebral veins
Axillary	**Cephalic**	palmar venous arch (hand), arm and forearm
Inferior mesenteric	**Colic (left)**	descending colon
Superior mesenteric	**Colic (middle)**	transverse colon
Superior mesenteric	**Colic (right)**	ascending colon
Cephalic or basilic	**Dorsal venous network**	digitals, dorsal hand
Great/small saphenous	**Dorsal venous arch**	digitals, dorsal foot
Anterior tibial vein	**Dorsalis pedis**	dorsal venous arch (foot)
Internal jugular	**Facial**	face
External iliac	**Femoral**	popliteal vein, thigh muscles
Hepatic portal	**Gastric**	lesser curvature of stomach
Internal iliac	**Gluteals**	gluteal muscles, coccyx, upper thigh
Left renal	**Gonadal (left)**	left gonad
Inferior vena cava	**Gonadal (right)**	right gonad

(Continued)

Table 18.3 Major Systemic Veins *(Cont'd)*

Azygos	**Hemiazygos (accessory)**	left posterior intercostal vein, superior left thorax
Azygos	**Hemiazygos**	left ascending lumbar vein, left posterior intercostal
Inferior vena cava	**Hepatic**	capillaries of liver
Capillaries of liver	**Hepatic portal**	superior mesenteric and splenic veins
Superior mesenteric	**Ileocolic**	ileum, cecum, appendix, lower ascending colon
Inferior vena cava	**Iliac (common)**	external and internal iliac veins
Common iliac	**Iliac (external)**	femoral vein, anterior abdominal wall
Common iliac	**Iliac (internal)**	gluteal/medial thigh muscles, bladder, pelvic organs
Internal thoracic	**Intercostals (anterior)**	anterior intercostal muscles and skin
Hemiazygos, accessory	**Intercostals (posterior)**	left, posterior intercostal muscles and skin
Azygos	**Intercostals (posterior)**	right, posterior intercostal muscles and skin
Subclavian	**Jugular (external)**	scalp and face
Brachiocephalic	**Jugular (internal)**	face and dural sinuses of brain
Inferior vena cava & ascending lumbar	**Lumbar**	posterior abdominal wall
Hemiazygos	**Lumbar (left ascending)**	left common iliac and left lumbar veins
Azygos	**Lumbar (right ascending)**	right common iliac and right lumbar veins
Basilic	**Median antebrachial**	anterior forearm
Basilic	**Median cubital**[1]	cephalic, anterior forearm
Splenic	**Mesenteric (inferior)**	colon (descending, sigmoid), rectum
Hepatic portal	**Mesenteric (superior)**	small intestine, cecum, appendix, ascending/transverse colon
External jugular	**Occipital**	posterior scalp
Cavernous sinus, facial	**Ophthalmic**	eye, orbit, forehead
Ulnar and median	**Palmar venous arch**	digital veins of fingers
Posterior tibial	**Plantar venous arch**	digital veins of toes, bottom of foot
Pudendal	**Penile** (deep)	dorsal penis
Posterior tibial	**Peroneal** (fibular)	deep lateral leg
Inferior vena cava	**Phrenic**	diaphragm
Posterior tibial	**Plantar arch**	sole of foot, digitals
Femoral	**Popliteal**	anterior/posterior tibial and small saphenous veins
Internal iliac	**Pudendal**	external genitalia, floor of pelvis, rectum

Table 18.3 Major Systemic Veins *(Cont'd)*

Brachial	**Radial**	palmar venous arch
Inferior mesenteric	**Rectal**	rectum
Inferior vena cava	**Renal**	kidney
Femoral	**Saphenous (great)**[2]	dorsal venous arch, superficial-anter/med leg
Popliteal	**Saphenous (small)**	dorsal venous arch, superficial posterior leg
Inferior mesenteric	**Sigmoidal**	sigmoid colon
Sigmoid sinus	**Sinus (cavernous)**	eye and inferior-anterior brain and cranial vault
Straight sinus	**Sinus (inferior sagittal)**	deep medial portion of cerebral hemispheres
Transverse sinus	**Sinus (occipital)**	posterior-inferior brain and cranial vault
Internal jugular	**Sinus (sigmoid)**	cavernous sinus, anterior cranium vault
Transverse sinus	**Sinus (straight)**	inferior sagittal sinus
Transverse sinus	**Sinus (superior sagittal)**	superior-anterior brain and cranial vault
Sigmoidal sinus	**Sinus (transverse)**	sinuses (superior sagittal, occipital, straight)
Hepatic portal	**Splenic**	spleen
Brachiocephalic	**Subclavian**	axillary vein (lymphatic duct also)
External jugular	**Superficial temporal**	most of scalp
Brachiocephalic	**Thoracic (internal)**	anterior intercostals
Popliteal	**Tibial (anterior)**	dorsal venous arch (foot)
Popliteal	**Tibial (posterior)**	fibular/plantar v, gastrocnemius, tibialis posterior
Brachial	**Ulnar**	palmar venous arch
Internal iliac	**Uterine**	uterus
Internal iliac	**Vaginal**	vagina
Right atrium	**Vena cava (inferior)**[3]	common iliac, ascending lumbars, lumbars, right gonadal, renal, right suprarenal, hepatic, phrenic veins
Right atrium	**Vena cava (superior)**	brachiocephalic and azygos vein
Small/great saphenous	**Venous arch (dorsal**-foot)	digital veins, dorsum of foot
Radial, cephalic, basilic	**Venous arch (dorsal**-hand)	digital veins, dorsum of hand
Subclavian	**Vertebral**	neck, vertebrae, spinal cord
Internal iliac	**Vesicular**	urinary bladder

[1]Vein most commonly used for withdrawing blood
[2]Longest vessel in the body
[3]Widest vessel in the body

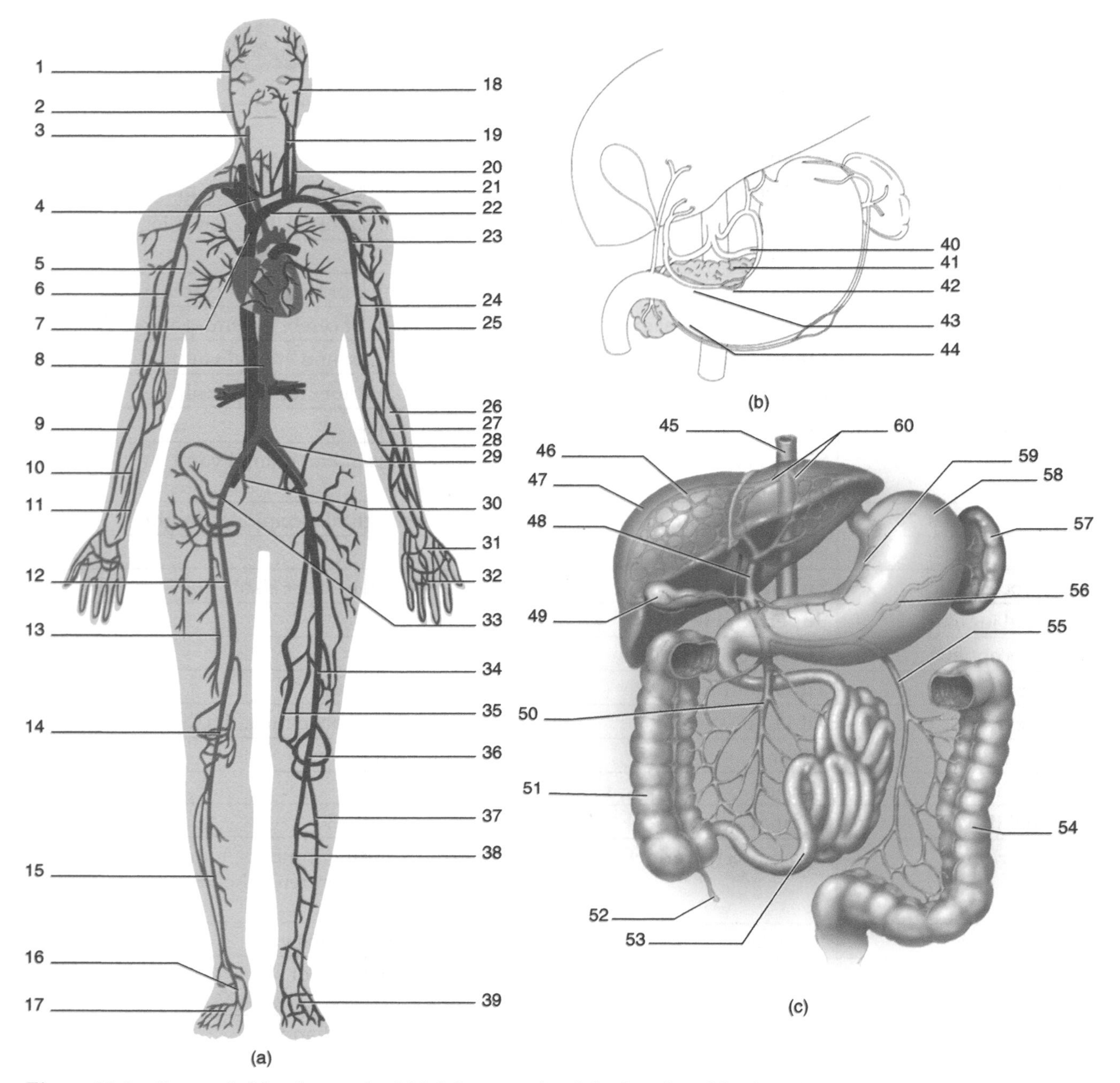

Figure 18-2 Systemic blood vessels: (a) Major vessels of the head and limbs., (b) branches of the celiac artery, (c) the hepatic portal system.

(a, c) © Kendall Hunt Publishing Company, (b) Illustrations by Jamey Garbett. © 2003 Mark Nielsen.

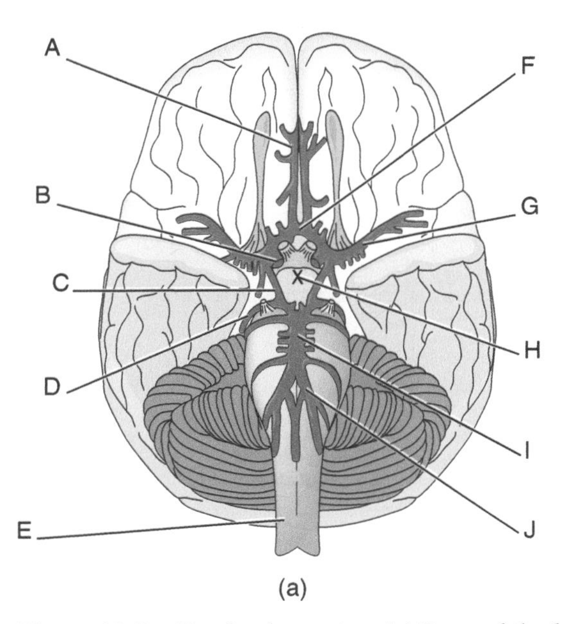

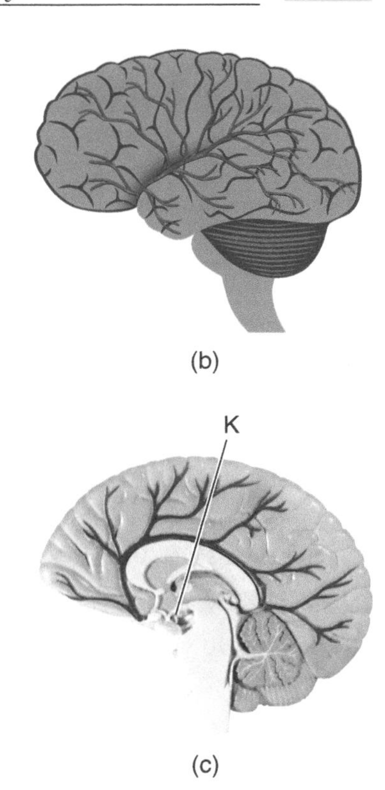

Figure 18-3 Cerebral arteries: (a) Base of the brain, (b) middle cerebral artery, (c) anterior cerebral artery.

Identifying Vessels In Preserved Specimen

You may be required to identify selected blood vessels in a preserved specimen, such as a fetal pig or cat. If so, and based on the quality of the preserved specimens, your instructor will provide a supplemental list of vessels to identify and visual references as a guide.

Measuring Blood Pressure

Blood pressure is the hydrostatic pressure exerted by blood on the inner walls of blood vessels. In the clinical profession, it usually refers to the pressure in the major arteries during ventricular systole. In this exercise, you can measure your own blood pressure and/or the blood pressure of a lab partner. Terms related to blood pressure and its measurement are shown in **Table 18.4**. Your instructor will demonstrate how to use the blood pressure equipment.

Table 18.4 Terms Related to Blood Pressure

Diastolic pressure	Pressure in a vessel when the ventricles are in diastole (relaxed)
Korotkoff sounds	Sounds heard in the vessels when they are constricted; constriction prevents blood from flowing in a laminar (straight line) manner and causes eddy (circular) currents in the vessels; turbulent flow as blood strikes the vessel walls are Korotkoff sounds.

(Continued)

Table 18.4 Terms Related to Blood Pressure *(Cont'd)*	
Mean arterial pressure (MAP)	Diastolic pressure + 1/3 pulse pressure; it is the average pressure in the vessels that transports blood to the tissues
Mercury manometer	Mercury-filled tube used to measure a person's blood pressure
Pulse	Expansion and recoiling of a vessel after each systole of the left ventricle
Pulse pressure	Systolic – diastolic pressure
Sphygmomanometer	Instrument used to restrict blood flow in the brachial artery when determining blood pressure
Stethoscope	Instrument used to listen to the heart, lung, or other sounds in ventral cavity
Systolic pressure	Pressure in a vessel when the ventricles contract; pressure when first Korotkoff sound can be heard as the pressure is relieved in a blood pressure cuff.

1. Determine your blood pressure as demonstrated by the instructor.
   Systolic Pressure = ________ Diastolic Pressure = ________
2. Calculate your pulse pressure: ______________________________
3. Calculate your MAP: ______________________________

Name ____________________________ Course Number: _________ Lab Section _______

Lab 18: Vessels Worksheet

1. Match the following ARTERIES with a correct choice:

A.	Middle colic	______	Receives blood from the brachiocephalic artery
B.	Vertebral	______	Receives blood from the popliteal artery
C.	Subclavian	______	Delivers blood to thigh muscles
D.	Popliteal	______	Delivers blood to the dorsal arterial arch
E.	Left common carotid	______	Delivers blood to the transverse colon
F.	Renal	______	Delivers blood to the plantar arterial arch
G.	Femoral	______	Delivers blood to the descending colon
H.	Brachial	______	Major artery behind the knee
I.	Occipital	______	Delivers blood to external genitalia
J.	Aorta	______	Unites anterior and posterior brain arteries
K.	Dorsalis pedis	______	Largest artery
L.	External carotid	______	Receives blood from the abdominal aorta
M.	Middle cerebral	______	Delivers blood to the gastric, splenic, and hepatic arteries
N.	Radial	______	Delivers blood to most of the brain's lobes
O.	Basilar	______	Delivers blood to the scalp
P.	Phrenic	______	Receives blood from the aortic arch
Q.	Celiac	______	Delivers blood to the kidney
R.	Right common carotid	______	Delivers blood to the appendix
S.	Ileocolic	______	Delivers blood to the facial artery
T.	External carotid	______	A deep artery in the forearm
U.	Tibial	______	Delivers blood to the urinary bladder
V.	Right colic	______	Delivers blood to the liver
W.	Pudendal	______	Receives blood from axillary artery
X.	Vesicular	______	Delivers blood to the diaphragm
Y.	Posterior tibial	______	Delivers blood to the scalp on the back of the head
Z.	Hepatic	______	Receives blood from vertebral arteries
AA.	Circle of Willis	______	Delivers blood to the ascending colon
AB.	Left colic	______	Delivers blood to the basilar artery
AC.	Common iliac	______	Major artery in the ankle

2. Match the following VEINS with a correct choice:

A.	Cephalic	______	Delivers blood to the external jugular
B.	Axillary	______	Receives blood from the small intestine
C.	External jugular	______	Delivers blood to the liver
D.	Ulnar	______	Receives blood from the common iliacs
E.	Inferior mesenteric	______	Delivers blood to the right atrium
F.	Internal jugular	______	Receives blood from anterior intercostals
G.	Superior sagittal sinus	______	Receives blood from the dorsal venous arch
H.	Brachiocephalic	______	Receives blood from descending colon
I.	Anterior tibial	______	Receives blood from scalp veins
J.	Hepatic portal	______	Superficial vein on lateral side of arm
K.	Superior vena cava	______	Receives blood from ascending lumbar veins and posterior intercostals
L.	Basilic	______	Receives blood from upper limb veins
M.	Superior mesenteric	______	Receives blood from the sigmoid sinus
N.	Azygos	______	Receives blood from most of the brain
O.	Great saphenous	______	Receives blood from the palmar venous arch
P.	Superficial temporal	______	Superficial vein on medial side of arm
Q.	Inferior vena cava	______	Delivers blood to the superior vena cava
R.	Internal thoracic	______	Longest blood vessel

Lab

19 Lymphatic System and Immunity

Overview of The Lymphatic System

The lymphatic system is an extension of the cardiovascular system in that it contains vessels that transport a clear fluid, *lymph*, into the blood. The lymphatic system also contains organs that house large numbers of *lymphocytes*, the second most abundant type of leukocyte (white blood cell). Lymph can ultimately trace its origin back to the blood. As blood courses through capillaries, some of the plasma is squeezed out into the surrounding tissues to become interstitial fluid. Most of this fluid is reabsorbed in the venule end of the capillary, but any excess is picked up by lymphatic capillaries to become lymph. Eventually, lymph makes its way back into the blood by entering the subclavian veins near the heart.

Vessels of The Lymphatic System

Just like in the cardiovascular system, there is a hierarchy of different-size vessels in the lymphatic system. The smallest vessels are **lymph capillaries**, which are *closed-ended* and collect excess interstitial fluid. Lymph capillaries have overlapping endothelial cells that connect to the surrounding tissues via small elastic threads called **anchoring filaments**. When tissues swell, the anchoring filaments pull the endothelial cells, which peel back like little flaps and allow more interstitial fluid to enter the capillary. **Lacteals** are lymph capillaries in the small intestine that pick up digested lipids from food.

Lymph flows from lymph capillaries into **afferent lymphatic vessels**, which ultimately enter bean-like filtering organs called *lymph nodes*. Lymph flows out of lymph nodes through **efferent lymphatic vessels** and into **lymphatic trunks**. There are five major lymphatic trunks, each named according to the region from which it receives lymph: *jugular* from the head and neck, *intestinal* from the intestines, *bronchomediastinal* from the thoracic cavity, *lumbar* from the lower abdomen, and *subclavian* from the shoulder region. Use the acronym JIBLS ("jibbles") to remember the lymphatic trunks.

After leaving a trunk, lymph passes into one of two lymphatic ducts. The **right lymphatic duct** collects lymph from the right side of the body above the abdomen and is only a few centimeters long. It empties lymph into the right subclavian vein. The **thoracic** (or **left lymphatic**) **duct** is much longer and courses upward through the abdomen and thorax to empty lymph into the left subclavian vein. At the inferior end of the thoracic duct is the sac-like **cisterna chyli**, which receives lymph from the intestines. Label the parts of the lymphatic system in **Figure 19-1**.

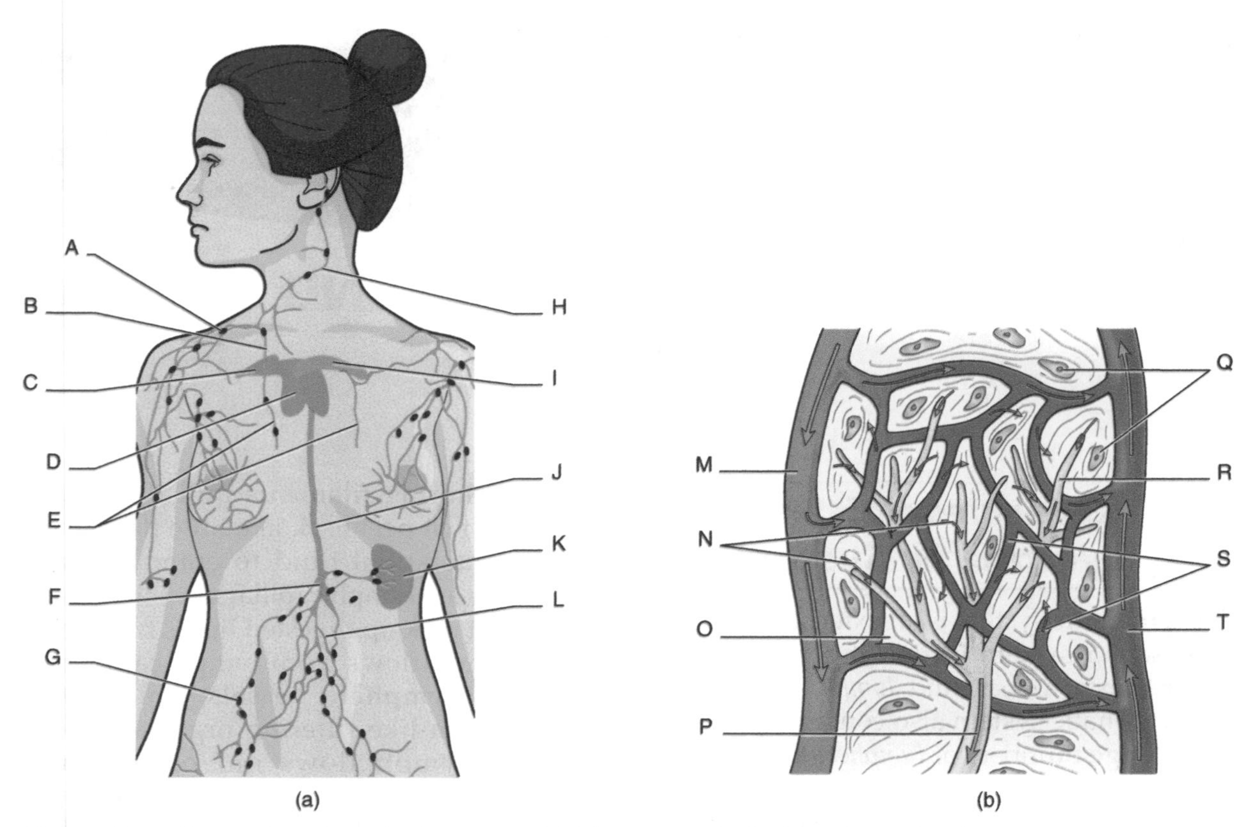

Figure 19-1 Parts of the lymphatic system: (a) Major vessels, (b) lymphatic capillaries.

Lymphatic Organs

In addition to vessels, the lymphatic system contains specialized lymphatic tissue containing abundant lymphocytes that can help defend the body against foreign items that might enter the lymph or blood. Lymphatic organs have a well defined shape and include the *spleen, thymus gland*, and *lymph nodes.* The **spleen** is in the left hypochondriac region of the abdomen and it filters blood. Its *red pulp* serves as the red blood cell "graveyard" where old RBCs can be broken down and their components recycled. Its *white pulp* houses numerous lymphocytes.

The **thymus gland** is in the mediastinum and houses numerous lymphocytes called T cells. It becomes less important in the adult and its size diminishes over time.

Lymph nodes are scattered throughout the body at sites along major lymphatic vessels. They filter lymph and house numerous lymphocytes. The node *cortex* is the outer portion of the node, while the *follicle* is the region containing abundant lymphocytes. The *trabeculae* are beam-like fibers within the node that trap foreign particles in the lymph. Label the structures in **Figure 19-2**.

There are five major groups of lymph nodes that receive lymph from specific regions: *intestinal* from the intestines; *lumbar* from the other abdominal organs; *inguinal* from the lower appendages; *axillary* from the upper appendages; and *cervical* from the head and neck. Remember the major

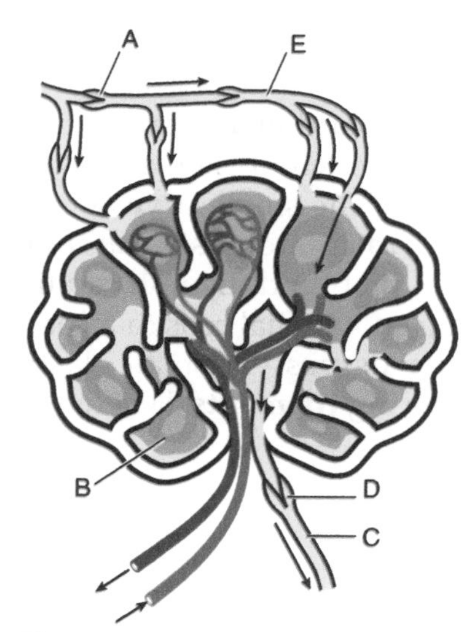

Figure 19-2 Structure of a lymph node.

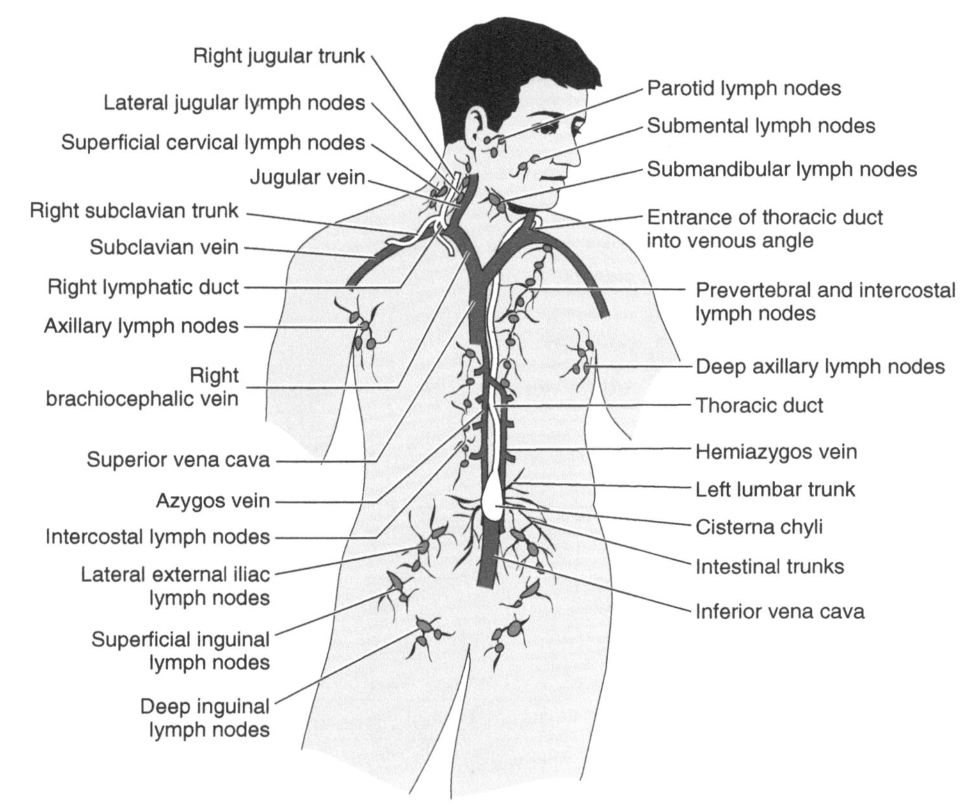

Figure 19-3 Major groups of lymph nodes.

groups of lymph nodes with the acronym *ILIAC*. See the lymphatic structures in **Figure 19-3**.

Diffuse Lymphatic Tissue

Whereas lymphatic organs have a well-defined outer border, or connective tissue capsule, **diffuse lymphatic tissue** does not. Since this tissue is found in contact with certain mucous membranes, it is collectively referred to as **MALT** (mucosa-associated lymphatic tissue). Diffuse lymphatic tissues include *tonsils*, *Peyers patches*, and the *appendix*. The tonsils, which are actually partially encapsulated, contain crevices called *crypts* that can trap foreign particles entering the body in food and air. The tonsils include the following: **lingual tonsils** on the posterior surface of the tongue, **palatine tonsils** in the arches alongside the posterior region of the tongue, **tubal tonsils** around the openings of the auditory (Eustachian) tubes in the nasopharynx; and the **pharyngeal tonsils** (also called **adenoids**) in the nasopharynx.

Peyer's patches are lymphatic nodules in the walls of the small intestine. The **appendix** is an extension of the large intestine in the lower right quadrant. Peyer's patches and the appendix help protect the blood from bacteria entering it from the intestines.

Immunoscenarios

In this part of the lab, you will participate in a game for the immune system, and your lab instructor will provide the directions. Since this game relates to how lymphocytes help defend your body against foreign pathogens (disease-causing agents), you should be able to identify the relevant components of lymphocytes shown in **Table 19.1** and **Figure 19-4**.

Diseases in the ImmunoScenarios game

The following diseases and adverse reactions may be encountered in the ImmunoScenarios game. For the next lab, you should have a general understanding of each of them.

Table 19.1 Terms Related to the Lymphocytes	
Antibody	Protein that can interact with an antigen to cause agglutination, neutralization, etc.
Antigen	Foreign body that causes the body to generate antibodies
Antigen-presenting cell	APC; cells capable of endocytosis of an antigen and presentation of the antigen to T lymphocytes; include macrophages, Kuppfer cells, Langerhans cells, B cells, and microglia
B lymphocyte	Cells whose ancestors developed in bone marrow; may become antibody-secreting plasma cells
CD4 marker	On helper T cells; interacts with MHC-2 surface markers on macrophages and B cells
CD8 marker	On cytotoxic T cells; interacts with MHC-1 markers on all body cells except RBCs
Cytotoxic T cell	Killer T cell; uses perforin and lymphotoxin to destroy antigen-infected body cells
MHC I receptor	Identifying marker on all body cells except RBCs
MHC II receptor	Additional identifying marker on macrophages and B cells
Plasma cell	Differentiated B cell that secretes antibodies
T cell receptor	TCR; located on T cells and recognizes specific antigen
T helper cell	Interacts with macrophages and B cells; secretes interleukins to stimulate production of lymphocytes
T lymphocyte	Cell whose ancestors developed within the thymus gland early in life; display T-cell receptors and CD markers; includes helper T and cytotoxic T cells

AIDS (Acquired Immune Deficiency Syndrome): The HIV (Human Immunodeficiency Virus) invades helper T cells after binding to a specific molecule ($CD4^+$) on the cell's surface. As the infection progresses, the number of T_H cells declines. Since these cells are necessary for stimulating the humoral (antibody) and cellular immune responses, a decline in their number can lead to increasing difficulty in resisting other infectious agents. AIDS results in the inability to resist even the organisms that grow naturally in the body. Spread of HIV occurs through contact with infected body fluids such as blood and semen, since these fluids carry free virus particles and infected cells. Due to the complexity of the virus, there is no effective immunization yet.

Allergy: Also called hypersensitivity; an immune response that goes beyond what is required to protect the body and, in the process, can cause harm to healthy tissues.

Blood transfusion reaction: Agglutination of RBCs when antibodies attach to specific agglutinogens on their surfaces.

Chickenpox: Chickenpox is a disease caused by a type of herpes virus called varicella, and its transmission is by infected airborne droplets. Although it initially infects the respiratory tract, this virus localizes in the skin after about two weeks to form vesicles filled with fluid. These can rupture and form a scab before healing. Activated B cells produce antibodies that attach to these specific virus particles causing their neutralization, thus, providing protection from developing chicken pox upon subsequent exposure to the virus. Cytotoxic T cells help eliminate the viruses by killing body cells infected with them.

Cold (common): Caused by a number of different viruses, including rhinoviruses and adenoviruses. These viruses enter the respiratory tract and can cause IgA antibodies to form. Since there are at least 200 different agents that can cause the common cold, repeated infections are common. As people age, they contract fewer colds due to accumulated immunity during their lifetime. Occasionally, a cold can lead to some complications such as laryngitis and middle ear infection.

Diphtheria: Diphtheria is a disease of the upper respiratory tract that is caused by the bacterium *Corynebacterium diphtheriae*. The bacteria grow in the throat and produce a grayish membrane that can block the passage of air to the lungs. The bacteria also produce a toxin that circulates in the blood and damages the heart and kidneys, often leading to death if the patient is not treated with antibiotics. Antitoxin (an IgG) can form that can neutralize the toxin. A vaccine consisting of an inactivated form of the toxin (toxoid) gives long-lasting immunity.

Hepatitis: Hepatitis B virus is transmitted by infected blood or other body fluids such as semen, breast milk, and saliva, and it can cause an acute form of liver disease that may or may not be noticeable. Although antibodies to the virus develop, the disease may become chronic and may lead to either cirrhosis or cancer of the liver. Now genetically engineered vaccines can prevent infection by inducing specific antibody production against this virus.

Influenza (flu): The "flu" is caused by a virus that is spread through airborne droplets or by contact with contaminated hands or surfaces. Influenza virus replicates in cells of the respiratory tract to cause damage that can lead to complications such as bacterial pneumonia. Specific IgA antibodies can block the binding of flu viruses to respiratory cells. However, since viruses can frequently change their surface structure, the body must make new IgA antibodies periodically. Receiving a vaccine made from the latest virus strains can induce production of new antibodies.

Measles (rubeola): Measles is an extremely infectious viral disease spread by infected respiratory droplets. Viruses replicate in the respiratory tract and then spread through the blood to the skin and mucous membranes, including the conjunctiva. IgG antibodies form to inactivate the virus. Cytotoxic T cells interacting with virus-infected cells in the small blood vessels are the cause of the typical measles rash. Protection against measles is possible through the administration of a highly effective vaccine containing attenuated, live measles virus.

Mumps: Mumps is a viral disease characterized by swelling of the parotid salivary glands and a general feeling of tiredness and loss of appetite. Spread of the virus occurs through respiratory droplets that enter the mouth and nose. During the disease various antibodies form. Cytotoxic T cells also are present but their function in recovery from the virus is unclear. After recovery, the person has life-long immunity. There is also an attenuated live-virus vaccine given to young children.

Pneumonia: A bacterium called *Streptococcus pneumoniae* causes **bacterial pneumonia** by entering the body through infected respiratory droplets. Although healthy persons frequently do not contract the disease even when they are exposed to the organism, those who have had a recent viral respiratory infection, such as influenza, or those who have lowered resistance, are more susceptible. Specific antibodies arise when a person contracts pneumonia, or after immunization with a vaccine made with antigens from the bacterium's capsule. **Aspiration pneumonia** is due to the entrance of foreign matter (e.g., food, saliva, etc.) into the bronchi. Certain viruses cause **viral pneumonia**.

Polio: Initially, polioviruses replicate in the cells of the throat and gastrointestinal tract where they rarely cause a serious problem. However, if they enter the blood and travel to the central nervous system, they can replicate and cause paralysis. Specific antibodies block the transmission of polioviruses and prevent damage to the central nervous system. Immunization consists of either a killed or a live-virus vaccine, both of which give long-lasting immunity.

Staph infection: Although staphylococci bacteria are common on human skin and in the respiratory and gastrointestinal tracts, certain species can cause a number of diseases. *Staphylococcus aureus* is a common cause of "staph" infections of the skin such as abscesses, acne, and vitiligo (loss of skin pigment). Antibodies that form against the organisms are not completely protective and a person may have repeated infections.

Strep throat: Pathogenic *Streptococcus pyogenes* living in the throat or nasal passages of an infected individual can infect a healthy individual through sneezing, kissing, sharing eating utensils, etc. They cause inflammation in the throat region known as "strep" throat or pharyngitis. Antibodies can form against a specific strain of *S. pyogenes*; however, there are many different strains, and a person may contract strep throat repeatedly.

Tetanus: **Pathologic**: If a person steps on a nail that has spores of the bacterium, *Clostridium tetani*, on its surface, the spores can germinate and multiply in the anaerobic environment of the injured tissue. The bacteria release a toxin that acts on the nervous system to cause severe muscle spasms that can be fatal. Immunization with an inactivated form of the toxin (toxoid) can prevent the spasms by inducing the formation of specific IgG molecules that neutralize the toxin. Spores may be abundant in feces-contaminated soil.

Tuberculosis: Disease of the lungs caused by a bacterium called *Mycobacterium tuberculosis,* which enters the lungs through sputum droplets from an infected patient. Once infected, a person develops both cellular immunity and delayed type hypersensitivity (DHR) responses. The DHR involves T_{DH} cells sensitized by the antigens from the TB organism. The T_{DTH} cells then cause the migration of macrophages to the infection site. In an otherwise healthy person, the T_{DTH} and cytotoxic T cells confine the organism to the original site of infection. However, if the person is immunocompromised, as would be the case in an HIV-infected person, the organisms may spread and cause a systemic infection that can be life threatening.

Whooping cough (**pertussis**): Whooping cough is caused by a bacterium called *Bordetella pertussis* that infects the respiratory tract from droplets spread when an infected patient coughs. The cough, called a "paroxysmal" cough, is so severe that it can lead to difficulty in breathing. A characteristic "whooping" sound occurs when the patient attempts to get his or her breath. After recovery, immunity is good due to the presence of IgG antibodies. A vaccine prepared from heat-inactivated whole bacteria is given regularly to young children to immunize them against this disease.

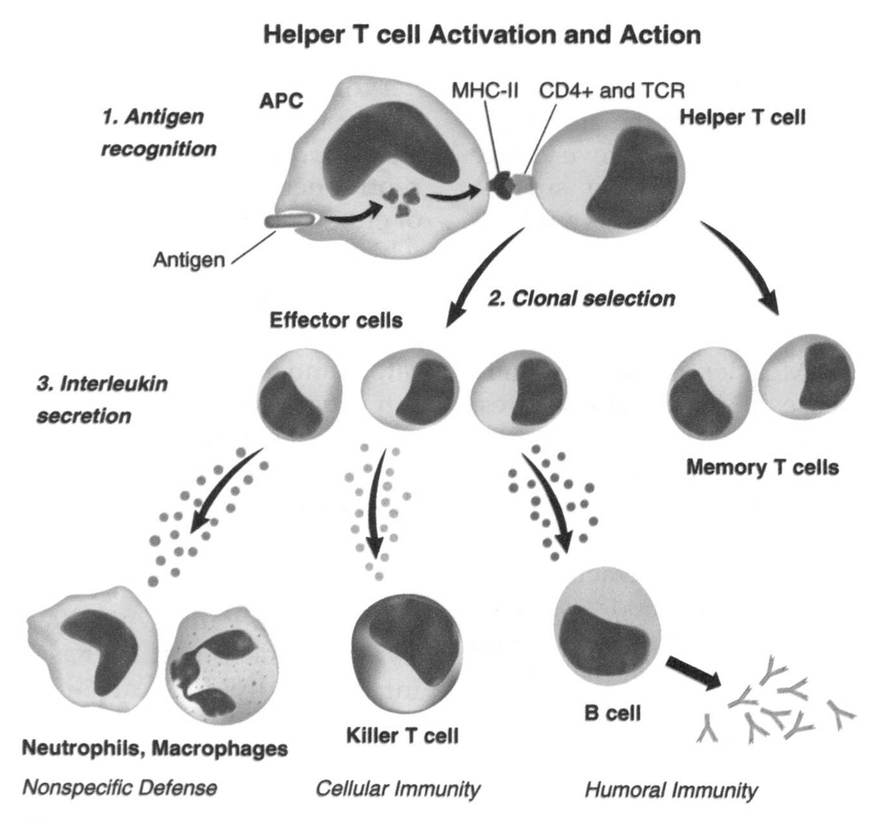

Figure 19-4 Activation and action of lymphocytes.

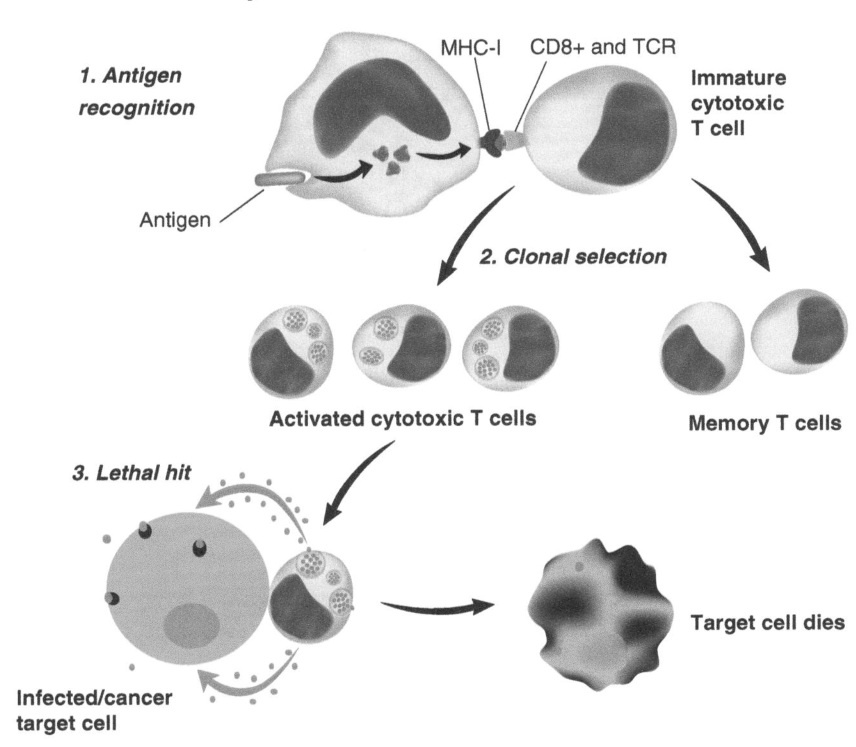

Figure 19-4 Continued

Name ______________________ Course Number: ________ Lab Section ______

Lab 19: Lymphatic System Worksheet

Write out the names of the labeled items in the selected figures.

Figure 19-1.

A. ______________________

B. ______________________

C. ______________________

D. ______________________

E. ______________________

F. ______________________

G. ______________________

H. ______________________

I. ______________________

J. ______________________

K. ______________________

L. ______________________

M. ______________________

N. ______________________

O. ______________________

P. ______________________

Q. ______________________

R. ______________________

S. ______________________

T. ______________________

Figure 19-2.

A. ______________________

B. ______________________

C. ______________________

D. ______________________

E. ______________________

LAB 20 RESPIRATORY SYSTEM

The **respiratory system** is responsible for the exchanging gases (oxygen and carbon dioxide) between the blood and the atmosphere. In this lab, you will identify the anatomical features of the upper and lower respiratory systems and also measure certain aspects of breathing. An overview of the respiratory system is shown in **Figure 20-1**, and you should be able to label its parts after learning material in the tables that follow.

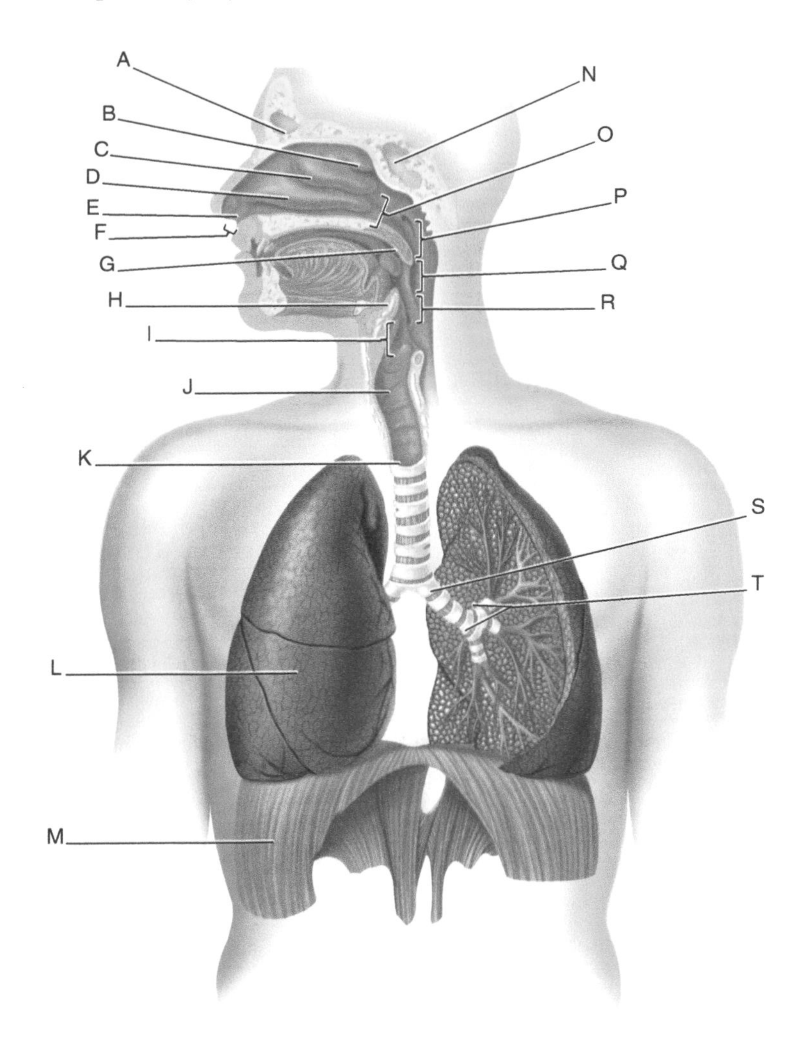

Figure 20-1 Overview of the respiratory system.

Upper Respiratory Tract

Most clinicians recognize the upper respiratory system to include structures and passageways located superior to the trachea (windpipe). These items include the nose, nasal cavity, and pharynx (throat). Look at **Table 20.1** to see a list of these items and their descriptions. Label the items in **Figures 20-1** and **20-2**.

Table 20.1	Features of the Upper Respiratory Tract
Alar cartilages	Plates of hyaline cartilages forming the most lateral regions of the nose (lateral to the lateral cartilages) and forming the shape of the nostrils; includes the larger **greater alar** and smaller **lesser alar cartilages**
Alar fibrofatty tissue	Fibrous connective tissue around the lesser alar cartilages; forms the "wings" of the nostrils
Arytenoid cartilages	Ladle-shaped hyaline cartilages in the larynx; site of muscle attachment
Arytenoid muscle	Skeletal muscle that closes the glottis
Corniculate cartilages	Horn-like hyaline cartilages attached to the superior end of the arytenoid cartilages
Cricoarytenoid muscles	Laryngeal muscles that open and close the glottis; **lateral cricoarytenoids** close the glottis; **posterior cricoarytenoids** open the glottis
Cricoid cartilage	Ring-like hyaline cartilage at bottom of larynx and top of trachea
Cricothyroid membrane	Fibrous membrane connecting the cricoid cartilage to the thyroid cartilage
Cricothyroid muscles	Laryngeal muscle connecting the cricoid and thyroid cartilages; stretches (tightens) the vocal cords to produce high pitch sounds
Cuneiform cartilage	Hyaline cartilages in the lateral wall of the laryngeal vestibule; prevents vestibule collapse
Epiglottis	Flap containing elastic cartilage that closes the laryngeal vestibule during swallowing
Ethmoid bone	Forms the roof of the nasal cavity; contains conchae and perpendicular plate
External nares	External nostrils; openings between the nasal vestibule and the outside
Frontal bone	Bone of the forehead; contains frontal sinus
Glottis	Opening between the vocal cords at the base of the larynx
Hard palate	Roof of the mouth (floor of nasal cavity) formed by maxillae and palatine bones
Hyoid bone	U-shaped bone in the neck, superior to the larynx; attachment site for muscles and membranes
Internal nares	Internal openings between the nasal cavity and nasopharynx
Laryngeal inlet	Opening to the laryngeal vestibule
Laryngeal prominence	Anterior projection on the thyroid cartilage; known as the Adam's apple
Laryngeal vestibule	Cavity between the epiglottis and glottis
Laryngopharynx	Muscular throat region around the epiglottis
Larynx	(Voice box); hyaline cartilage structure containing vocal cords; entrance to trachea
Lateral cartilages	Large plates of hyaline cartilages on the sides of the nose
Maxillary bones	Top jaw bones; forms the base of the nasal cavity
Nasal bones	Two narrow bones that forms the bridge of the nose; connect to frontal bone
Nasal conchae	Ridges inside the nasal cavity that increase surface area for warming and moistening air; includes **inferior**, **middle**, and **superior nasal conchae** (last two are on the ethmoid bone)

(Continued)

Table 20.1 Features of the Upper Respiratory Tract *(Cont'd)*

Nasal septum	Partition that divides the nasal cavity; made of perpendicular plate of ethmoid and vomer
Nasal vestibule	Space between the external nares and nasal cavity
Nasopharynx	Muscular throat region at the posterior end of the nasal cavity
Oropharynx	Muscular throat region at the back of the oral cavity between nasopharynx and laryngopharynx
Palatine bones	Two bones that form the posterior one-third of the hard palate
Paranasal sinuses	Cavities within cranial bones around the nasal cavity; includes **ethmoid air cells**, **frontal sinus**, **maxillary sinuses** and **sphenoid sinus**
Perpendicular plate	Vertical projection of ethmoid bone that forms most of the nasal septum
Septal cartilage	Central, vertical hyaline cartilage of the nose; inferior to the nasal bones
Soft palate	Soft tissue region posterior region of oral cavity roof; posterior to the hard palate
Sphenoid bone	Bird-shaped bone forming the back of the nasal cavity
Thyroarytenoid muscle	Muscles that connect the thyroid cartilage to the arytenoid cartilages; relaxes (shortens) vocal cords to produce low pitch sounds
Thyrohyoid membrane	Fibrous membrane connecting the thyroid cartilage to the hyoid bone
Thyroid cartilage	Largest hyaline cartilage of the larynx
Uvula	Flap of tissue extending from the soft palate; closes the nasopharynx during swallowing
Vestibular folds	(**Ventricular** or **false vocal folds**); superior to the true cords; closes the glottis
Vocal folds	Elastic tissue bands on either side of the glottis; produce sound when vibrating
Vomer bone	Plow-shaped bone that forms the inferior one-third of the nasal septum

Lower Respiratory Tract

The lower respiratory tract will be considered those components located inferior to the glottis. These items include the trachea, bronchi, bronchioles and smaller tubes, alveoli, and the respiratory membrane. **Table 20.2** lists the features of the lower respiratory tract. After you become familiar with these features, label them on the figures that follow.

Muscles of Breathing

Breathing (inhaling and exhaling air) requires pressure changes within the lungs and this is brought about by muscle contractions changing the volume of the thoracic cavity. (Read about Boyles' Law in the lecture notes). The skeletal muscles used for breathing are listed in **Table 20.3**. Refer back to Lab 9 to see illustrations of these muscles. Also, be able to identify them on models.

Preserved Lungs

Observe the real lungs on demonstration. You can cause them to inflate and deflate using the manual air pump. You should also be able to identify the various parts, including visceral peritoneum on the lung surface.

Demonstration of Boyle's Law

Several demonstrations are available for Boyle's Law, which states that the pressure inside a cavity is inversely proportional to the cavity's volume. We can change the volume of the thoracic

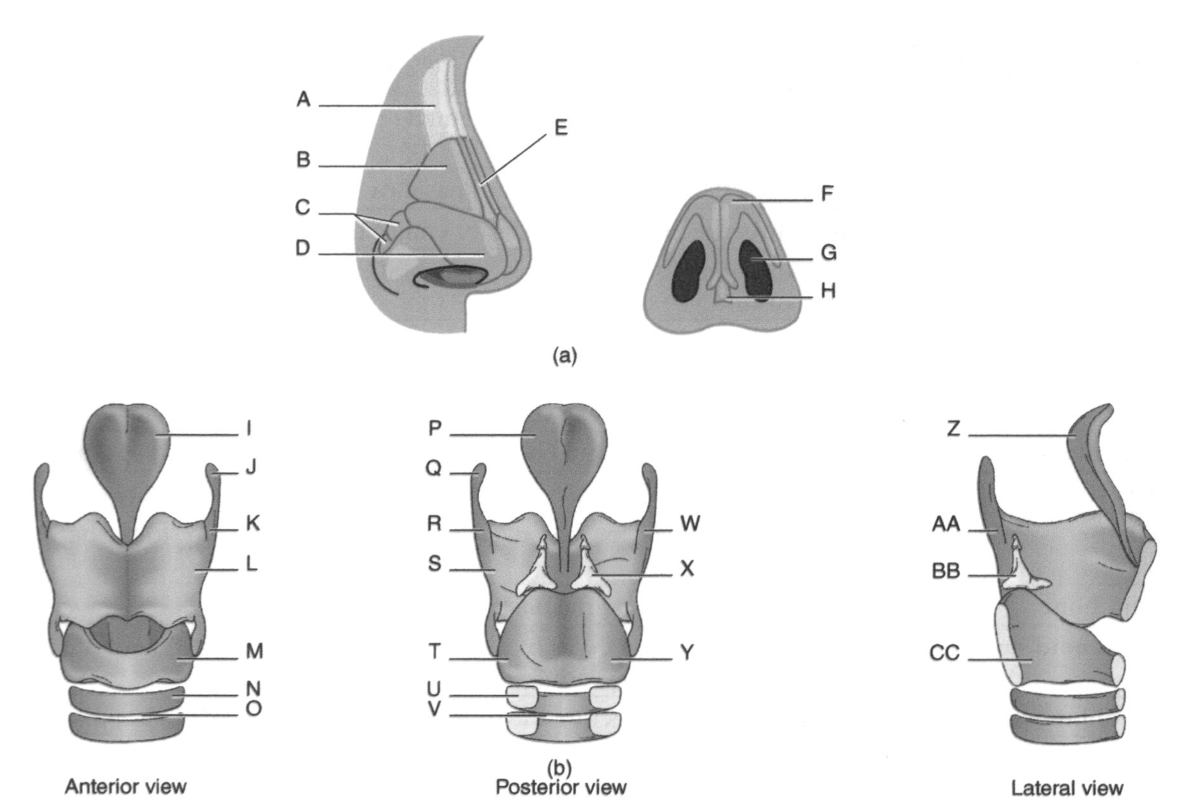

Figure 20-2 Cartilages in the upper respiratory tract: (a) Nose, (b) larynx.

cavity both vertically, using the diaphragm, and horizontally, using the intercostals and other muscles that pull on the ribs. Both demonstrations utilize rubber balloons to mimic the lungs. One demonstration has an elastic diaphragm that you can pull downward to increase the size of the cavity around the balloons. The other demonstration requires pressing inward on the sides of a plastic soda bottle from the sides to change the cavity's volume horizontally. Be able to explain the processes that cause inflation and deflation of the balloons.

Respiratory Volumes and Capacities

Clinicians often measure the amount of air that can be inhaled and/or exhaled to determine the relative health of the respiratory system. A **respiratory volume** is the total amount of air that can be measured directly, whereas a **respiratory capacity** is a summation of various lung volumes. **Anatomical dead space** is the amount of air existing between the external nares and the alveolar ducts at any one time. It is a difficult volume to measure, but other volumes are more easily measured.

A **spirometer** is a device used to measure respiratory volumes. In this lab, you can use a computerized spirometer that measures volumes electronically and/or a *wet* spirometer, one that measures the amount of water displaced when air is exhaled into a plastic bucket. Your lab instructor will provide directions on how to use the spirometers. **Table 20.4** and **Figure 20-4** show the respiratory volumes and capacities.

Table 20.2 Features of the Lower Respiratory Tract	
Adventitia	Outer areolar connective tissue covering of the trachea
Alveolar duct	Small tube between a respiratory bronchiole and an alveolar sac
Alveolar macrophages	(Dust cells); Large phagocytes that remove debris from inside of alveoli
Alveolar sac	Cluster of alveoli around an alveolar duct
Alveolus	Grape-like sac where gas exchange occurs between blood and the atmosphere
Bronchi	Tubes extending from the trachea toward the bronchioles; contain hyaline cartilage
Bronchiole	Muscular tube between a bronchus and alveolar duct; no cartilage present
Carina	Keel-like internal ridge where bronchi branch from the trachea; sensitive to foreign objects
Costal cartilage	Hyaline cartilage extending from the anterior end of a rib
Lumen	Cavity of a hollow organ, such as in a lung's tubes
Lung lobes	Three exist on the right and two exist on the left
Pleura	Serous membranes associated with the lungs: **parietal pleura** lines each pleural cavity and is continuous with the **visceral pleura** that covers the surface of each lung
Pleural cavity	Virtual space between a lung and the thoracic wall
Respiratory membrane	Site of gas exchange between the blood and atmosphere; includes alveolar cells and their basement membrane and the capillary wall and its basement membrane
Respiratory mucosa	Ciliated, pseudostratified columnar epithelium in respiratory tract
Ribs	Enclose the thoracic cavity and are important in breathing mechanism; includes true ribs (**vertebrosternal** or pairs 1-7) with cartilage attached to sternum; **vertebrochondral** (false ribs or pairs 8-10) with cartilages attached to other cartilages; **vertebral** (false ribs or pairs 11-12) with cartilage not attached to anything else
Stroma	Elastic tissue matrix within a lung
Trachea	(Windpipe) Large tube between the larynx and bronchi; contains tracheal cartilages
Tracheal cartilage	Semicircular rings of hyaline cartilage around the trachea
Type I cells	Simple squamous cells lining the inside of alveoli; site of gas exchange
Type II (septal) cells	Surfactant-secreting cells inside alveoli; surfactant decreases water surface tension

Table 20.3 Muscles Used in Breathing	
MUSCLES OF INHALATION	
Sternocleidomastoid	Muscle attaching mastoid process to sternum: used in forced inhalation
Diaphragm	Muscular partition between thorax and abdomen; muscle fibers pull on the ***central tendon*** located within the diaphragm; a contracting diaphragm moves inferiorly to increase thoracic volume vertically; relaxing the diaphragm moves superiorly
External intercostals	Between the ribs; contraction elevates and spreads ribs to increase thoracic volume
Scalenus	Muscles in the neck that connects cervical vertebrae to the sternum: elevates sternum
Serratus anterior	Attaches the scapula to the ribs
MUSCLES OF EXHALATION	
Transversus abdominis	Muscle in abdomen; compresses abdomen during forced exhalation
External obliques	Abdominal muscle used during forced exhalation; compresses abdomen
Internal intercostals	Between ribs; pull ribs down and together—exhalation
Internal obliques	Middle layer in abdomen; compresses abdomen in forced exhalation
Rectus abdominis	Abdominal muscle; compresses abdomen during forced exhalation
Trachealis	Smooth muscle layer on posterior side of trachea

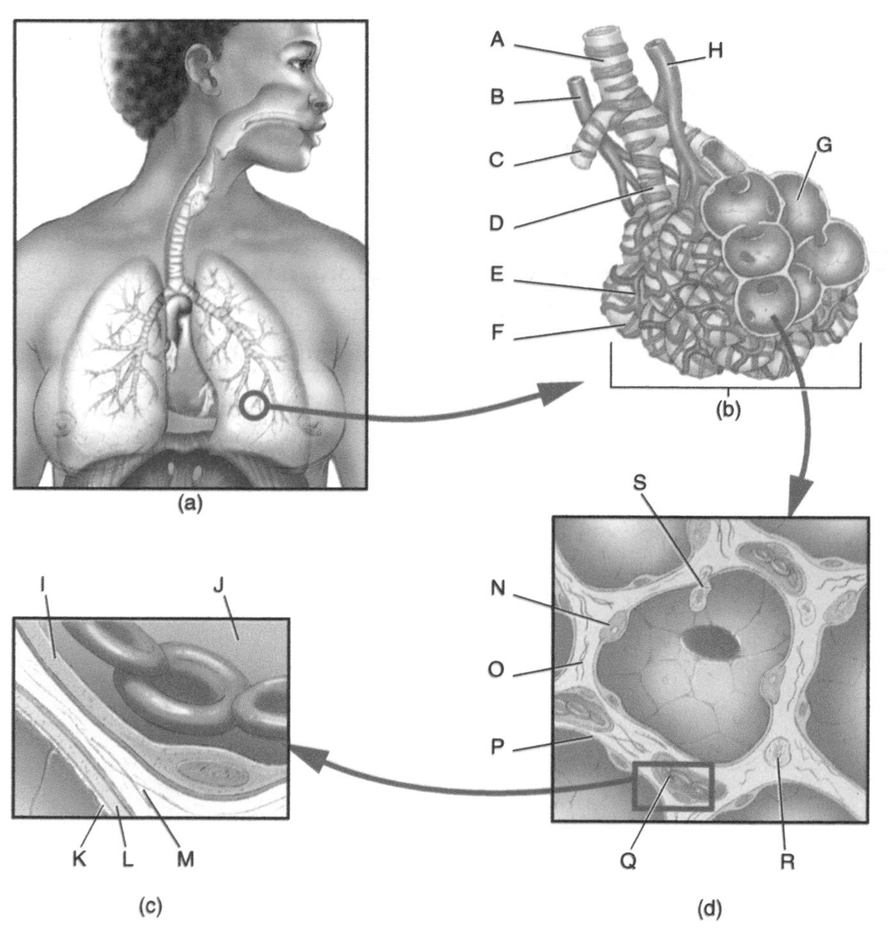

Figure 20-3 Site of gas exchange in the lungs. (a) Lungs; (b) Bronchioles and alveoli; (c) Respiratory membrane; (d) Inside an alveolus.

Table 20.4 Respiratory Volumes and Capacities

RESPIRATORY VOLUMES			
Anatomical dead space	Total amount of air between the external nares and alveolar ducts at any one time		
Expiratory reserve volume	(**ERV**) Amount of air exhaled forcefully after normal exhalation		
Forced expiratory volume	(**FEV**) Maximum amount of air exhaled in one second after maximum inhalation; healthy individual's FEV should be about 80% of their vital capacity		
Inspiratory reserve volume	(**IRV**) Maximum amount of air inhaled forcefully after normal inhalation		
Minute respiratory volume	(**MRV**) Amount of air breathed in or out in one minute: TV x Number of breaths/min.		
Residual volume	(**RV**) Amount of air remaining in *all* alveoli after maximum exhalation		
Tidal volume	(**TV**) Total amount of air either inhaled or exhaled during normal breathing		
RESPIRATORY CAPACITIES			
Functional residual capacity	(**FRC**) Amount of air left in the lungs at the end of normal exhalation: ERV+RV		
Inspiratory capacity	(**IC**) Amount of air inhalable beginning at rest; IC = TV+IRV		
Total lung capacity	(**TLC**) Amount of air in the lungs after maximum inhalation; TV+IRV+ERV+RV (ERV and RV because they exist in the lungs at the beginning of tidal inhalation)		
Vital capacity	(**VC**) Maximum amount of air exhaled after maximum inhalation; TV+IRV+ERV		
AVERAGE VOLUMES AND CAPACITIES			
	Males	**Females**	**My Values**
ERV	~1200	~950	Measured =
FEV	~3750	~3000	Measured =
IRV	~3000	~2500	
RV	~1200	~950	
TV	~500	~400	
IC	Calculated =	Calculated =	
TLC	Calculated =	Calculated =	Calculated =
VC	Calculated =	Calculated =	Calculated =

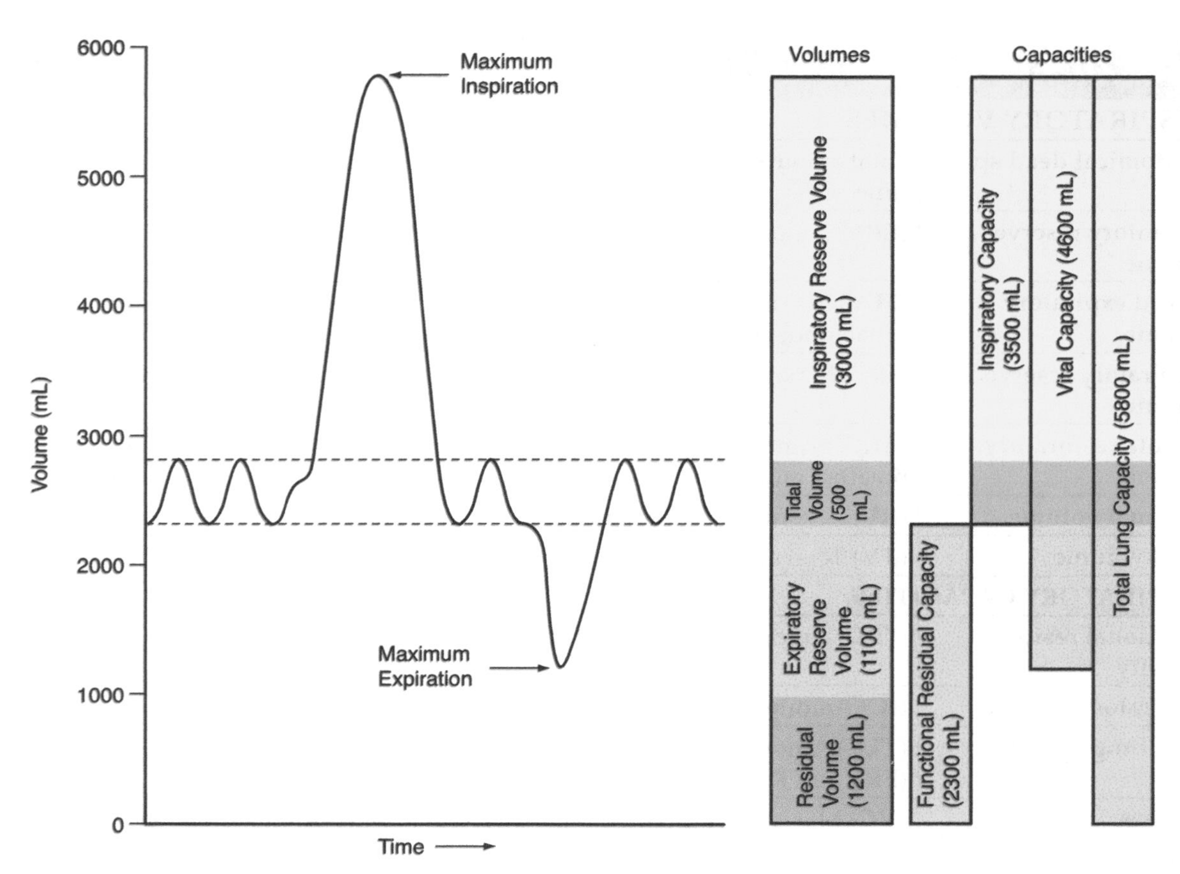

Figure 20-4 Respiratory volumes and capacities.

Name ____________________ Course Number: ________ Lab Section ______

Lab 20: Respiratory System Worksheet

Write out the names of the labeled items in the selected figures.

Figure 20-1.

A. ____________________
B. ____________________
C. ____________________
D. ____________________
E. ____________________
F. ____________________
G. ____________________
H. ____________________
I. ____________________
J. ____________________
K. ____________________
L. ____________________
M. ____________________
N. ____________________
O. ____________________
P. ____________________
Q. ____________________
R. ____________________
S. ____________________
T. ____________________

Figure 20-2.

A. ____________________
B. ____________________
C. ____________________
D. ____________________
E. ____________________
F. ____________________
G. ____________________
H. ____________________
I. ____________________
J. ____________________
K. ____________________
L. ____________________
M. ____________________
N. ____________________
O. ____________________
P. ____________________
Q. ____________________
R. ____________________
S. ____________________
T. ____________________
U. ____________________
V. ____________________
W. ____________________
X. ____________________
Y. ____________________
Z. ____________________
AA. ____________________
AB. ____________________
AC. ____________________

Figure 20-3.

A. ____________________
B. ____________________
C. ____________________
D. ____________________
E. ____________________
F. ____________________
G. ____________________
H. ____________________
I. ____________________
J. ____________________
K. ____________________
L. ____________________
M. ____________________
N. ____________________
O. ____________________
P. ____________________
Q. ____________________
R. ____________________

Lab

21 Digestive System

Overview of the Digestive System

The digestive system is responsible for breaking apart food items and converting them into components that can be absorbed (assimilated) into the blood. It consists of a long tube called the **alimentary canal** that extends from the mouth to the anus, and a variety of accessory organs. The **accessory organs** are found outside of the alimentary canal and secrete substances that aid in the digestion (break down) of food. For the study of digestive system anatomy we present structures by regions, beginning with the oral cavity. After you become familiar with the structures, label them on the figures. **Figure 21-1** provides an overview of the digestive system.

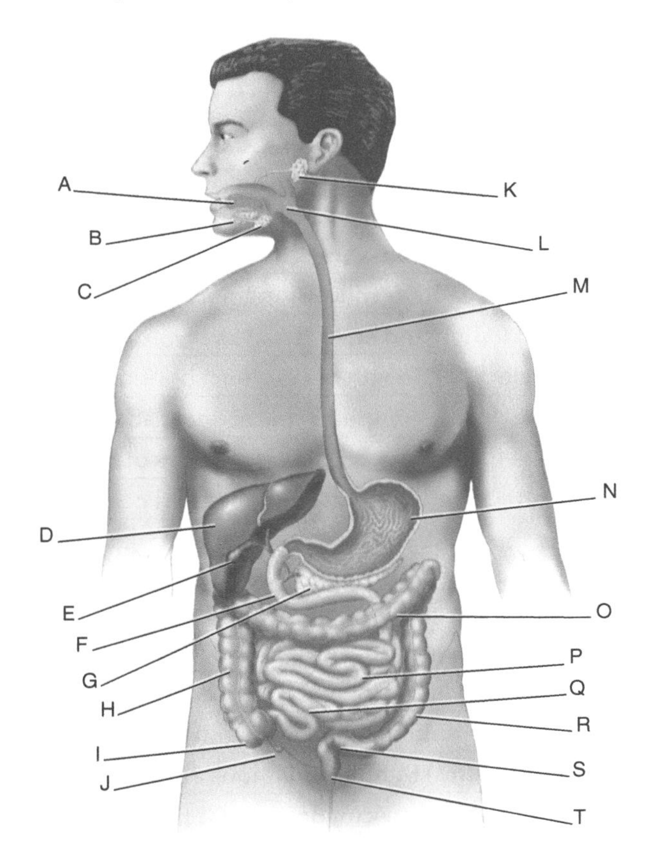

Figure 21-1 Overview of the digestive system.

Tissues of the Alimentary Canal

Although different regions of the alimentary canal may have different overall shapes, the basic makeup of the wall in those regions is very similar. At your table you will find a microscope slide containing a section of the small intestine. View this slide on low and high power and be able to identify the components specified by your lab instructor; you can mark these items in **Table 21.1**. In addition, on **Figure 21-2** you should be able to label all relevant items listed in the table.

Mouth, Oral Cavity, and Pharynx

The path through which food passes through the alimentary canal begins at the opening called the **mouth**. Food then passes into the **oral cavity** where it is chewed and formed into a soft mass called a *bolus*. The bolus then passes into the **pharynx**, or throat, before moving to the stomach. **Table 21.2** lists the features of the mouth, oral cavity, and pharynx. After you become familiar with these items, identify them on the lab models and label them on **Figure 21-3**.

Esophagus and Stomach

After leaving the pharynx, the bolus passes through the esophagus and then enters the stomach. The major anatomical features of the stomach can be seen in **Figure 21-1**, while the anatomy of the stomach lining is shown in **Figure 21-4**. Learn to recognize the items in **Table 21.3** on lab models and figures.

The Intestines

After experiencing physical and some chemical digestion in the stomach, chyme enter the small intestine where chemical digestion is completed and most absorption of nutrients occurs. The

Table 21.1 Components of the Alimentary Canal Walls

Component	Description
Goblet cells	Scattered among the epithelial cells of the mucous membrane; secretes mucus
Greater omentum	Connects stomach to transverse colon
Lamina propria	Areolar tissue beneath the mucous membrane lining the inside of the alimentary canal
Lesser omentum	Connects duodenum and stomach to liver and diaphragm
Longitudinal muscle	Smooth muscle along length of GI tract; shortens segments of GI tract
Lymphatic nodule	Peyer's patch; diffuse lymphatic tissue most abundant in wall of large intestine
Mesentery	Double serous membrane between segments of intestine
Mesocolon	Membrane attaching transverse and sigmoid colon to posterior abdominal wall
Mucosa	Includes the mucous membrane (epithelium), lamina propria, and muscularis mucosae
Mucous membrane	Epithelial tissue covering the inner surface of the alimentary canal
Muscularis mucosae	Smooth muscle between mucosa and submucosa; contraction produces wrinkles
Myenteric plexus	(**Auerbach plexus**) Nerve network of the longitudinal and circular muscles
Parietal peritoneum	Serous membrane lining wall of abdominopelvic cavity
Serosa	(Visceral peritoneum) Simple squamous epithelium on the outer surface of the wall
Submucosa	Between mucosa and muscle tunics in GI tract; contains vessels and nerves
Submucosal gland	(**Brunner's glands**) In the small intestine; secrete bicarbonate that neutralizes acid
Submucosal plexus	(**Meissner's plexus**) Nerve network of blood vessels and the muscularis mucosa
Visceral peritoneum	Serosa; covers the surface of abdominopelvic organs

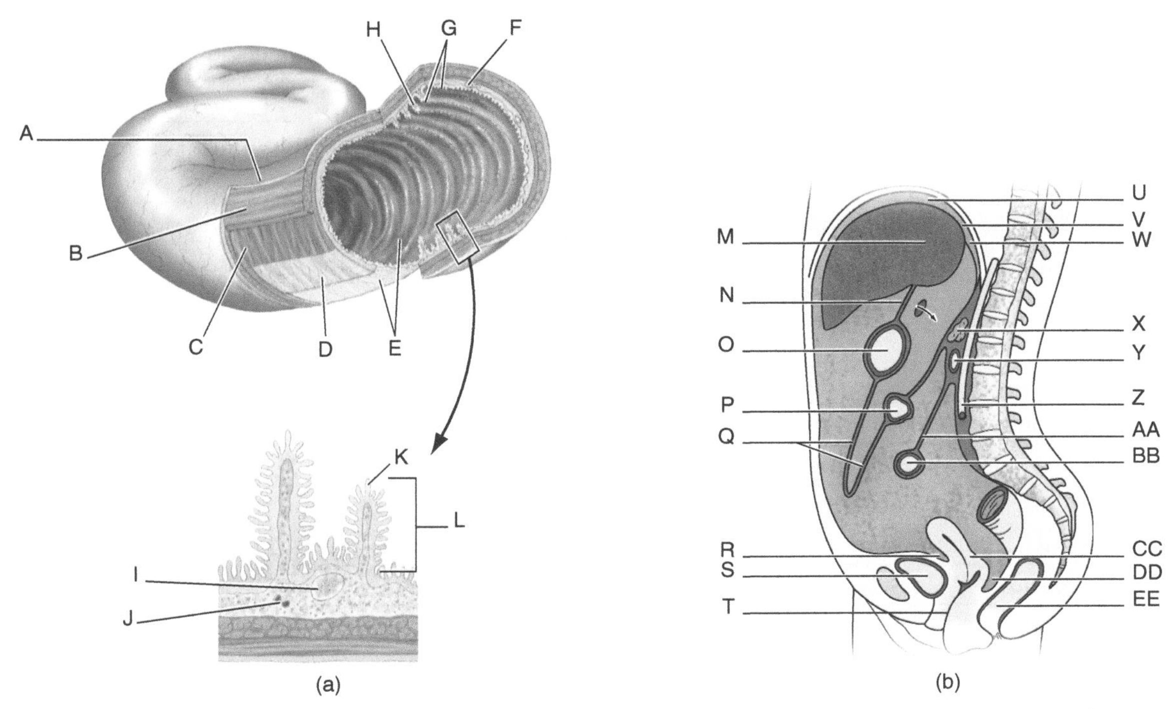

Figure 21-2 Tissue layers of the digestive system: (a) Wall of the small intestine, (b) serous membranes.

Table 21.2 Components of the Mouth, Oral Cavity, and Pharynx

Component	Description
Alveolar socket	Cavity in which tooth fits into mandible or maxilla
Apical foramen	Opening in tooth root through which vessels and nerves pass
Bicuspid teeth	1st and 2nd premolars; between cuspid and first molar; absent in deciduous dentition
Cementum	Periosteum around a tooth's root
Cuspids	(Canine or eye teeth) Located between the lateral incisor and first premolar
Dentin	Between enamel and pulp cavity of a tooth
Dentition	Set of teeth: Permanent dentitions have 32 teeth; Deciduous dentitions have 20 teeth
Enamel	Hard outer layer of a tooth
Frenulum	Flaps connecting lips and tongue to the gums (labial, lingual)
Gingiva	Gums
Hard palate	Roof of mouth and floor of nasal cavity; maxilla and palatine bones
Incisor (Medial/Lateral)	Most anterior teeth four teeth
Labium	Lips; superior and inferior
Laryngopharynx	Throat region posterior to the larynx
Mandible	Lower jaw bone
Maxilla	Upper jaw bone
Molar (1st, 2nd, 3rd)	Most posterior "grinding" teeth; 3rd molar (wisdom tooth) absent in deciduous dentition
Oral cavity	(**Buccal cavity**) Located between the mouth and oropharynx

(*Continued*)

Table 21.2 Components of the Mouth, Oral Cavity, and Pharynx *(Cont'd)*	
Oral vestibule	Between lip and gum
Oropharynx	Back of oral cavity; between nasopharynx and laryngopharynx
Palatine bone	Posterior portion of hard palate
Palatine raphe	Ridge in the hard palate where right and left maxillae fuse
Palatoglossal arch	(**Glossopharyngeal arch**) Most anterior arch in the oropharynx
Palatopharyngeal arch	(**Pharyngopalatine arch**) Most posterior arch in the oropharynx
Parotid duct	Stenson's duct; transports saliva from parotid gland to oral cavity
Parotid glands	Large salivary glands near the ears
Periodontal ligaments	Microscopic dense connective tissue bands holding tooth in alveolar socket
Philtrum	Depression between upper lip and nose
Root canal	Long projections of a tooth that are anchored to the alveolar sockets
Sublingual duct	Rivinus' duct; transports saliva from sublingual gland to oral cavity
Sublingual gland	Salivary gland beneath tongue
Submandibular duct	Wharton's duct; transports saliva from submandibular gland to oral cavity
Submandibular gland	Salivary gland under mandible
Tongue	Skeletal muscle organ in oral cavity that helps forms bolus for swallowing
Tongue papillae	"Bumps" on the tongue's surface that hold taste buds
Tooth crown	Part of tooth above the gum line
Tooth pulp cavity	Internal cavity of a tooth; contains vessels and nerves
Tooth root	Part of the tooth below gum line
Uvula	Flap of the soft palate that closes nasopharynx when swallowing
Vermillion	Pink part of lips between skin and oral vestibule

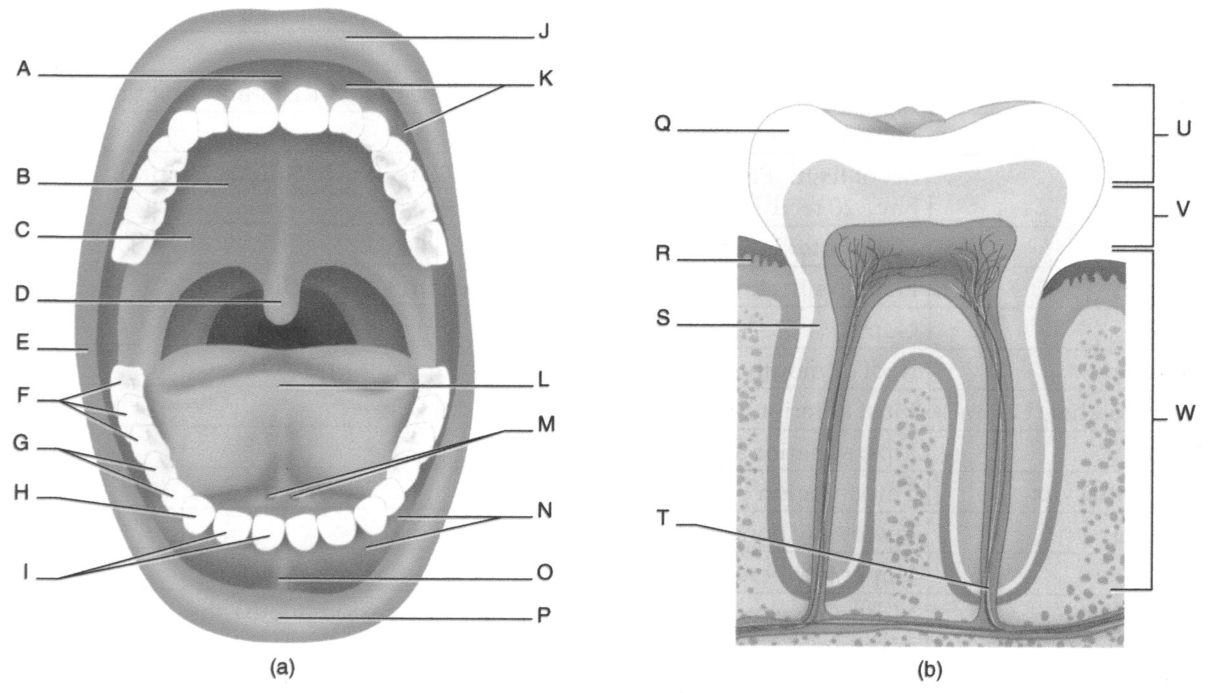

Figure 21-3 Features of the oral cavity: (a) Mouth and oral cavity, (b) anatomy of a tooth.

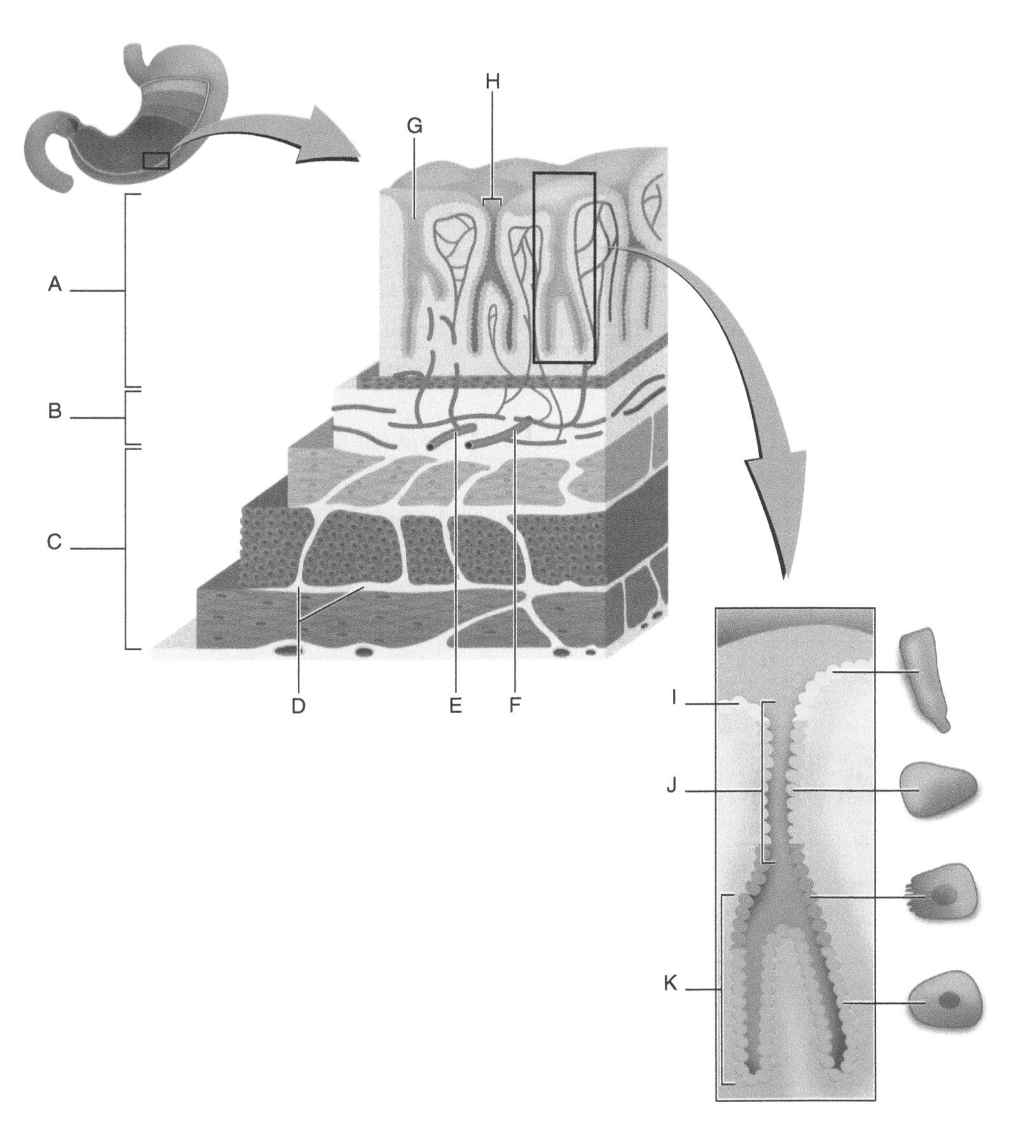

Figure 21-4 Anatomy of the stomach lining.

Table 21.3 Components of the Esophagus and Stomach

Component	Description
Adventitia	Loose (areolar) connective tissue on the outer surface of the esophagus
Cardia	Part of stomach attached to the esophagus
Circular muscle	Smooth muscle wrapped like rings around the tubes of the alimentary canal; when contracting it squeezes sections of the canal to cause elongation of that section
Esophageal hiatus	Opening in the diaphragm through which the esophagus passes
Esophagus	Muscular tube connecting the pharynx to the stomach
Fundus	Domed, superior part of the stomach
Gall bladder	Bag-like organ attached to the liver that stores bile made in the liver

(*Continued*)

Table 21.3 Components of the Esophagus and Stomach *(Cont'd)*

Gastric pit	Microscopic tube in the stomach's mucosa; lined with parietal and chief cells
Gastroesophageal sphincter	(**Cardiac** or **lower esophageal sphincter**); between stomach and esophagus
Greater curvature	Outer long edge of the stomach
Lesser curvature	Inner curved region of the stomach between the cardia and pylorus
Oblique muscle layer	Located between the circular muscle and submucosa of the stomach
Oxyntic cell	(**Parietal cell**); secretes HCl and intrinsic factor in stomach
Pyloric sphincter	Smooth muscle band that regulates chyme from the stomach into the small intestine
Pylorus	(**Antrum**) Distal end of the stomach that is attached to duodenum
Rugae	Wrinkles in the stomach's mucosa; allows the stomach to distend (expand)
Stomach	Major organ of physical digestion between the esophagus and small intestine
Upper esophageal sphincter	Not a true sphincter, but it is the region where the laryngopharynx and esophagus connect.
Zymogenic cell	**(Chief cell)**; secretes pepsinogen into stomach lumen

Table 21.4 Components of the Small and Large Intestines

Component	**Description**
Anal canal	Smooth muscular tube between rectum and anus
Anus	Distal opening of the alimentary canal
Appendix	Wormlike projection extending from the cecum
Brush border	Microvilli on the surface of intestine mucosa
Colon	Large intestine (ascending, descending, transverse, sigmoid sections)
Duodenal papilla	Projection where bile and pancreas juice enter duodenum
Duodenum	First 10-12 inches of small intestine
Epiploic appendages	Projections of fat from the outer surface of the colon
External anal sphincter	Skeletal muscle ring around anal opening; controlled voluntarily
Haustrum	Pouch in colon
Hepatic flexure	Bend in the colon between the ascending and transverse colon
Ileocecal (ileocolic) valve	Between ileum and cecum
Ileum	Longest part of small intestine; connects to colon
Internal anal sphincter	Smooth muscle ring around anal canal; controlled involuntarily
Intestinal crypt	**Crypt of Lieberkuhn**; pits in intestinal mucosa; secretes intestinal juices
Jejunum	Between duodenum and ileum
Lacteal	Lymph capillary inside villus
Microvilli	Extensions of cell membrane; on intestinal mucosa cells
Plicae circulares	Circular folds in intestine formed when muscularis mucosae contracts
Rectum	Straight portion of colon between sigmoid colon and anal canal
Splenic flexure	Bend in the colon between the transverse and descending colon
Taenia coli	Thin longitudinal bands of smooth muscle on colon
Villi	Projections of mucosa into intestinal lumen; covered with mucous epithelium

small intestine is a highly coiled, muscular tube consisting of three major regions: *duodenum*, *jejunum*, and *ileum*. Material remaining in the chyme at the end of the ileum is not digestible and passes into the large intestine (colon) to become a major component of feces (fecal material).

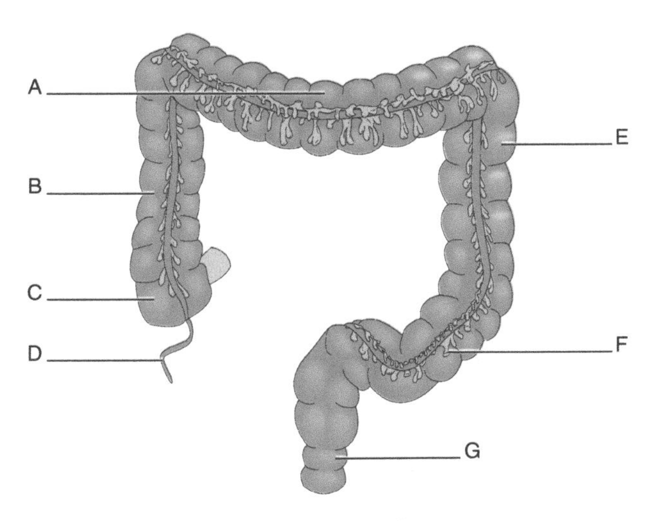

Figure 21-5 Anatomy of the colon.

In the clinical area, certain parts of the alimentary canal are classified under different headings. Some sources specify the **upper gastrointestinal (GI) tract** to include the esophagus, stomach, and duodenum, and the **lower GI tract** to include the jejunum, ileum, and colon. Learn the material in **Table 21.4** and be able to identify features of the intestines on lab models and figures. Use **Figures 21-1**, **21-2**, and **21-5** to identify regions of the intestine.

Liver, Gall Bladder, and Pancreas

The liver, gall bladder, and pancreas are accessory digestive organs in the abdomen that enhance the alimentary canal's ability to process food. The liver is the largest visceral organ and holds the gall bladder. The pancreas is located in the space between the stomach and small intestine. Learn the features in **Table 21.5** and identify them on lab models and in **Figure 21-6**.

Table 21.5 Features of the Liver, Gall Bladder, and Pancreas

Component	Description
Acinar cells	Most abundant cells of the pancreas; secrete exocrine substances
Bile canaliculi	Tiny tubes in a liver lobule for transporting bile
Central vein	In the center of a liver lobule
Common bile duct	Receives bile from hepatic duct and cystic duct
Cystic duct	Drains bile from gall bladder
Falciform ligament	Extension of visceral peritoneum into the liver; separates right/left lobes
Hepatic duct	Drains bile from liver to common bile duct
Hepatocyte	Liver cell
Hepatopancreatic ampulla	(**Ampulla of Vater**) Collects bile and pancreas juice; empties into duodenum
Hepatopancreatic sphincter	(**Sphincter of Oddi**) Controls the release of fluids from the ampulla of Vater
Islets of Langerhans	(**Pancreatic islets**) Pancreatic cells that secrete hormones
Kuppfer cells	Macrophages in the liver; also called *reticuloendothelial cells*; reside in liver sinusoids
Liver	Large organ in right hypochondriac region; secretes bile; processes nutrients
Liver lobes	Includes the caudate, left, quadrate, and right lobes
Liver lobules	Functional units within the lobes of the liver
Pancreatic duct	Transports pancreatic enzymes to the ampulla of Vater
Round ligament	Remnant of the umbilical vein; connects the falciform ligament to the umbilicus

Enzymes and Hormones

There are numerous enzymes and hormones synthesized in different parts of the digestive system. Now that you have had the opportunity to learn the locations of various digestive system organs, it is important to know the functions of those organs. **Table 21.6** lists the chemicals synthesized by the digestive system and how those chemicals are important to digestion.

Preserved Specimens

You may have the opportunity to identify different components of the digestive system in a preserved specimen, either a fetal pig or cat. Your lab instructor will provide a list of items that can be readily seen and also handouts to identify the structures.

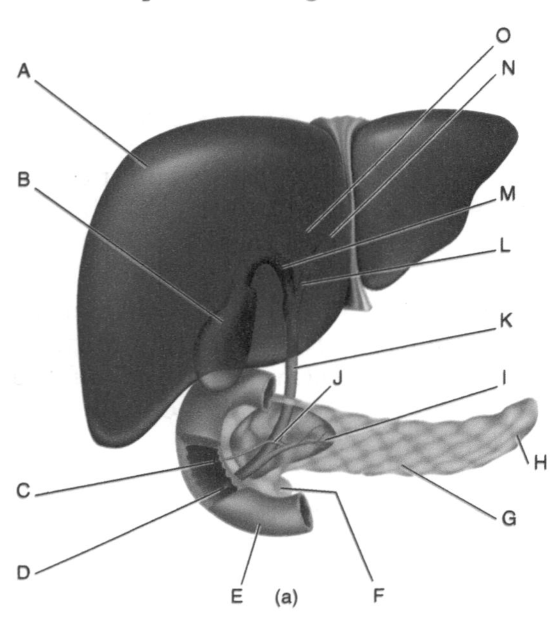

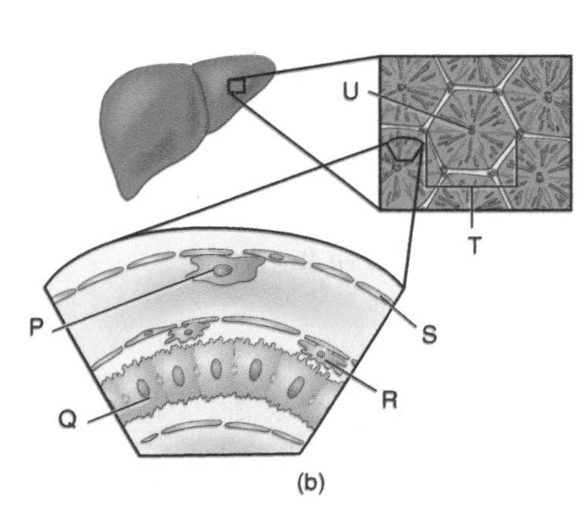

Figure 21-6 Accessory organs of the abdomen: (a) Liver, gall bladder, and pancreas; (b) liver's internal structure.

Table 21.6 Enzymes and Hormones of the Digestive System

Amylase	Enzyme of the salivary glands and pancreas; hydrolyzes starch to dextrins and maltose
Bicarbonate	Alkaline substance that neutralizes acidic chyme; secreted from the pancreas and duodenum
Bile	Secreted from the liver and stored in the gall bladder; it is a mixture of bile salts (used in fat emulsification) and bilirubin (byproduct of hemoglobin breakdown)
Bile salts	Amphipathic molecules derived from cholesterol and secreted by the liver into bile; emulsify fats
Carboxypeptidase	Pancreatic enzyme derived from *procarboxypeptidase* through the action of trypsin in the small intestine; hydrolyzes proteins to free amino acids
Chymotrypsin	Pancreatic enzyme derived from *chymotrypsinogen* through the action of trypsin in the small intestine; hydrolyzes proteins into smaller polypeptides
Dipeptidase	Enzyme within the small intestine's mucosal cells; hydrolyzes dipeptides to two amino acids
Enterokinase	Enzyme bound to the small intestine's mucosa; converts trypsinogen to trypsin

(Continued)

Table 21.6 Enzymes and Hormones of the Digestive System *(Cont'd)*

Hydrochloric acid	(HCl) Strong acid secreted from the stomach; converts pepsinogen to pepsin
Lactase	Enzyme bound to the small intestine's mucosa; hydrolyzes lactose to glucose and galactose
Lipase	Enzyme secreted from the tongue (lingual lipase), stomach (gastric lipase), and pancreas (pancreatic lipase); hydrolyzes lipids to monoglycerides and fatty acids
Maltase	Enzyme bound to the small intestine's mucosa; hydrolyzes maltose to two glucose molecules
Mucus	Viscous fluid secreted from goblet cells; lubricates and protects the mucosa
Nuclease	Pancreatic enzyme that hydrolyzes nucleic acids to nucleotides
Nucleosidase	Enzyme bound to the small intestine's mucosa; hydrolyzes nucleotides to sugars, bases, and PO_4
Pepsin	Stomach enzyme derived from *pepsinogen* through the action of stomach acid; hydrolyzes proteins to smaller polypeptides
Sucrase	Enzyme bound to the small intestine's mucosa; hydrolyzes sucrose to glucose and fructose
Tripeptidase	Enzyme in the small intestine's mucosal cells; hydrolyzes tripeptides to dipeptides and amino acid
Trypsin	Pancreatic enzyme derived from *trypsinogen* through the action of enterokinase in the small intestine; hydrolyzes proteins to smaller polypeptides

Name ______________________ Course Number: _________ Lab Section _______

Lab 21: Digestive System Worksheet

Write out the names of the labeled items in the selected figures.

Figure 21-1.

A. ______________________
B. ______________________
C. ______________________
D. ______________________
E. ______________________
F. ______________________
G. ______________________
H. ______________________
I. ______________________
J. ______________________
K. ______________________
L. ______________________
M. ______________________
N. ______________________
O. ______________________
P. ______________________
Q. ______________________
R. ______________________
S. ______________________
T. ______________________

Figure 21-2.

A. ______________________
B. ______________________
C. ______________________
D. ______________________
E. ______________________
F. ______________________
G. ______________________
H. ______________________
I. ______________________
J. ______________________
K. ______________________
L. ______________________
M. ______________________
N. ______________________
O. ______________________
P. ______________________
Q. ______________________
R. ______________________
S. ______________________
T. ______________________
U. ______________________
V. ______________________
W. ______________________
X. ______________________
Y. ______________________
Z. ______________________
AA. ______________________
AB. ______________________
AC. ______________________
AD. ______________________
AE. ______________________

Figure 21-3.

A. ______________________
B. ______________________
C. ______________________
D. ______________________
E. ______________________
F. ______________________
G. ______________________
H. ______________________
I. ______________________
J. ______________________
K. ______________________
L. ______________________
M. ______________________
N. ______________________
O. ______________________
P. ______________________
Q. ______________________
R. ______________________
S. ______________________
T. ______________________
U. ______________________
V. ______________________
W. ______________________

Figure 21-4.

A. ______________________
B. ______________________
C. ______________________
D. ______________________
E. ______________________
F. ______________________
G. ______________________
H. ______________________
I. ______________________
J. ______________________
K. ______________________

Figure 21-5.

A. ______________________
B. ______________________
C. ______________________
D. ______________________
E. ______________________
F. ______________________
G. ______________________

Figure 21-6.

A. ______________________
B. ______________________
C. ______________________
D. ______________________
E. ______________________
F. ______________________
G. ______________________
H. ______________________
I. ______________________
J. ______________________
K. ______________________
L. ______________________
M. ______________________
N. ______________________
O. ______________________
P. ______________________
Q. ______________________
R. ______________________

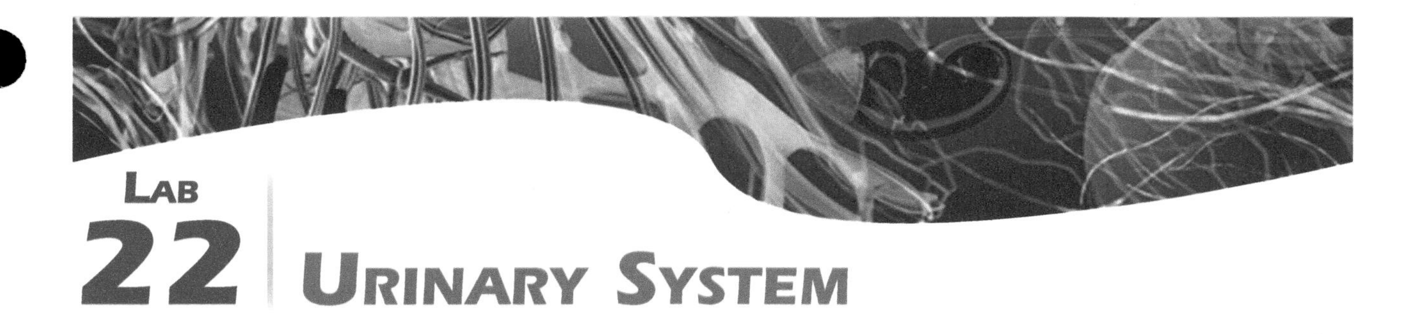

Lab 22 Urinary System

Overview of the Urinary System

The urinary system is responsible for filtering the blood, and it is important as a producer of chemicals that affect blood cell production and blood pressure. The major organs of the urinary system include the *kidneys, ureters, urinary bladder,* and *urethra.* First, learn the major features of the urinary system listed in **Table 22.1** and

Table 22.1 Major Features of the Urinary System

Detrusor muscle	Smooth muscle of urinary bladder
External urethral orifice	Opening of the urethra to the outside of the body
External urethral sphincter	Skeletal muscle ring in the urogenital diaphragm; under voluntary control
Hilus	Indentation on medial side of kidney; attachment site of ureter and vessels
Internal urethral orifice	Opening from the urinary bladder into the urethra
Internal urethral sphincter	Smooth muscle ring at the base of the urinary bladder; under autonomic control
Kidney	Major blood filtering organ; right kidney is slightly more inferior than left kidney
Major calyx	Located between the minor calyx and renal pelvis
Membranous urethra	Passes through the urogenital diaphragm
Minor calyx	Receives urine from a renal papilla
Prostatic urethra	Only in the male; passes through the prostate gland
Renal capsule	Dense connective tissue covering of a kidney
Renal cortex	Outer portion of kidney
Renal medulla	Deeper portion of kidney
Renal papilla	Projection into a minor calyx; receives urine from collecting ducts
Renal pelvis	Cavity in kidney that collects urine from all nephrons
Renal pyramid	Triangular region in kidney; site of collecting ducts
Spongy urethra	**(Penile urethra)** Only in the male; passes through the penis
Trigone	Region at the base of the urinary bladder; includes openings of the ureters and urethra
Ureters	Smooth muscle tubes that transport urine from the kidneys to the urinary bladder
Urethra	Tube that transports urine from the bladder to the outside
Urinary bladder	Smooth muscle organ that stores urine
Urogenital diaphragm	Skeletal muscles that form base of pelvic cavity

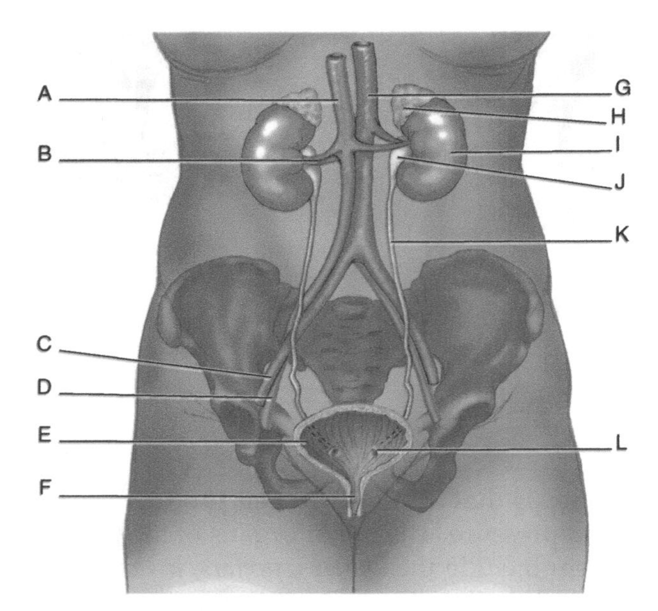

Figure 22-1 Organs of the urinary system.

then label the items in **Figure 22-1**. Also, be able to identify these items on lab models and preserved specimens, if available.

Blood Flow Through the Kidney

As a blood filtering organ, it is important to understand the pathway along which blood takes as it passes through the kidney. Inside each kidney are functional units called *nephrons*, and around each nephron is a blood vessel network with which the nephron can exchange materials. Learn the kidney's blood vessels listed in **Table 22.2** and label them in **Figure 22-2**.

Anatomy of the Nephron

The ways in which nephrons act as the functional units of the kidneys involve three processes: *filtration*, *reabsorption*, and *secretion*. **Filtration** is the removal of fluids and solutes from the blood and it takes place in the **renal corpuscle**. Fluid that enters the nephron at the renal corpuscle is called filtrate. As filtrate passes through the nephron, most of the fluid and solutes are put back into the blood in the process of **reabsorption**. If materials do not enter the nephron in the renal corpuscle, they may still enter it but at different locations in a process called **secretion**. Learn the parts of the nephron in **Table 22.3** and label them in **Figure 22-3**.

Trace a water molecule from the GLOMERULUS to the EXTERNAL URETHRAL ORIFICE, including all relevant structures and regions listed in this lab.

Trace a water molecule from the AORTA through relevant blood vessels in the kidney and back to the INFERIOR VENA CAVA.

Table 22.2 Blood Vessels of the Kidney

Afferent arteriole	Transports blood from an interlobular artery into the glomerulus
Arcuate artery	Transports blood from an interlobar artery to a cortical radiate artery
Arcuate vein	Transports blood from a cortical radiate vein to an interlobar vein
Cortical radiate artery	(**Interlobular arteries**) Transports blood from an arcuate artery to an afferent arteriole
Cortical radiate vein	(**Interlobular veins**) Transports blood from a peritubular capillary to an arcuate vein
Efferent arteriole	Transports blood from the glomerulus to peritubular capillaries
Interlobar artery	Transports blood from a segmental artery to an arcuate artery
Interlobar vein	Transports blood from an arcuate vein to the renal vein
Peritubular capillary	Transports blood around a nephron and to a cortical radiate vein
Renal artery	Transports blood from the aorta to a segmental artery
Renal vein	Transports blood from an interlobar vein to the inferior vena cava
Segmental artery	Transports blood from the renal artery to an interlobar artery
Vasa recta	Peritubular capillaries that surround the nephron loop of a juxtamedullary nephron; transports blood from peritubular capillaries into cortical radiate veins

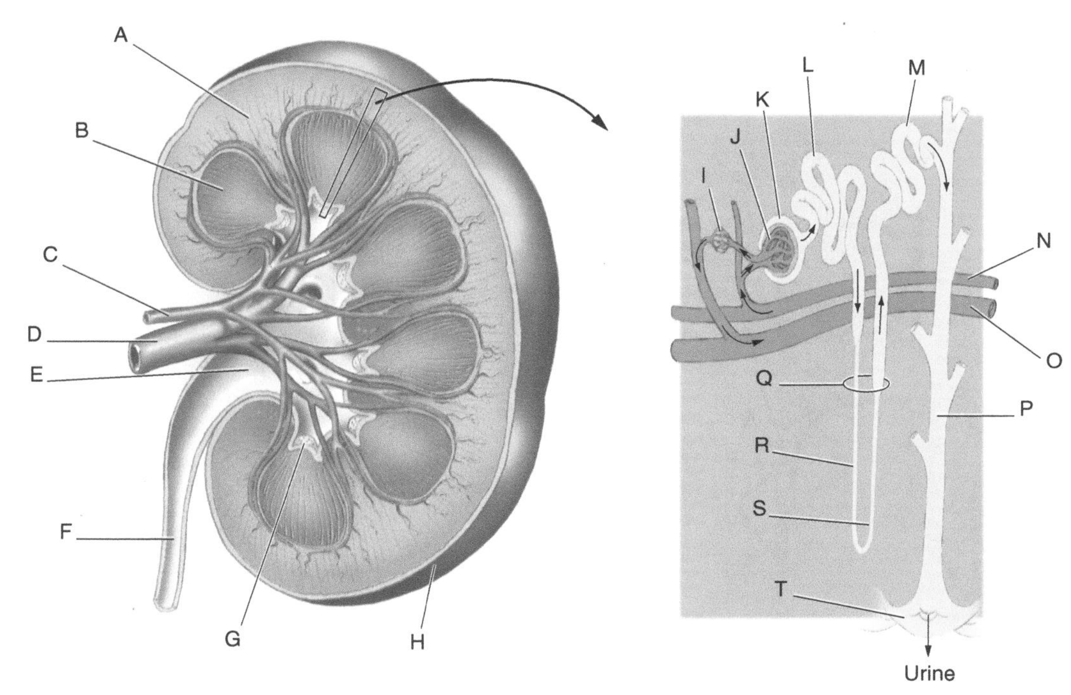

Figure 22-2 Blood vessels of the kidney.

Table 22.3 Features of a Nephron	
Ascending limb	Transports filtrate toward the renal cortex from the nephron loop
Collecting duct	Drains urine from nephrons; between DCT and minor calyx
Convoluted tubules	Proximal: between glomerular capsule and nephron loop Distal: between nephron loop and collecting duct
Cortical nephron	In renal cortex
Descending limb	Transports filtrate toward the renal pelvis from the PCT
Glomerular capsule	Bowman's capsule; collects filtrate from glomerulus
Glomerulus	Ball of capillaries within renal corpuscle
Juxtaglomerular apparatus	Union of afferent arteriole and distal convoluted tubule; releases renin and erythropoietin
Juxtamedullary nephron	Nephron with nephron loops deep in the medulla
Nephron loop	(**Loop of Henle**); Connection between the descending and ascending limbs
Pedicel	Fingerlike extension of a podocyte wrapping partly around the glomerular capillary
Podocyte	Spider shaped cell around a glomerular capillary
Renal corpuscle	Glomerular capsule and the glomerulus together

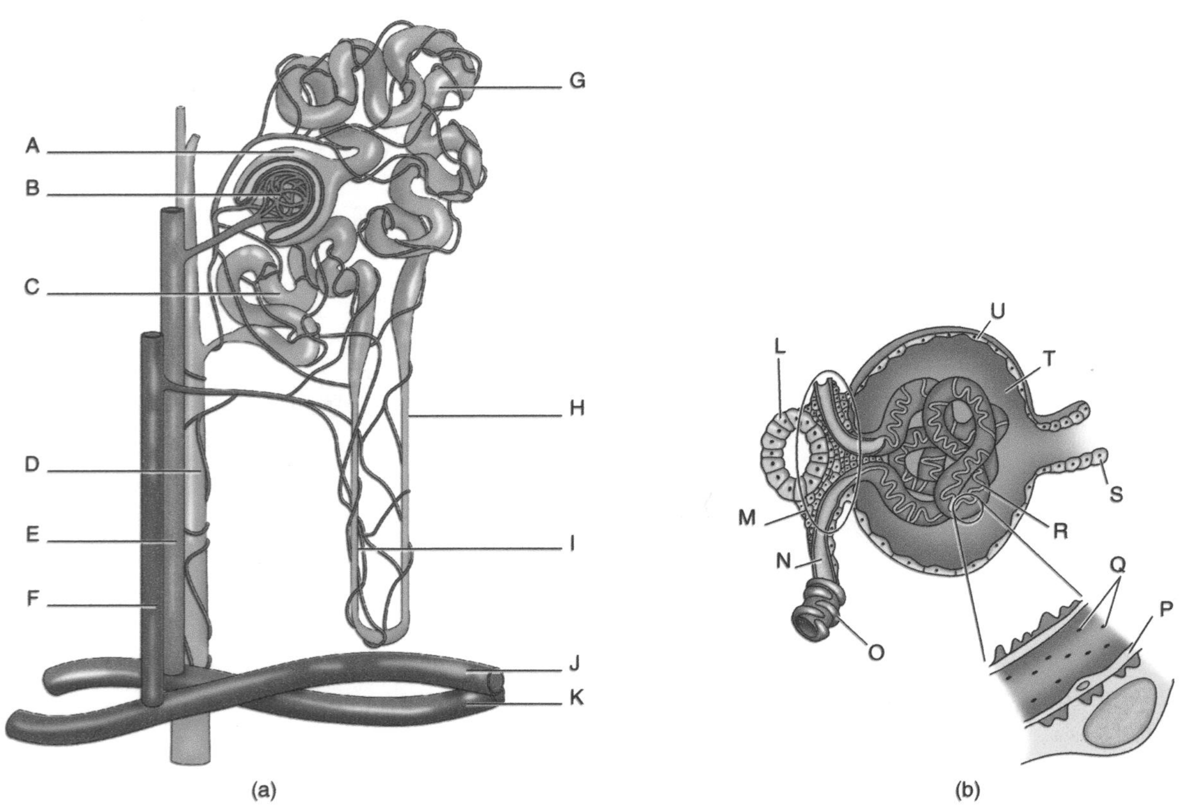

Figure 22-3 Anatomy of a nephron: (a) Vessels around a nephron, (b) anatomy of the glomerulus.

Name ______________________ Course Number: ________ Lab Section _______

Lab 22: Urinary System Worksheet

Write out the names of the labeled items in the selected figures.

Figure 22-1.

A. ______________________
B. ______________________
C. ______________________
D. ______________________
E. ______________________
F. ______________________
G. ______________________
H. ______________________
I. ______________________
J. ______________________
K. ______________________
L. ______________________

Figure 22-2.

A. ______________________
B. ______________________
C. ______________________
D. ______________________
E. ______________________
F. ______________________
G. ______________________
H. ______________________
I. ______________________
J. ______________________
K. ______________________
L. ______________________
M. ______________________
N. ______________________
O. ______________________
P. ______________________
Q. ______________________
R. ______________________
S. ______________________
T. ______________________

Figure 22-3.

A. ______________________
B. ______________________
C. ______________________
D. ______________________
E. ______________________
F. ______________________
G. ______________________
H. ______________________
I. ______________________
J. ______________________
K. ______________________
L. ______________________
M. ______________________
N. ______________________
O. ______________________
P. ______________________
Q. ______________________
R. ______________________
S. ______________________
T. ______________________
U. ______________________

LAB 23 REPRODUCTIVE SYSTEMS

Overview of the Reproductive Systems

The reproductive systems are responsible for making and transporting sex cells (gametes) that can be involved in fertilization. In addition, the female reproductive system nourishes a developing individual until birth. The production of gametes is called *gametogenesis* and it will be covered in the next lab. In this lab you will concentrate on learning the anatomical features of the female and male reproductive systems. We will begin with the female system. However, since the anatomy of the breast is more related to human development, it will be covered in the next lab. Become familiar with the female items listed in **Table 23.1** and then label them in **Figures 23-1** through **23-3**. Then learn the parts of the male reproductive system in **Table 23.2** and label them in **Figures 23-4** and **23-5**. Also, be able to identify all relevant structures on lab models and photographs that might be provided.

Table 23.1 Features of the Female Reproductive System

Ampulla of oviduct	Wider portion of the oviduct near an ovary
Broad ligament	Continuation of the peritoneum covering the uterus
Cervix	Narrow region of the uterus where it meets the vagina
Clitoris	Erectile organ in females anterior to the urethral orifice
Corpora cavernosa	Pair of erectile tissue tubes within the clitoris
Corpus albicans	Degenerated corpus luteum in an ovary
Corpus luteum	Differentiated ruptured follicle; secretes progesterone
Endometrium	Inner lining of the uterus; consists of a stratum basalis and stratum functionalis
External os	Opening between the vagina and cervical canal
Fimbriae	Fingerlike projections around the infundibulum of an oviduct
Fundus of uterus	Dome-shaped superior portion of the uterus
Glans of clitoris	Distal end of the clitoris; it is not enlarged like that of the penis
Graafian follicle	Mature follicle that becomes a corpus luteum after ovulation
Hymen	Thin membrane covering the vaginal orifice
Infundibulum	Funnel-shaped opening of the oviduct near an ovary; surrounded by fimbriae

(Continued)

Table 23.1 Features of the Female Reproductive System *(Cont'd)*	
Internal os	Opening between the uterus and cervical canal
Isthmus of oviduct	Narrow portion of the oviduct near the uterus
Labia majora	Lateral folds adjacent to the vestibule; part of the vulva
Labia minora	Medial folds that directly enclose the vestibule; part of the vulva
Mesovarium	Membrane that attaches the ovary to the oviduct
Mons pubis	Thick layer of fat covering the pubic bone in females; cushions the pubis during intercourse
Myometrium	Smooth muscle tissue of the uterus
Ovarian ligament	Attaches the ovary to the uterus
Ovaries	Primary sex organs in females
Oviduct	Tube between an ovary and the uterus
Paraurethral (Skenes) glands	Several mucus-secreting glands around the urethra; homologous to the male's prostate gland
Perimetrium	Visceral peritoneum that covers the outer surface of the uterus
Perineum	Region between the anus and external genitalia
Prepuce of clitoris	(**Foreskin**) Covering the distal end of clitoris
Rectouterine pouch	Space between the uterus and rectum
Round ligament	Fibrous cord that connects the uterus to the labia majora through the inguinal canal
Rugae of vagina	Ridges along internal surface of vagina that stimulate the penis during intercourse
Stratum basalis	Layer of endometrium that generates the stratum functionalis
Stratum functionalis	Innermost layer of endometrium generated from the stratum basalis
Tunica albuginea	Tough, fibrous outer covering of an ovary
Uterine tubes	(**Fallopian tubes** or **oviducts**) Transports developing stages to the uterus
Uterus	(**Womb**) Site of fetal development
Vagina	Female copulatory organ and birth canal
Vaginal fornix	Recess formed around the short projection of the cervix into the vagina
Vaginal orifice	Opening of the vagina into the vestibule
Vesicouterine pouch	Space between the uterus and urinary bladder
Vestibular gland	(**Bartholin's gland**) Located in the vestibule; secretes fluid during intercourse
Vestibule	Region between the labia minora; contains urethral and vaginal orifices
Vulva	(**Pudendum**) External genitalia of the female

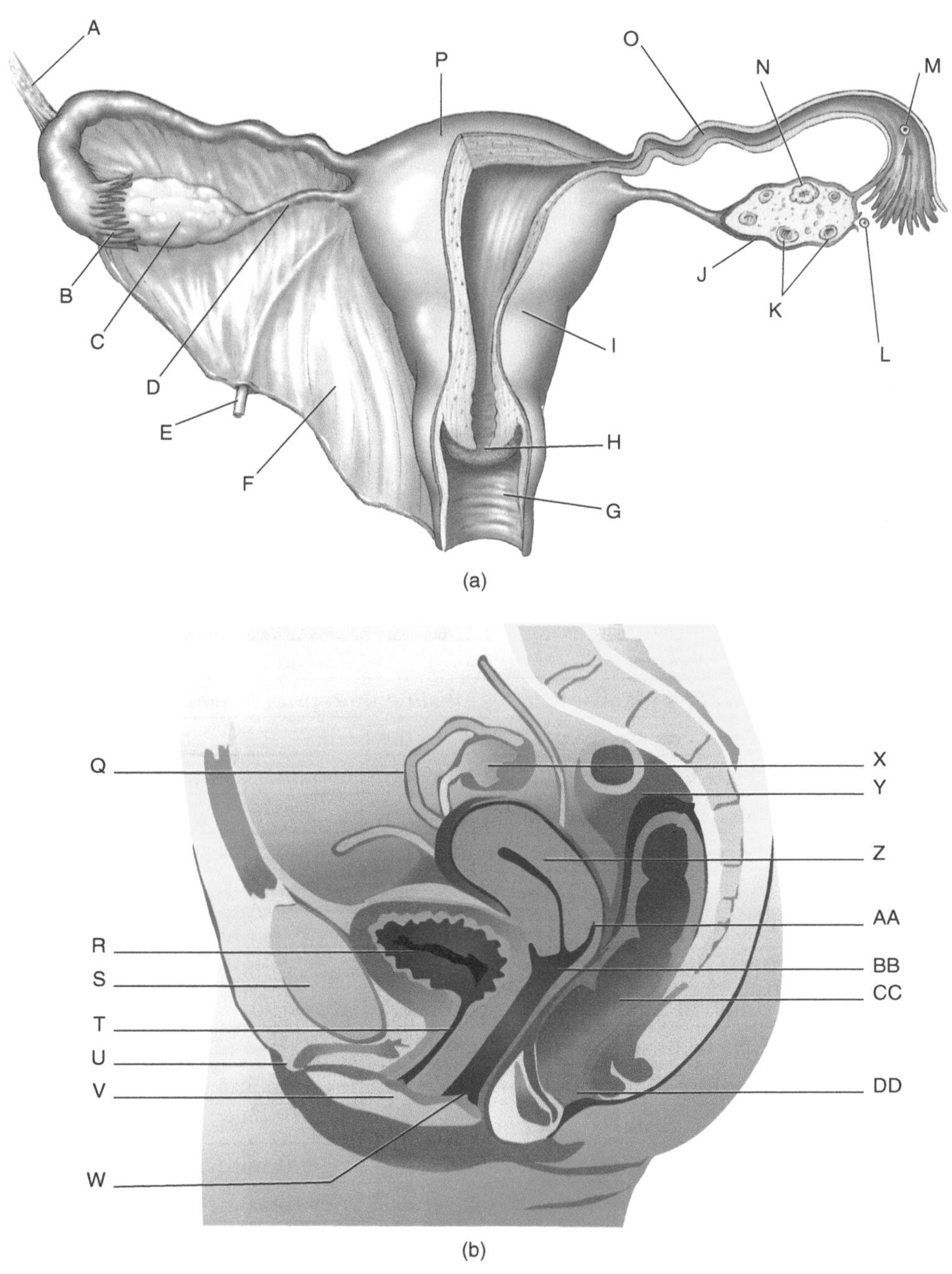

Figure 23-1 Female reproductive system: (a) Anterior view, (b) midsagittal view.

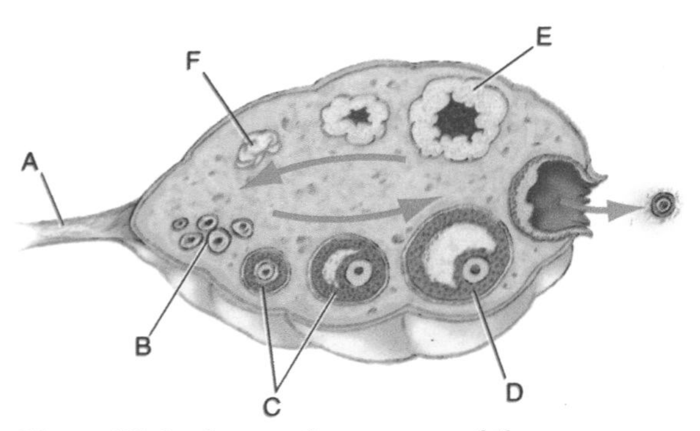

Figure 23-2 Internal anatomy of the ovary.

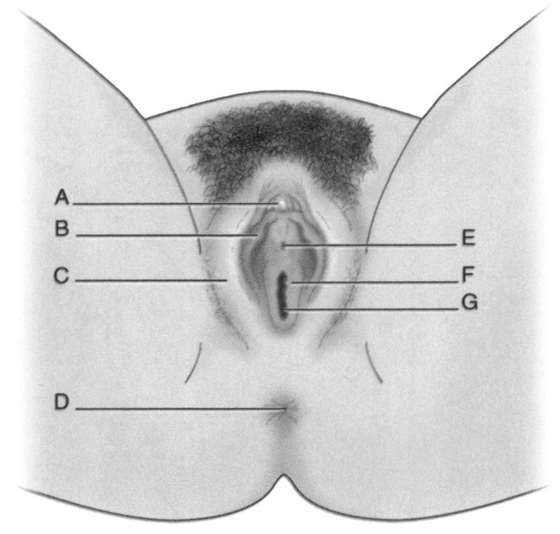

Figure 23-3 External female genitalia.

Table 23.2 Features of the Male Reproductive System

Bulb of penis	Base of penis
Bulbourethral glands	Cowper's glands; secretes alkaline fluid prior to ejaculation
Corpora cavernosa	Pair of erectile tissue "tubes" in the penis
Corpus spongiosum	Single tube of erectile tissue in penis; contains the urethra
Corona of penis	Raised ridge of the glans formed by the corpus spongiosum
Cremaster muscle	Suspends testis; extension of internal oblique muscle; thermoregulation of testis
Crus of penis	Projections of the corpora cavernosa; anchors penis to pubis
Dartos muscle	Smooth muscle layer in scrotum; affects scrotum's surface area for thermoregulation
Efferent ductules	Transports sperm between rete testis and epididymis
Ejaculatory duct	Union of ampulla and ducts of seminal vesicles
Epididymis	Tube testis where sperm mature; between efferent ductule and vas deferens
External urethral orifice	Opening of urethra to outside; provides for passage of sperm
Glans of penis	Swollen distal end of penis
Gubernaculum	Fibrous cord along which the testes descend in the abdominal cavity
Interstitial cells	Cells of Leydig between the seminiferous tubules; secrete testosterone
Inguinal canal	Opening between in pelvic cavity through which testes descend
Penile urethra	Urethra inside penis
Penis	Male copulatory organ
Prepuce of penis	Foreskin covering the distal end of penis
Prostate gland	Secretes clotting factors, plasminogen, and motility factors in semen
Prostatic urethra	Urethra inside prostate gland
Rete testis	Network of tubes between seminiferous tubules and efferent ductules of testis
Scrotum	Sac that contains the testes
Seminal vesicles	Makes 60% of seminal fluid; fructose, prostaglandins, fibrinogen, alkaline materials
Seminiferous tubules	Site of spermatogenesis in testes
Spermatic cord	Fibrous cord containing vas deferens, vessels, nerves, and cremaster muscle

(Continued)

Table 23.2 Features of the Male Reproductive System *(Cont'd)*	
Sustenacular cells	Nurse cells or Sertoli cells; protect developing spermatocytes in testes
Testes	Singular—testis or testicle; primary sex organs of male
Tunica albuginea	Tough fibrous covering of testis
Tunica vaginalis	Thin covering of testes derived from parietal peritoneum
Vas deferens	Tube between epididymis and ejaculatory duct

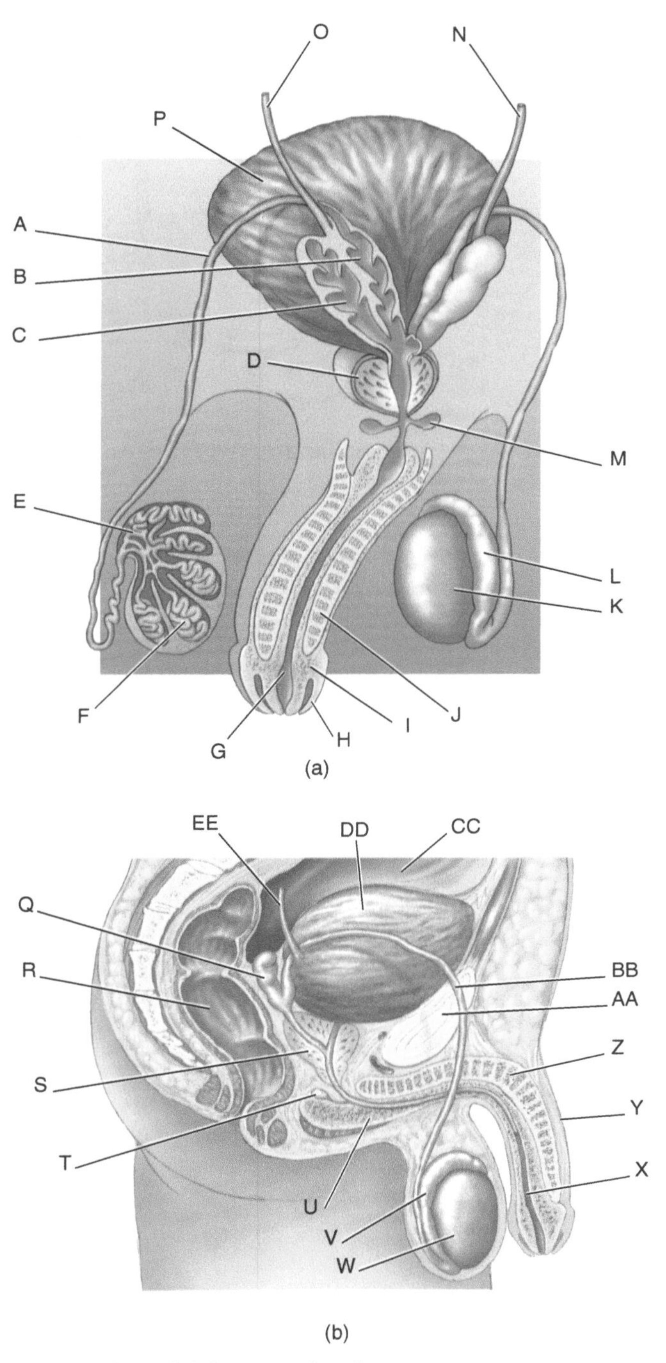

Figure 23-4 Male reproductive system: (a) Anterior view, (b) midsagittal view.

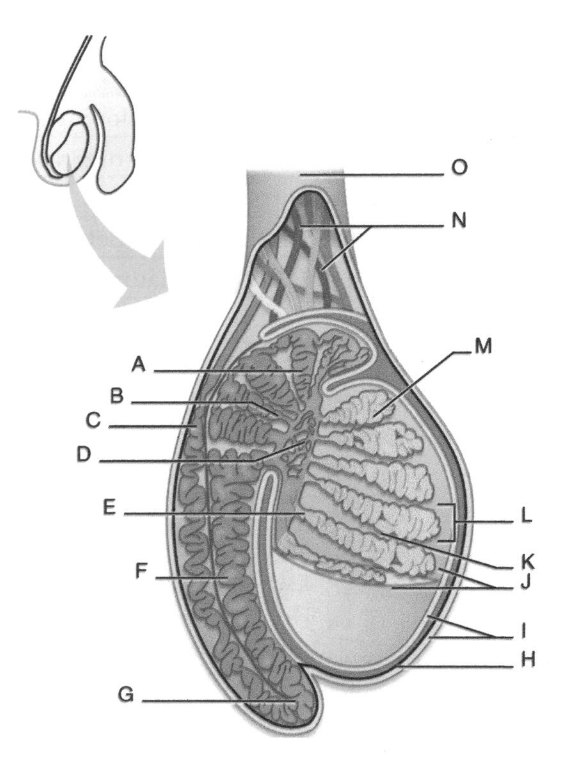

Figure 23-5 Internal anatomy of the testis.

Name ____________________ Course Number: __________ Lab Section ________

Lab 23: Reproductive System Worksheet

A. Trace the pathway through the female reproductive trace from the ovary to the outside of the body, including all relevant structures listed in Table 23-1.

1. Ovarian follicle
2. ____________
3. ____________
4. ____________
5. ____________
6. ____________
7. ____________
8. ____________
9. ____________
10. ____________
11. ____________
12. ____________

B. Trace the pathway through the male reproductive tract from the seminiferous tubule to the external urethral orifice, including all relevant structures listed in Table 23-2.

1. Seminiferous tubule
2. ____________
3. ____________
4. ____________
5. ____________
6. ____________
7. ____________ (surrounded by urogenital diaphragm)
8. ____________
9. ____________
10. ____________
11. ____________

Write out the names of the labeled items in the selected figures.

Figure 23-1.

A. ____________
B. ____________
C. ____________
D. ____________
E. ____________
F. ____________
G. ____________
H. ____________
I. ____________
J. ____________
K. ____________
L. ____________
M. ____________
N. ____________
O. ____________
P. ____________
Q. ____________
R. ____________
S. ____________
T. ____________
U. ____________
V. ____________
W. ____________
X. ____________
Y. ____________
Z. ____________
AA. ____________
AB. ____________
AC. ____________
AD. ____________

Figure 23-2.

A. ____________
B. ____________
C. ____________
D. ____________

Figure 23-3.

A. ____________
B. ____________
C. ____________
D. ____________
E. ____________
F. ____________
G. ____________

Figure 23-4.

A. ____________
B. ____________
C. ____________
D. ____________
E. ____________

F. ______
G. ______
H. ______
I. ______
J. ______
K. ______
L. ______
M. ______
N. ______
O. ______
P. ______
Q. ______
R. ______
S. ______
T. ______
U. ______
V. ______
W. ______
X. ______

Figure 23-5.

A. ______
B. ______
C. ______
D. ______
E. ______
F. ______
G. ______
H. ______
I. ______
J. ______
K. ______
L. ______
M. ______
N. ______
O. ______

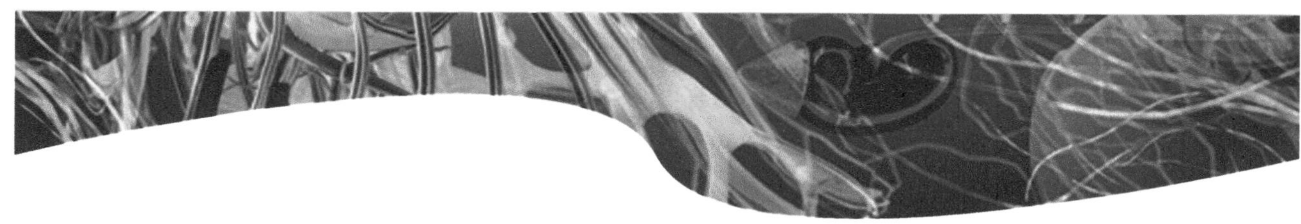

Lab 24 Gametogenesis, Development, and Inheritance

Overview of Development

Your development from a single cell was influenced by the expression of *genes* in your chromosomes and this lab helps reveal, in part, how that happens. When you studied cell division in an earlier lab, you learned that half of your 46 chromosomes came from your father (these are your *paternal chromosomes*) and half came from your mother (these are your *maternal chromosomes*). When the reproductive systems make gametes (sex cells), it is necessary that these cells contain only 23 chromosomes so that when the gametes unite there will be only 46 chromosomes in the new cell. We will begin with a look at how gametes are made.

Gametogenesis

The process of making gametes is called **gametogenesis** and it is called **oogenesis** in the female and **spermatogenesis** in the male. Similar to the cell cycle of a somatic cell, gametogenesis involves division of a cell's nuclear material and division of its cytoplasm. The nuclear division is called **meiosis** and it has some aspects in common with mitosis. However, unlike the cell cycle, which has one mitotic division and one cytokinesis and produces two genetically identical cells, gametogenesis has two meiotic divisions and two cytoplasmic divisions and produces four cells, each having only half the number of nuclear chromosomes as the original cell. Review the terms in **Table 24.1** and then see the comparison of mitosis with meiosis in **Figure 24-1**. In addition, be able to recognize these components on drawings and models.

Demonstrating Gametogenesis

You should be able to demonstrate gametogenesis using chromosome models provided at your table. The different stages of meiosis are described below and your lab instructor will explain how to demonstrate these stages,

Table 24.1 Terms Related to Gametogenesis

Autosome	Chromosome that does not determine the sex of the individual; human somatic cells have 44 autosomes
Centromere	Region where two identical double-helix molecules (chromatids) are held together during mitosis or meiosis
Chromatid	One of two identical strands held together by a centromere; each chromatid contains a kinetochore

(Continued)

Table 24.1 Terms Related to Gametogenesis *(Cont'd)*

Chromatin	DNA (double helix) coupled with histone; coils to make a visible chromosome
Chromosome	DNA molecule containing genes that provide information for protein synthesis; become more visible when stained; usually refers to the highly coiled (condensed) form of chromatin seen during mitosis or meiosis
Crossing over	Process by which homologous maternal and paternal chromosomes exchange genetic information while they are synapsed during prophase I of meiosis
Cytokinesis I	Cytoplasmic division of a primary oocyte or primary spermatocyte
Cytokinesis II	Cytoplasmic division of a secondary oocyte or secondary spermatocyte
Daughter cells	Cells formed when the parent cell divides
Diploid (2N)	Having two sets of chromosomes, i.e., having a homologous pair for each kind of chromosome; the diploid number for humans is 46
Gamete	Sex cell (sperm or egg); haploid in humans
Gene	Segment of DNA containing information for sequencing amino acids in a protein
Genome	Complete set of genes in an organism; i.e., all genes in a human's 46 chromosomes
Germinal cell	Cell that will undergo meiosis and cytokinesis to form gametes
Haploid (1N)	Number of different kinds of chromosomes in a species; different kinds can be identified by (1) length, (2) centromere location, (3) banding patterns; humans 1N=23
Homologous chromosomes	Chromosomes with information that codes for the same traits; humans have 23 pairs of homologous chromosomes (23 from each parent)
Independent assortment	Process by which tetrads align themselves independently of other tetrads along the metaphase plate during meiosis I
Karyotype	A picture of an organism's chromosomes
Kinetochore	Protein-component of each chromatid; the site where a spindle fiber attaches
Meiosis I	Nuclear division in a primary oocyte or primary spermatocyte
Meiosis II	Nuclear division in a secondary oocyte or secondary spermatocyte
Parent cell	Cell that will undergo division
Replicated chromosome	Chromosome having two chromatids (molecules) joined at a centromere
S phase	"Synthesis" phase; time when the DNA is replicates; may last up to 8 hours
Sex chromosomes	Responsible for determining the sex of the individual; "x" and "y" in humans
Spindle fibers	Microtubules that pull chromatids apart during nuclear division
Synapsis	Homologous chromosomes come together in prophase I; allows for crossing over.
Tetrad	Formed when synapsis occurs; refers to the 4 chromatids in the group.

including how to replicate the chromosomes. For any hypothetical cell, you should be able to determine the following:

- Number of homologous chromosomes
- Number of molecules present in the cell
- Number of chromosomes
- Number of chromatids (if any)
- Number of centromeres and kinetochores (if any)
- The 1N and 2N number for the organism
- The stage and process taking place in the cell
- Whether the cell is 1N or 2N

Interphase During S phase the DNA replicates and the two chromatids are held together by a centromere. The chromatin has not condensed, so chromosomes are not visible. Spindle fibers form and attach to the kinetochores of the chromatids.

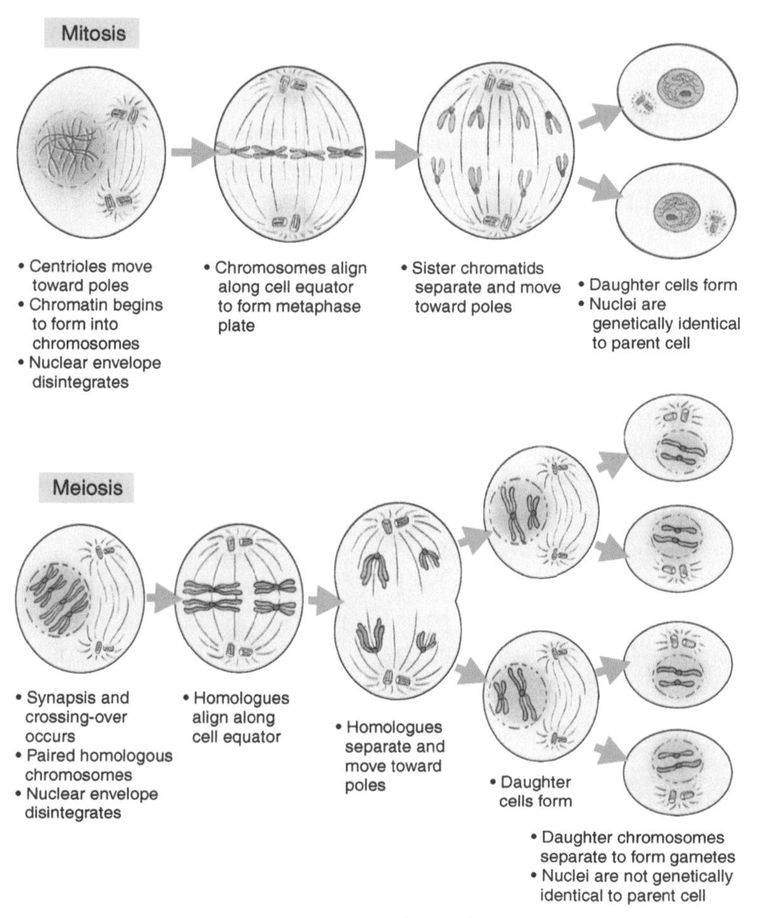

Figure 24-1 Comparison of mitosis and meiosis.

Meiosis I

Prophase I The events during this time are similar to prophase of mitosis, except that homologous chromosomes come together (synapse) resulting in the formation of tetrads. The number of tetrads equals the organism's haploid number. Crossing over occurs during this stage.

Metaphase I The tetrads align themselves independently along the cell's equator; this process is called independent assortment.

Anaphase I The tetrads separate, but the centromeres continue to hold sister chromatids together. The replicated maternal and paternal chromosomes within a homologous pair move toward opposite poles.

Telophase I During this stage, 23 replicated chromosomes exist at opposite poles of the cell (each chromosome still has two chromatids). If the cell did not begin to undergo cytoplasmic division (cytokinesis) during late anaphase I, it will happen during this stage.

Cytokinesis I

A cleavage furrow pinches inward to separate the parent cell into two daughter cells. Each daughter cell is haploid (1N), having only one of each kind of chromosome, but because each chromosome still has two chromatids, there are still 46 linear DNA molecules present in each cell.

Meiosis II

The events in meiosis II are identical to those in mitosis except that the cell is haploid; i.e., there are no homologous chromosomes present.

Prophase II The chromosomes are randomly arranged.

Metaphase II The chromosomes align on the equator.

Anaphase II Sister chromatids separate and move toward opposite poles.

Telophase II A nuclear membrane forms around the chromosomes at the opposite poles.

Cytokinesis II

A cleavage furrow pinches inward to separate each daughter cell into two more daughter cells. Each daughter cell is haploid (1N), having only one of each kind of chromosome and only 23 linear DNA molecules present. Keep in mind that it is theoretically possible for half of a man's sperm cells to contain only the man's maternal chromosomes (including the "X"), while the other half would contain only the man's paternal chromosomes (including the "Y"). However, due to crossing over and the law of independent assortment, the chance of this happening is not likely.

Stages of gametogenesis
Each cell during meiosis has a special name in the male and female reproductive systems. In addition, there are certain structures that are closely associated with various stages. Review these terms in **Table 24.2** and then label them in **Figure 24-2.**

Developmental Structures

Fertilization (or syngamy) refers to the union of a sperm cell with a secondary oocyte. Shortly thereafter, the secondary oocyte nucleus unites with the sperm's nucleus and the cell is a **zygote**. The zygote then experiences repeated mitotic and cytokinetic divisions to produce a multicellular mass. Cilia in the oviduct then sweep this mass toward the uterus. You should become familiar with the terms related to development in **Table 24-3** and be able to recognize the stages in **Figure 24-3** drawings and models.

Table 24.2 Terms Associated with Gametogenesis

Term	Description
Acrosome	Enzyme-containing package in the head of sperm; its enzymes help dissolve the oolemma
Corona radiata	Follicular cells around an ovulated secondary oocyte
Corpus luteum	Yellow body derived from the ruptured ovarian follicle; secretes progesterone
Graafian follicle	Mature follicle in the ovary; after rupturing it becomes a corpus luteum
Oogonium	2N parent cell prior to replication
Ootid	1N cell immediately after cytokinesis II
Ovum	1N cell just prior to its nucleus uniting with a sperm nucleus
Polar bodies	Small 1N daughter cells at the end of cytokinesis I and II
Primary follicle	Contains a primary oocyte inside the ovary
Primary oocyte	2N cell during meiosis I inside the ovary
Primary spermatocyte	2N cell during meiosis I inside the testis
Secondary follicle	Contains a secondary oocyte inside the ovary
Secondary oocyte	1N cell during meiosis II; expelled from the ovary during ovulation; becomes an ootid if fertilized by a sperm cell
Secondary spermatocyte	1N cell during meiosis II; gives rise to spermatids
Spermatids	1N cells after cytokinesis II; not yet flagellated
Spermatogonium	2N parent cell prior to replication
Spermatozoa	Flagellated sperm cells; mature in the epididymis
Spermiogenesis	Process of converting a spermatid into a spermatozoan; completes spermatogenesis
Zona pellucida	**(Oolemma)** A thick membrane surrounding a secondary oocyte

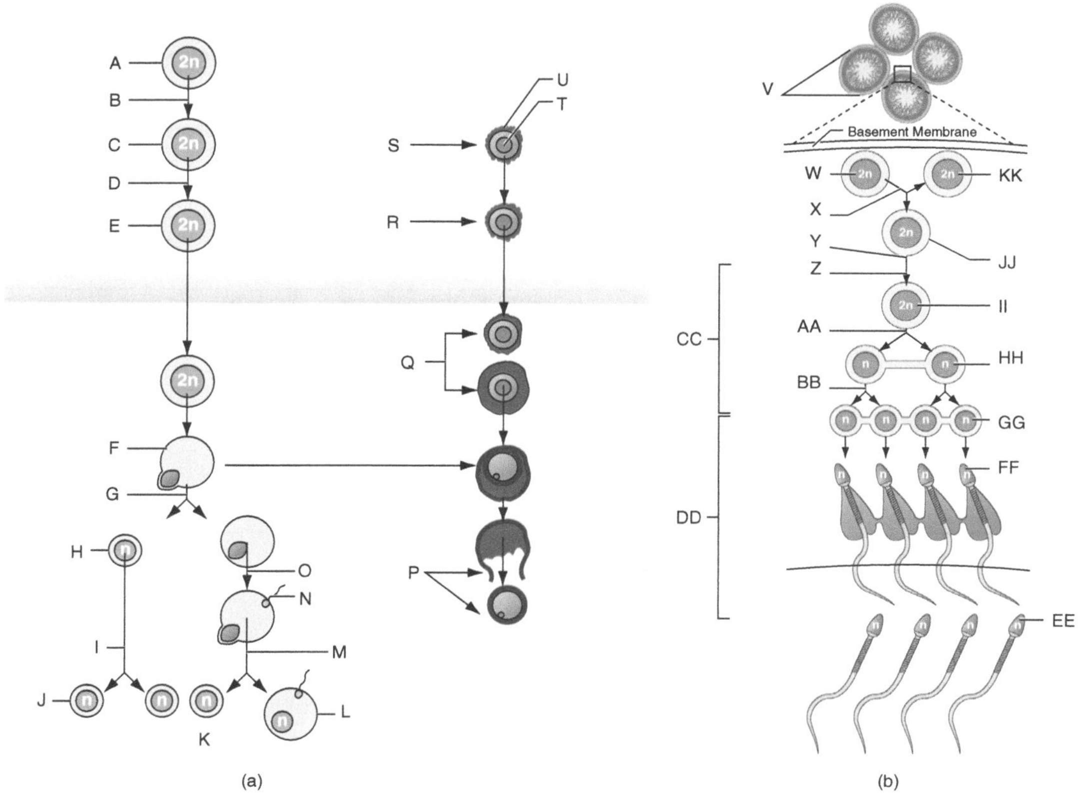

Figure 24-2 Gametogenesis: (a) oogenesis, (b) spermatogenesis.

Table 24.3 Terms Related to Developmental Stages	
Allantois	Projection of the yolk sac; produces blood cells and umbilical vessels for the embryo
Amnion	Membranous sac surrounding the amniotic cavity
Amniotic cavity	Fluid-filled cavity around the developing fetus
Areola	Pigmented region around the nipple; serves as point of focus for a nursing infant
Autosome	Any chromosome not classified as a sex chromosome
Blastocoel	Hollow cavity inside a blastocyst
Blastocyst	Hollow ball of cells developed from a zygote; stage that implants in the uterus
Cellular differentiation	Change in developing cells by activation and/or inactivation of genes
Chorion	Baby's contribution to the placenta
Chorionic villi	Fingerlike projections of the chorion; may be sampled to generate a karyotype
Decidua basalis	Mother's contribution to the placenta
Ductus arteriosus	Bypass from the pulmonary trunk to the aorta in a fetus
Ductus venosus	Bypass from the umbilical vein to the inferior vena cava in a fetus
Ectoderm	First germ layer to form; it is between the amniotic cavity and the blastocoel
Embryonic disc	The three germ layers between the amniotic sac and blastocoel
Endoderm	Second germ layer to form; between ectoderm and blastocoel
Extraembryonic coelom	Cavity around amniotic cavity and developing embryo

(Continued)

Table 24.3 Terms Related to Developmental Stages *(Cont'd)*	
Foramen ovale	Opening in the interatrial septum of a fetus
Lactiferous ducts	Transport milk from mammary glands to the nipple; smooth muscle fibers encircling it contract in response to oxytocin
Mammary glands	Milk-producing organs in the breasts of a female; respond to prolactin
Mesoderm	Last germ layer to form; it arises from ectoderm cells passing through the primitive streak
Morula	Solid ball of cells that develop from the zygote; usually 16-32 cells
Nipple	Projection in the middle of the areola through which milk is ejected
Notochord	Flexible cord along the back of developing fetus; in all chordates
Placenta	Site of nutrient and waste exchange between mother and fetus
Primitive knot	Site in ectoderm where brain will develop
Primitive streak	Site in ectoderm where spine will develop
Trophoblast	Outer layer of cells in a blastocyst
Umbilical arteries	Arise from each internal iliac artery in a fetus; deliver deoxygenated blood to the placenta
Umbilical cord	Contains two arteries and one vein; filled with Wharton's jelly
Umbilical vein	Carries oxygenated blood from placenta to fetus
Yolk sac	Baglike structure derived from endoderm; produces blood cells for embryo
Zygote	2N cell after the union of sperm and ovum nuclei

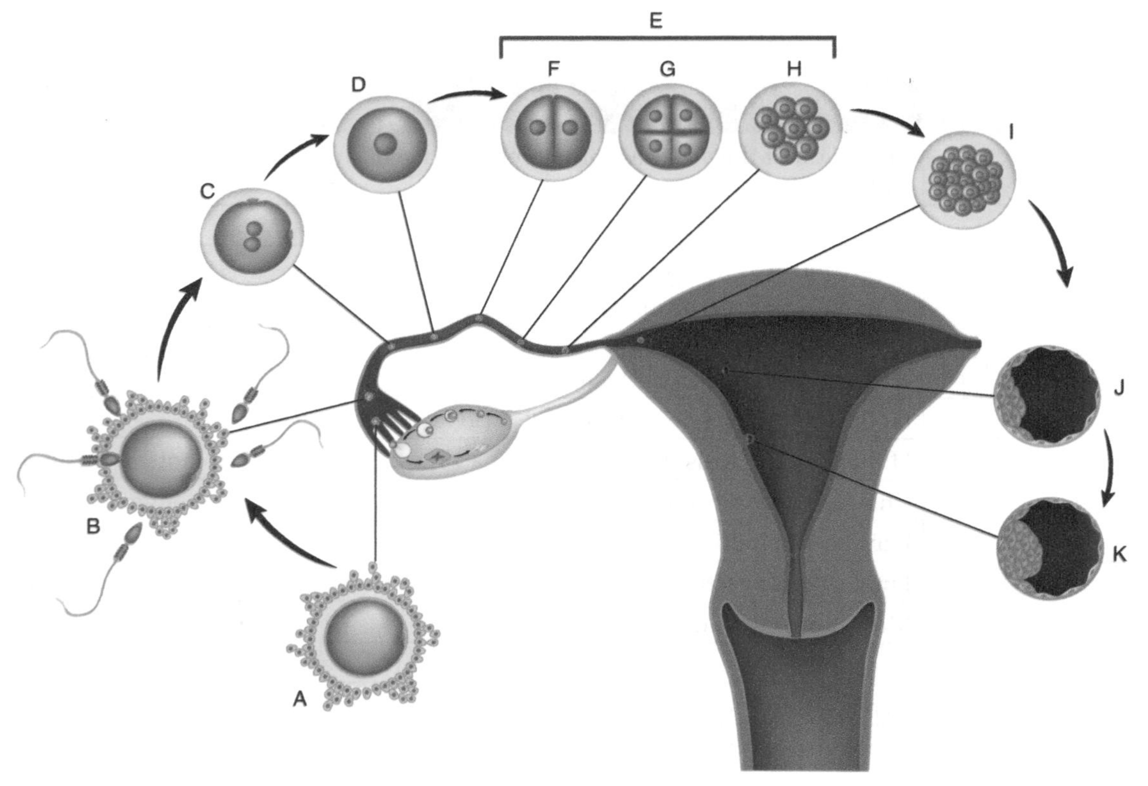

Figure 24-3 Stages of development from fertilization to implantation.

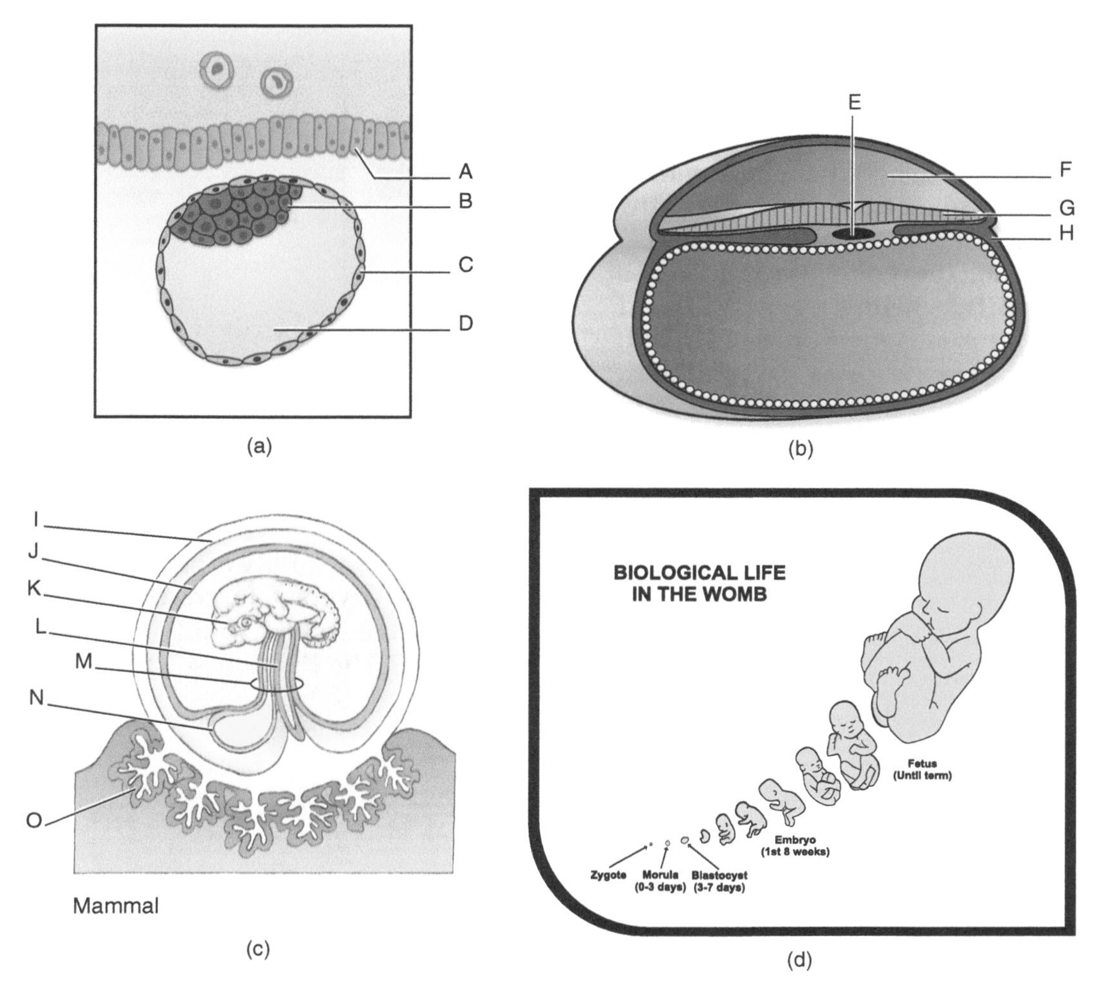

Figure 24-4 Stages of development: (a) Parts of the blastocyst, (b) parts of the embryonic disc, (c) Early structures of the fetus (d) Stages from the zygote to the fetus.

Breast Anatomy

A newborn infant's nourishment comes from the mother's milk, which is produced by mammary glands in the breasts. Using the appropriate terms from **Table 24.3**, label the structures in **Figure 24-5**.

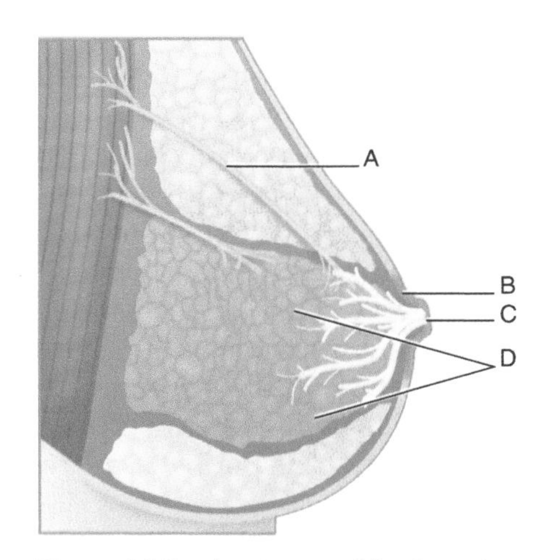

Figure 24-5 Anatomy of the breast.

Genetics and Probability

This exercise will show how an independent assortment of tetrads during metaphase I of meiosis can give rise to different combinations of genes in a zygote. In addition, you will learn how to determine the probability of getting certain combinations of genes in a zygote when you know the genetic makeup of the parents. You learned about dominant and recessive genes in Lab 16, but we will provide a review here. You may need to refer back to Lab 16 for information needed to answer some of the problems at the end of this lab.

Traits controlled by genes on autosomes are **autosomal traits**, whereas traits controlled by genes on sex chromosomes are **sex-linked**. Some genes have alternate forms, called **alleles.** Your **genotype** refers to the types of genes in your genome, whereas, your **phenotype** refers to the way those genes are expressed. Any allele expressed whenever it is inherited is a **dominant allele**. If an allele is not expressed unless there are two copies (i.e., maternal and paternal copies) is a **recessive allele**.

For example, the allele for having six digits on each hand and foot is caused by dominant; let's call this "P". In contrast, the allele for having only five digits on each hand and foot is recessive; let's call this "p". If either the mother or father donates a P, the child will have six digits on each hand and foot. If both parents donate a P, then the offspring is **homozygous** for the dominant allele, or PP. If one parent donates a "P" and the other parent donates a "p", then the offspring is **heterozygous**, or Pp. In either case, the offspring expresses the dominant trait. The only way a child will have five digits on each hand and foot is for him or her to be homozygous recessive, or pp.

Using a Punnett Square

A Punnett square is a tool that shows all possible combinations of alleles in offspring when the genetic makeup of the parents is known. In our example, let's assume Billy and Betty are both heterozygous, Pp. What is the probability they will have a child who has only five digits on each hand and foot? First, you must determine all types of gametes possible with respect to the alleles, P and p. Half of Billy's gametes will contain a "P" and half will contain a "p", and the same is true for Betty. The Punnett Square is then set up as follows:

According to the Punnett Square, the chances are .25 (or 25%) for PP; .5 (or 50%) for Pp; and .25 (or 25%) for pp. Therefore, since PP and Pp yield six digits and pp yields five digits, the chance is 25%.

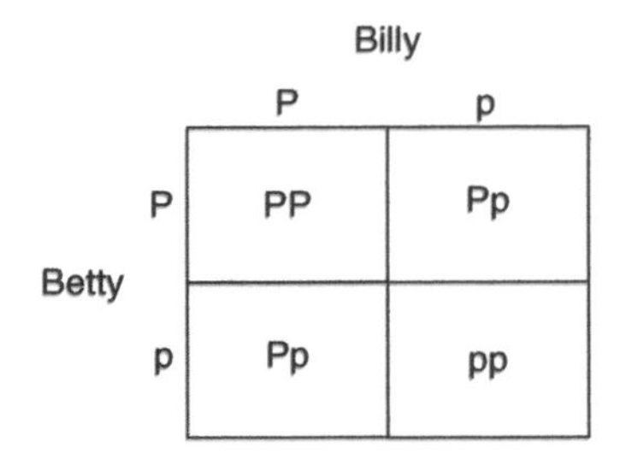

The trait just described is an autosomal trait. How then do we make a Punnett Square for a sex-linked trait? Most sex-linked traits are coded for by genes on the "X" chromosome, so we will consider the "X-linked" trait *colorblindness*. This is a recessive trait, written X^c, while the dominant allele codes for normal vision and will be written X^N. If Billy is normal and Betty is a carrier (heterozygous) for the colorblind allele, what is the chance they will have a child who is colorblind? The Punnett square is drawn as follows:

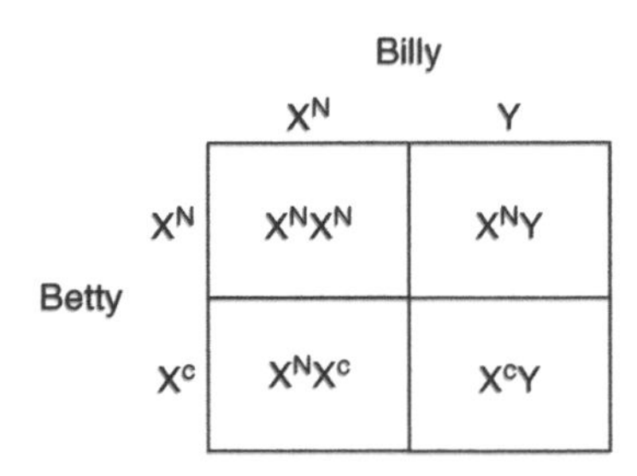

According to this Punnett Square, the chance of having a colorblind child is .25 (or 25%). What is the chance that a boy in this family will be born colorblind? The answer is .5 (or 50%).

Product and Addition Rules

The last aspect of probability you need to know is the product and addition rules. The **product rule** involves multiplying probabilities of separate events that yield a particular sequence that can happen only one way. For example, what is the chance of a couple having two boys as offspring? This sequence can happen only one way; the first child must be a boy and the second child must be a boy. The probability of this happening is 0.5 (1st boy) x 0.5 (2nd boy) = 0.25 or 25%. The chance of having a girl first and a boy second, in that order, is also 0.5 x 0.5 = 0.25 or 25%.

The **addition rule** involves adding probabilities of more than one sequence of events that can yield the same outcome; that is, it is the probability of getting a particular combination of events when considering all sequences possible. For example, what is the probability of a couple having one boy and one girl as their two children? The chance that the 1st child will be a boy = 0.5 or 50%. To have a boy and girl, the 2nd child would have to be a girl, which is 0.5 or 50% chance. Therefore, the chance of having a boy and girl in this order is 0.5 x 0.5 = 0.25 or 25%. However, the children could be born in the reverse order: 1st child girl = 0.5 and 2nd child boy = 0.5. The chance of having a girl and boy in that order is also 0.5 x 0.5 = 0.25 or 25%. Therefore, according to the addition rule, the chance of having a boy and girl, when the order is not important, is 0.25 + 0.25 = 0.5 or 50%.

Name ____________________ Course Number: ________ Lab Section ______

Lab 24: Worksheet

For the following problems, be sure to show your work.

1. What is the chance of having two boys in a family?

2. What is the chance of having a boy first and a girl second?

3. What is the chance of having a boy and a girl, if the order is not important?

4. What is the chance of having two boys and one girl, if the order is not important?

5. A man who has blood type AB- marries a woman who has blood type O-. What is the probability that this couple will have a child with type A- blood?

6. A man who has Rh positive blood (heterozygous) marries a woman who has Rh negative blood. What is the probability that this couple will have two children with Rh negative blood?

7. A man with AB+ blood (his father has A-; his mother has AB+) marries a woman with B+ (her mother has type O-; her father has B+). The probability that this couple will have two boys with AB- is what?

8. A child has type A+ blood. The child's mother has type O-. The child's father could have what blood type?

9. A man who has type AB- blood and is a carrier for an autosomal recessive trait marries a woman who has type B+ blood (she is heterozygous type B and also for the Rh factor) and has the autosomal recessive trait. What is the probability that this couple will have two girls with type A- blood and the recessive trait?

10. A man (XY) who is colorblind (a recessive sex-linked trait, i.e., the allele is located only on the X chromosome) marries a woman (XX) who has normal vision but is heterozygous for the color-blind allele. What is the chance that this couple will have a colorblind boy and a girl who is a carrier?

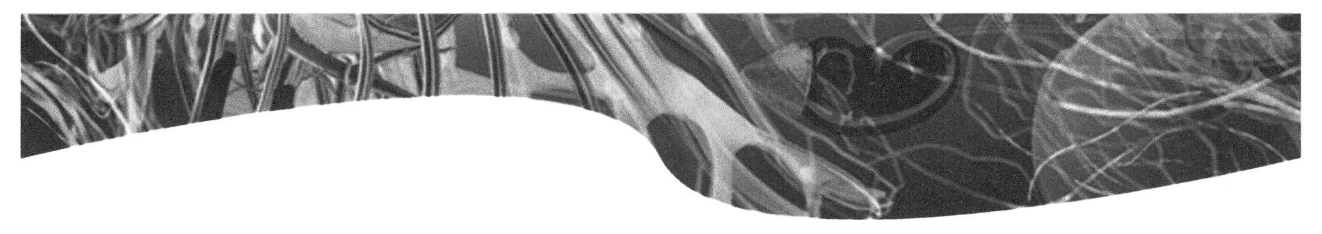

Taylor Lab Manual Keys for Figures

Figure 1-1

A. Frontal
B. Buccal
C. Oral
D. Acromial
E. Axillary
F. Brachial
G. Antecubital
H. Antebrachial
I. Carpal
J. Palmar
K. Femoral
L. Patellar
M. Fibular (peroneal)
N. Tarsal
O. Digital
P. Orbital
Q. Nasal
R. Mental
S. Sternal
T. Pectoral
U. Abdominal
V. Umbilical
W. Pelvic
X. Pubic
Y. Crural
Z. Occipital
AA. Otic
BB. Cervical
CC. Natal cleft
DD. Gluteal fold
EE. Calcaneal
FF. Cephalic
GG. Nuchal
HH. Scapular
II. Vertebral (spinal)
JJ. Olecranal
KK. Lumbar
LL. Pollex
MM. Popliteal
NN. Sural
OO. Fibular (peroneal)
PP. Plantar

Figure 1-3

A. Cranial cavity
B. Vertebral or spinal cavity
C. Larynx
D. Thyroid gland
E. Trachea
F. Aorta
G. Lung
H. Heart
I. Liver
J. Spleen
K. Gallbladder
L. Transverse colon
M. Small intestine
N. Ascending colon
O. Greater omentum (left half; yellow)
P. Thoracic cavity
Q. Abdominopelvic cavity
R. Stomach
S. Descending colon

Figure 1-5

A. Visceral peritoneum
B. Greater omentum
C. Parietal peritoneum
D. Lesser omentum
E. mesentery

Figure 1-6

A. Upper right
B. Upper left
C. Lower right
D. Lower left
E. Right hypochondriac
F. Epigastric
G. Left hypochondriac
H. Right lumbar
I. Umbilical
J. Left lumbar
K. Right inguinal (iliac)
L. Hypogastric (pubic)
M. Left inguinal (iliac)

Figure 5-1

A. Reticular dermis
B. Epidermis
C. Papillary dermis
D. Hair follicle
E. Eccrine sweat gland
F. Sebaceous gland
G. Arrector pili muscle
H. Hair root

Figure 5-2

A. Epithelial root sheath
B. Hair
C. Hair matrix
D. Hair papilla
E. Adipose tissue of hypodermis
F. Hair bulb
G. Hair papilla
H. Hair matrix
I. Henle's layer of internal epithelial root sheath
J. Hair cuticle
K. Border between internal and external epithelial root sheaths
L. Dermal root sheath
M. Huxley's layer of internal epithelial root sheath
N. Hair medulla
O. Hair cortex
P. External epithelial root sheath
Q. Hair cuticle
R. Hair medulla
S. Hair cortex

Figure 6-1

A. Cranial bones
B. Facial bones
C. Scapula
D. Sternum
E. Ulna
F. Metacarpal
G. Phalanx
H. Patella
I. Tibia
J. Metatarsal
K. Phalanx
L. Frontal bone
M. Zygomatic bone
N. Maxilla (maxillary bone)
O. Mandible
P. Rib
Q. Ilium (part of pelvic girdle and pelvis)
R. Scapula
S. Humerus
T. Radius
U. Carpal bone
V. Femur
W. Fibula
X. Tarsal
Y. Parietal bone
Z. Occipital bone
AA. Temporal bone
BB. Cervical vertebrae
CC. Thoracic vertebrae
DD. Lumbar vertebrae
EE. Sacral vertebrae (sacrum)
FF. Coccygeal vertebrae (coccyx)

Figure 6-2

1. Frontal bone
2. Lacrimal bone
3. Ethmoid bone
4. Parietal bone
5. Superior orbital (supraorbital) margin
6. Sphenoid bone
7. Temporal bone
8. Sphenoid bone
9. Zygomatic bone
10. Zygomatic bone
11. Maxilla
12. Zygomaxillary suture
13. Infraorbital foramen
14. Ramus of mandible
15. Nasal spine on maxilla

16. Angle of mandible
17. Mandibular symphysis
18. Glabella
19. Coronal suture
20. Supraorbital foramen
21. Optic foramen
22. Frontozygomatic suture
23. Superior orbital fissure
24. Nasal bone
25. Inferior orbital fissure
26. Middle nasal concha (turbinate)
27. Inferior nasal concha (turbinate)
28. Maxilla
29. Vomer
30. Alveolar process
31. Mental foramen
32. Body of mandible
33. Frontal bone
34. Coronal suture
35. Greater wing of sphenoid bone
36. Frontozygomatic suture
37. Lacrimal bone
38. Nasolacrimal fossa
39. Nasal bone
40. Zygomatic bone
41. Infraorbital foramen
42. Maxilla
43. Nasal spine of maxilla
44. Mental foramen
45. Parietal bone
46. Lambdoidal suture
47. Squamosal suture
48. Occipital bone
49. Temporal bone
50. Zygomatic process of temporal bone
51. Mandibular condyle
52. External auditory meatus
53. Mastoid process of temporal bone
54. Styloid process of temporal bone
55. Mandibular notch
56. Angle of mandible

Figure 6-3

1. Incisive foramen
2. Intermaxillary suture
3. Temporal process of zygomatic bone
4. Palatine bone
5. Zygomatic process of temporal bone
6. Basilar (basioccipital) process of occipital bone
7. Foramen ovale
8. Styloid process of temporal bone
9. Carotid canal
10. Stylomastoid foramen
11. Occipital condyle
12. Foramen magnum
13. External occipital protuberance
14. Temporal process of zygomatic bone
15. Zygomatic process of temporal bone
16. Mandibular fossa of temporal bone
17. Foramen lacerum
18. Condylar canal
19. Crista galli of ethmoid bone
20. Cribriform plate of ethmoid bone
21. Lesser wing of sphenoid bone
22. Greater wing of sphenoid bone
23. Sella turcica of sphenoid bone
24. Foramen ovale
25. Middle cranial fossa
26. Internal auditory meatus
27. Hypoglossal canal
28. Anterior cranial fossa (above frontal bone)
29. Foramen rotundum
30. Clinoid process of sphenoid's lesser wing
31. Foramen lacerum
32. Clivus of occipital bone
33. Petrous portion of temporal bone
34. Jugular foramen
35. Foramen magnum
36. Posterior cranial fossa (above occipital bone)
37. Sagittal suture
38. Parietal bone
39. Lambdoidal suture
40. Squamosal suture
41. External occipital protuberance
42. Sagittal suture
43. Coronal suture
44. Lambdoidal suture
45. Lambdoidal suture
46. Occipital bone
47. Posterior fontanel
48. Anterior fontanel
49. Frontal suture
50. Frontal bones
51. Coronal suture
52. Squamosal suture
53. Lambdoidal suture
54. Mastoid (posterolateral) fontanel
55. Sphenoid (anterolateral) fontanel
56. Coronal suture

Figure 6-4

A. Frontal bone
B. Coronal suture
C. Frontal sinus
D. Crista galli of ethmoid bone
E. Sella turcica of sphenoid bone
F. Nasal bone
G. Perpendicular plate of ethmoid bone
H. Vomer
I. Nasal spine of maxilla
J. Intermaxillary suture
K. Mandibular symphysis
L. Parietal bone
M. Squamosal suture
N. Lambdoidal suture
O. Sphenoid sinus
P. Internal auditory meatus
Q. Hypoglossal canal
R. Styloid process of temporal bone
S. Mandibular foramen
T. Palatine bone

Figure 6-5

A. Greater cornu (horn)
B. Lesser cornu (horn)
C. Body of hyoid bone

Figure 6-6

A. Vertebral foramen
B. Articular facet
C. Dens (odontoid process) of C2 (axis)
D. Spinous process
E. Lamina of vertebra
F. Transverse foramen
G. Pedicle of vertebra
H. Costal facet of thoracic vertebra
I. Vertebral lamina
J. Body (centrum) of vertebra
K. Spinous process
L. Sacral foramen
M. Coccyx
N. Coccyx
O. Sacrum
P. Lumbar vertebrae
Q. Thoracic vertebrae
R. Cervical vertebrae

Figure 6-7

A. Jugular notch
B. Manubrium
C. Body (gladiolus) of sternum
D. Xiphoid process
E. Costal cartilage
F. Vertebrosternal ribs
G. Vertebrochondral ribs
H. Vertebral (floating) ribs

Figure 7-1

A. Acromial end of clavicle
B. Manubrial end of clavicle
C. Superior border of scapula; scapular notch to immediate left
D. Acromion process
E. Coracoid process
F. Glenoid fossa
G. Subscapular fossa
H. Axillary (lateral) border
I. Vertebral (medial) border
J. Supraspinous fossa
K. Spine of scapula
L. Infraspinous fossa

Figure 7-2

A. Greater tubercle
B. Intertubercular groove
C. Deltoid tuberosity
D. Coronoid process of ulna
E. Radial fossa
F. Capitulum
G. Trochlea
H. Medial epicondyle
I. Lesser tubercle
J. Head of humerus
K. Middle of diaphysis
L. Olecranon fossa
M. Olecranon process
N. Coronoid process of ulna
O. Trochlear notch of ulna
P. Radial notch
Q. Diaphysis
R. Head of ulna
S. Styloid process of ulna
T. Head of radius
U. Neck of radius
V. Radial tuberosity
W. Diaphysis
X. Styloid process of radius
Y. Scaphoid
Z. Trapezium
AA. Trapezoid
BB. 1st metacarpal

CC. Proximal phalanx of digit 1
DD. Distal phalanx of digit 1
EE. Lunate
FF. Pisiform
GG. Triquetrum
HH. Hamate
II. Capitate
JJ. Proximal phalanx of digit 5
KK. Middle phalanx of digit 5
LL. Distal phalanx of digit 5

Figure 7-3

A. Iliac crest
B. Posterior superior iliac spine
C. Posterior inferior iliac spine
D. Greater sciatic notch
E. Ischial spine
F. Obturator foramen
G. Ischial tuberosity
H. Anterior superior iliac spine
I. Anterior inferior iliac spine
J. Acetabulum
K. Pubis (pubic bone)
L. False pelvis
M. Inlet of true pelvis
N. Outlet of true pelvis

Figure 7-4

A. Neck of femur
B. Fovea capitis
C. Head of femur
D. Lesser trochanter
E. Lateral epicondyle
F. Patellar surface
G. Medial epicondyle
H. Greater trochanter
I. Linea aspera
J. Lateral epicondyle
K. Lateral condyle
L. Intercondylar notch
M. Head of fibula
N. Right fibula
O. Lateral malleolus
P. Intercondylar eminence
Q. Median condyle
R. Medial malleolus
S. Lateral condyle
T. Calcaneus
U. Talus
V. Navicular
W. Medial cuneiform
X. 1st metatarsal
Y. Proximal phalanx
Z. Distal phalanx
AA. Medial cuneiform
BB. Intermediate cuneiform
CC. Navicular
DD. Talus
EE. 5th metatarsal
FF. Lateral cuneiform
GG. Cuboid
HH. Calcaneus
II. Lateral malleolus

Figure 8-1

1st column boxes:

1. 29
2. 8
3. 6
4. 15
5. 9
6. 23
7. 19
8. 7
9. 31
10. 30
11. 14

2nd column boxes:

1. 6
2. 24
3. 25
4. 34
5. 22
6. 11
7. 13
8. 28
9. 18
10. 12

3rd column boxes:

1. 26
2. 3
3. 1
4. 34
5. 10
6. 21
7. 17
8. 17
9. 32
10. 27

4th column boxes:

1. 2
2. 22

3. 20
4. 4
5. 5

Figure 8-2

1st column boxes:

1. 3-6-13
2. 3-6-16
3. 2-5-11
4. 2-5-11
5. 3-6-12
6. 2-5-11
7. 3-6-12
8. 3-6-12
9. 3-6-13
10. 3-6-12

2nd column boxes:

1. 1-4-9
2. 3-6-15
3. 3-6-12
4. 2-5-11
5. 3-6-17
6. 3-6-12
7. 3-6-13
8. 3-6-12
9. 3-6-14
10. 3-6-13

3rd column boxes:

1. 1-4-7
2. 3-6-14
3. 3-6-16
4. 2-5-11
5. 3-6-13
6. 3-6-17
7. 3-6-14
8. 3-6-14
9. 3-6-13
10. 3-6-13

4th column boxes:

1. 3-6-17
2. 3-6-17
3. 3-6-14
4. 3-6-12
5. 3-6-16

Figure 8-3

1. Dorsiflexion
2. Forearm pronation
3. Head lateral flexion
4. Hand flexion
5. Arm extension
6. Thigh medial rotation
7. Leg extension
8. Trunk extension
9. Thigh adduction
10. Thigh lateral rotation
11. Forearm flexion
12. Thigh adduction
13. Hand abduction
14. Arm extension
15. Head rotation
16. Thigh flexion
17. Trunk rotation
18. Head flexion
19. Forearm extension
20. Thigh abduction
21. Thigh extension
22. Foot inversion
23. Plantar flexion
24. Arm abduction
25. Hand extension
26. Arm circumduction
27. Head extension
28. Trunk extension
29. Hand adduction
30. Arm adduction
31. Head rotation
32. Leg flexion
33. Arm medial rotation
34. Supination
35. Trunk flexion
36. Leg hyperextension
37. Head extension
38. Foot eversion
39. Forearm hyperextension
40. Arm flexion
41. Hand extension
42. Arm lateral rotation
43. Arm adduction
44. Variety of motions

Figure 8-4

A. Femur lateral condyle
B. Anterior cruciate ligament
C. Lateral meniscus
D. Head of fibula
E. Femur medial condyle
F. Posterior cruciate ligament
G. Medial meniscus
H. Medial (tibial) collateral ligament
I. Tibial tuberosity

Figure 9-2

A. Frontalis
B. Sternocleidomastoid
C. Deltoid
D. Pectoralis major
E. Serratus anterior
F. Tendinous intersection
G. Rectus abdominis
H. Temporalis
I. Linea alba
J. External oblique
K. Trapezius
L. Infraspinatus
M. Latissimus dorsi
N. Epicranial aponeurosis
O. Occipitalis
P. Teres minor
Q. Teres major
R. Lumbodorsal fascia

Figure 9-3

A. Corrugator supercilia
B. Temporalis
C. Levator labii superioris
D. Zygomaticus major
E. Masseter
F. Buccinator
G. Platysma (extends over neck)
H. Depressor anguli oris
I. Frontalis
J. Procerus
K. Orbicularis oculi
L. Nasalis
M. Levator labii superioris alaeque nasi
N. Zygomaticus minor
O. Orbicularis oris
P. Risorius
Q. Depressor labii
R. Mentalis

Figure 9-5

A. Longus colli
B. Sternalis
C. Internal intercostal (on back wall)
D. Diaphragm
E. Psoas minor
F. Quadratus lumborum
G. Iliacus
H. Psoas major
I. Scalene muscles
J. Omohyoid
K. Internal intercostal
L. External intercostal
M. Rectus abdominis
N. External oblique
O. Rectus sheath

Figure 9-6

A. Supraspinatus
B. Infraspinatus
C. Teres minor
D. Teres major

Figure 9-7

A. Sternocleidomastoid
B. Semispinalis
C. Splenius
D. Levator scapulae
E. Rhomboid
F. Serratus anterior
G. Iliocostalis
H. Longissimus
I. Spinalis
J. Serratus posterior (inferior)
K. Semispinalis
L. Spinalis
M. Longissimus (not delineated from the iliocostalis-n)
N. Iliocostalis

Figure 10-1

A. Deltoid
B. Biceps brachii
C. Brachioradialis
D. Palmaris longus
E. Flexor retinaculum
F. Sartorius
G. Rectus femoris
H. Vastus lateralis
I. Tibialis anterior
J. Superior extensor retinaculum
K. Adductor longus
L. Vastus medialis
M. Patellar tendon
N. Gastrocnemius
O. Deltoid
P. Triceps brachii
Q. Flexor carpi ulnaris
R. Extensor carpi ulnaris
S. Extensor retinaculum
T. Vastus lateralis

U. Semitendinosus
V. Popliteus
W. Gastrocnemius
X. Soleus
Y. Extensor digitorum
Z. Gluteus medius
AA. Gluteus maximus
BB. Adductor magnus
CC. Biceps femoris
DD. Semimembranosus
EE. Calcaneal (Achilles) tendon

Figure 10-3

A. Brachioradialis
B. Pronator teres
C. Palmaris longus
D. Flexor carpi ulnaris
E. Flexor carpi radialis
F. Pronator teres
G. Palmaris longus
H. Flexor carpi ulnaris
I. Flexor carpi radialis
J. Flexor pollicis
K. Pronator quadratus
L. Palmar aponeurosis
M. Flexor digitorum
N. Flexor digitorum
O. Anconeus
P. Extensor digitorum
Q. Extensor carpi ulnaris
R. Extensor digiti minimi
S. Brachioradialis
T. Extensor carpi radialis
U. Abductor pollicis
V. Extensor pollicis brevis and longus
W. Extensor pollicis longus
X. Extensor pollicis tendon
Y. Supinator

Figure 10-4

A. Abductor pollicis
B. Flexor pollicis
C. Palmar aponeurosis
D. Adductor pollicis
E. Lumbrical muscles
F. Opponens digiti minimi
G. Flexor digiti minimi
H. Palmar interossei
I. Dorsal interossei
J. Lumbrical muscles
K. Opponens digiti minimi
L. Opponens pollicis
M. Adductor pollicis

Figure 10-5

A. Gluteus medius
B. Tensor fasciae latae
C. Gluteus maximus
D. Iliacus
E. Gluteus minimus
F. Piriformis
G. Gemellus superior
H. External obturator
I. Gemellus inferior
J. Quadratus femoris
K. Internal obturator
L. Piriformis
M. Internal obturator
N. Quadratus femoris

Figure 10-6

A. Psoas major
B. Iliacus
C. Tensor fasciae latae
D. Sartorius
E. Adductor magnus
F. Iliotibial tract
G. Vastus lateralis
H. Tibialis anterior
I. Gastrocnemius
J. Extensor digitorum
K. Piriformis
L. Pectineus
M. Adductor longus
N. Gracilis
O. Rectus femoris
P. Vastus medialis
Q. Gastrocnemius
R. Soleus
S. Psoas major
T. Gluteus maximus
U. Adductor magnus
V. Semitendinosus
W. Biceps femoris
X. Semimembranosus
Y. Gastrocnemius
Z. Soleus
AA. Calcaneal (Achilles) tendon
BB. Gluteus medius
CC. Tensor fasciae latae
DD. Gracilis
EE. Iliotibial tract

FF. Popliteus
GG. Iliotibial tract
HH. Gluteus maximus
II. Adductor longus
JJ. Gracilis
KK. Adductor magnus
LL. Vastus medialis (line crosses gracilis and sartorius)
MM. Vastus medialis
NN. Gastrocnemius
OO. Tibia (line crosses soleus)
PP. Tensor fasciae latae
QQ. Vastus lateralis
RR. Vastus lateralis
SS. Biceps femoris
TT. Extensor digitorum
UU. Fibularis (peroneus) longus
VV. Soleus
WW. Calcaneal (Achilles) tendon

Figure 10-7

A. Adductor brevis
B. Pectineus
C. Adductor longus
D. Gracilis
E. Adductor magnus

Figure 10-8

A. Tibialis anterior
B. Extensor digitorum
C. Fibularis (peroneus) brevis
D. Fibularis (peroneus) longus
E. Fibularis (peroneus) tertius
F. Soleus
G. Extensor hallucis
H. Popliteus
I. Tibialis posterior
J. Flexor hallicus
K. Flexor digitorum
L. Popliteus
M. Gastrocnemius
N. Calcaneal (Achilles) tendon

Figure 10-9

A. Adductor hallucis
B. Lumbrical muscles
C. Abductor digiti minimi
D. Flexor digitorum brevis
E. Abductor hallucis
F. Dorsal interossei
G. Extensor digitorum brevis
H. Extensor hallucis brevis
I. Lumbrical muscles
J. Quadratus plantae
K. Adductor hallucis (transverse part)
L. Flexor hallucis brevis
M. Adductor hallucis (oblique part)
N. Flexor digiti minimi brevis
O. Dorsal interossei
P. Plantar interossei

Figure 11-1

A. Brain
B. Spinal cord
C. Nerves

Figure 11-3

A. Schwann cells
B. Nodes of Ranvier
C. Axon
D. Schwann cell cytoplasm
E. Schwann cell nucleus
F. Axons
G. Schwann cell membrane
H. Schwann cell nucleus

Figure 11-4

A. Cerebrum
B. Cerebellum
C. Brain stem
D. Spinal cord
E. Lateral ventricles
F. Insula
G. 3rd ventricle
H. Cerebral white matter
I. Cerebral cortex
J. Thalamus
K. Gyri
L. Longitudinal fissure
M. Sulci
N. Frontal lobe
O. Parietal lobe
P. Occipital lobe
Q. Cerebellum
R. Temporal lobe

Figure 11-5

A. Cingulate gyrus
B. Parietal lobe
C. Fornix
D. Genu of corpus callosum
E. Diencephalon
F. Transverse fissure

G. Cerebellum
H. Body of corpus callosum
I. Frontal lobe
J. Callosal sulcus
K. Septum pellucidum
L. Location of 3rd ventricle's choroid plexus
M. 3rd ventricle (cavity) or diencephalon tissue
N. Hypothalamus
O. Pons
P. Medulla oblongata
Q. Mammillary bodies
R. Medullary pyramid
S. Thalamus
T. Cerebral peduncles of mesencephalon
U. Pons
V. Cranial nerves
W. Cerebral cortex
X. Cerebral white matter
Y. Longitudinal fissure
Z. Arachnoid villus

Figure 11-8

A. Cervical enlargement
B. Lumbar enlargement
C. Conus medullaris
D. Cauda equina
E. Dorsal column
F. Dorsal horn
G. Lateral horn
H. Lateral column
I. Ventral horn
J. Ventral column

Figure 11-9

A. Lateral ventricle
B. Choroid plexus
C. Choroid plexus
D. 3rd ventricle
E. Choroid plexus
F. 4th ventricle
G. Dura mater
H. Arachnoid mater
I. Subarachnoid space
J. Cerebrospinal fluid
K. Mesencephalic (cerebral) aqueduct
L. Pia mater
M. Spinal cord
N. Arachnoid villus
O. Superior sagittal sinus
P. Skull bone
Q. Dura mater (periosteal part)
R. Dura mater (meningeal part)
S. Subdural space
T. Arachnoid mater
U. Subarachnoid space
V. Pia mater

Figure 12-1

A. Dorsal root
B. Dorsal root ganglion
C. Dorsal ramus
D. Ventral root
E. Ramus communicans
F. Ventral ramus
G. Sympathetic chain
H. Paravertebral ganglion (sympathetic)
I. Splanchnic nerve
J. Prevertebral ganglion
K. Ganglion
L. Epineurium around nerve
M. Fascicle
N. Perineurium
O. Axon
P. Schwann cells
Q. Axon

Figure 12-3

A. Cervical plexus
B. Brachial plexus
C. Axillary nerve
D. Musculocutaneous nerve
E. Median nerve
F. Ulnar nerve
G. Radial nerve
H. Femoral nerve
I. Tibial nerve
J. Intercostal nerves
K. Lumbar plexus
L. Sacral plexus
M. Sciatic nerve
N. Fibular nerve

Figure 13-2

A. Glomerulus
B. Axon
C. Cribriform plate
D. Olfactory foramen
E. Basal cell
F. Olfactory neuron
G. Supporting cell
H. Dendrites

Figure 13-3

A. Supporting cell
B. Gustatory cell
C. Taste bud
D. Lingual tonsil
E. Circumvallate papillae
F. Foliate papilla
G. Fungiform papillae
H. Filiform papillae

Figure 13-4

A. Ora serrata
B. Ciliary body
C. Attachment site of suspensory ligament
D. Cornea
E. Pupil
F. Anterior aqueous chamber
G. Optic disc
H. Iris
I. Superior rectus muscle
J. Sclera
K. Choroid coat
L. Retina
M. Vitreous chamber
N. Fovea centralis (within macula lutea)
O. Optic nerve
P. Inferior rectus muscle
Q. Fovea centralis
R. Optic disc
S. Axons of ganglion cells
T. Ganglion cells
U. Amacrine cell
V. Bipolar cell
W. Horizontal cell
X. Cone
Y. Rod
Z. Melanocyte

Figure 13-5

A. Medial rectus
B. Superior rectus
C. Superior oblique
D. Lateral rectus
E. Inferior oblique
F. Inferior rectus
G. Pupil
H. Lacrimal gland
I. Lacrimal duct
J. Lateral canthus
K. Sclera
L. Lacrimal punctum
M. Superior palpebra
N. Iris
O. Scleral venous sinus
P. Superior lacrimal canal
Q. Lacrimal caruncle
R. Lacrimal sac
S. Nasolacrimal duct

Figure 13-6

A. Pinna (auricle)
B. Temporal bone
C. Malleus (hammer)
D. Incus (anvil)
E. Stapes (stirrup)
F. Semicircular canal
G. Cochlea
H. Vestibular nerve
I. Cochlear nerve
J. Round window
K. Tympanum
L. Styloid process of temporal bone
M. Internal carotid artery
N. Auditory (Eustachian) tube
O. Opening to nasopharynx
P. Pinna
Q. Scapha
R. Antihelix
S. Concha
T. Tragus
U. Antitragus
V. Lobule

Figure 13-7

A. Vestibule
B. Vestibulocochlear nerve
C. Oval window
D. Round window
E. Scala vestibuli
F. Vestibular membrane
G. Scala media (cochlear duct)
H. Spiral organ (of Corti)
I. Basilar membrane
J. Scala vestibuli
K. Spiral ganglion
L. Cochlear nerve fibers
M. Scala tympani
N. Attachment of semicircular duct to ampulla
O. Semicircular duct
P. Ampulla
Q. Utricle
R. Saccule

S. Cochlear nerve
T. Otolith
U. Otolithic membrane
V. Macula
W. Semicircular ducts
X. Cupula
Y. Hair cells of crista ampullaris

Figure 15-1

A. Skull
B. Hypothalamus
C. Anterior pituitary
D. Posterior pituitary
E. Optic chiasm
F. Anterior pituitary
G. Posterior pituitary
H. Infundibulum
I. Hypothalamus
J. Thyroid gland
K. Right lobe
L. Trachea
M. Isthmus
N. Left lobe

Figure 15-2

A. Thyroid gland
B. Thymus gland
C. Heart
D. Stomach
E. Small intestine
F. Parathyroid glands
G. Adrenal gland
H. Kidney
I. Pancreas

Figure 17-1

1. Right subclavian artery/vein
2. Right/left brachiocephalic veins
3. Superior vena cava
4. Right pulmonary arteries
5. Right pulmonary veins
6. Right coronary artery (within adipose tissue)
7. Right atrium
8. Anterior cardiac veins
9. Inferior vena cava
10. Right ventricle
11. Thoracic aorta
12. Left ventricle
13. Anterior interventricular artery
14. Left pulmonary veins
15. Left atrium
16. Left pulmonary arteries
17. Left subclavian artery/vein
18. Left internal jugular vein
19. Left common carotid artery
20. Brachiocephalic artery
21. Right common carotid artery
22. Right internal jugular vein
23. Left subclavian artery/vein
24. Left pulmonary arteries
25. Left pulmonary veins
26. Circumflex artery (within adipose tissue)
27. Branches of posterior vein
28. Middle cardiac vein
29. Right coronary artery
30. Inferior vena cava
31. Coronary sinus
32. Right coronary veins and right atrium
33. Right pulmonary arteries
34. Superior vena cava
35. Right brachiocephalic vein
36. Right subclavian artery/vein
37. Right internal jugular vein
38. Right common carotid artery
39. Left common carotid artery
40. Left internal jugular vein

Figure 17-2

A. Right atrium
B. Tricuspid valve
C. Right ventricle
D. Papillary muscle
E. Interventricular septum
F. Trabeculae carneae in left ventricle
G. Path of oxygenated blood
H. Bicuspid valve
I. Aortic semilunar valve
J. Oxygenated blood in left atrium
K. Pulmonary semilunar valve

Figure 17-3

A. Superior vena cava
B. Anastomosis of right/left coronary arteries
C. Right coronary artery
D. Right atrium
E. Right marginal artery
F. Posterior interventricular artery
G. Anterior interventricular artery

H. Left ventricle
I. Right ventricle
J. Circumflex artery
K. Left atrium
L. Left coronary artery
M. Pulmonary trunk
N. Aortic arch
O. Right coronary artery
P. Posterior cardiac vein
Q. Middle cardiac vein
R. Coronary sinus
S. Anterior interventricular artery

Figure 17-4

A. SA node
B. Internodal pathway
C. AV bundle (of His)
D. Bundle branches
E. Purkinje fibers

Figure 18-2

1. Superficial temporal artery
2. External carotid artery
3. Internal carotid artery
4. Brachiocephalic artery
5. Deep artery of arm
6. Brachial artery
7. Superior vena cava
8. Abdominal aorta
9. Radial artery
10. Interosseous artery
11. Ulnar artery
12. Femoral artery
13. Deep femoral artery
14. Popliteal artery
15. Posterior tibial artery
16. Dorsalis pedis artery
17. Metatarsal artery
18. Superficial temporal vein
19. Internal jugular vein
20. External jugular vein
21. Left subclavian vein
22. Left brachiocephalic vein
23. Axillary vein
24. Brachial vein
25. Cephalic vein
26. Median antecubital vein
27. Median antebrachial vein
28. Basilic vein
29. Common iliac artery
30. Internal iliac artery
31. Palmar venous arch
32. Dorsal venous network
33. External iliac artery
34. Femoral vein
35. Great saphenous vein (extends from ankle)
36. Popliteal vein
37. Small saphenous vein
38. Posterior tibial vein
39. Small saphenous vein
40. Splenic vein
41. Pancreas
42. Anastomosis of right/left gastric veins
43. Right gastric vein
44. Gastroepiploic vein
45. Inferior vena cava
46. Branches of hepatic veins
47. Liver
48. Hepatic portal vein
49. Gallbladder
50. Superior mesenteric vein
51. Ascending colon
52. Appendix
53. Ileum
54. Descending colon
55. Inferior mesenteric vein
56. Splenic vein
57. Spleen
58. Stomach
59. Left gastric vein
60. Hepatic veins

Figure 18-3

A. Anterior cerebral artery
B. Internal carotid artery (cut)
C. Circle of Willis
D. Posterior cerebral artery
E. Spinal cord
F. Circle of Willis
G. Middle cerebral artery
H. Pituitary gland
I. Basilar artery
J. Vertebral artery
K. Pituitary gland

Figure 19-1

A. Right subclavian trunk
B. Right lymphatic duct

C. Right subclavian vein
D. Thymus gland
E. Bronchomediastinal trunks
F. Cisterna chyli
G. Intestinal node
H. Jugular trunk
I. Left subclavian vein
J. Thoracic duct
K. Spleen
L. Intestinal trunk
M. Metarteriole
N. Lymph capillaries
O. Interstitial fluid
P. Lymphatic vessel
Q. Body cells
R. Lymph capillary
S. Blood capillaries
T. Thoroughfare channel

Figure 19-2

A. Lymphatic vessel valve
B. Node follicle
C. Efferent lymphatic vessel
D. Lymphatic vessel valve
E. Afferent lymphatic vessel

Figure 20-1

A. Frontal sinus
B. Superior nasal concha
C. Middle nasal concha
D. Inferior nasal concha
E. Nasal vestibule
F. External naris
G. Uvula
H. Epiglottis
I. Laryngeal vestibule
J. Glottis
K. Trachea
L. Right lung
M. Diaphragm
N. Sphenoid sinus
O. Internal naris
P. Nasopharynx
Q. Oropharynx
R. Laryngopharynx
S. Primary bronchus
T. Secondary bronchi

Figure 20-2

A. Nasal bone
B. Lateral cartilage
C. Lesser alar cartilages
D. Greater alar cartilage
E. Septal cartilage
F. Greater alar cartilage
G. External naris
H. Septal cartilage
I. Epiglottis
J. Superior horn of thyroid cartilage
K. Thyroid tubercle of thyroid cartilage
L. Lamina of thyroid cartilage
M. Cricoid cartilage
N. Tracheal cartilage
O. Connective tissue
P. Epiglottis
Q. Superior horn of thyroid cartilage
R. Thyroid tubercle of thyroid cartilage
S. Lamina of thyroid cartilage
T. Cricoid cartilage
U. Tracheal cartilage
V. Connective tissue
W. Superior horn of thyroid cartilage
X. Arytenoid cartilage
Y. Cricoid cartilage
Z. Epiglottis
AA. Superior horn of thyroid cartilage
BB. Arytenoid cartilage
CC. Cricoid cartilage

Figure 20-3

A. Bronchiole
B. Branch of pulmonary vein
C. Terminal bronchiole
D. Alveolar duct
E. Pulmonary capillary
F. Branch of pulmonary artery
G. Alveolus
H. Branch of pulmonary vein
I. Endothelial cell
J. Blood plasma
K. Type I cell
L. Basement membrane
M. Lung stroma
N. Septal (type II) cell
O. Lung stroma
P. Type I cell
Q. Erythrocyte
R. Fibroblast
S. Alveolar macrophage

Figure 21-1

A. Oral cavity

B. Sublingual gland
C. Submandibular gland
D. Liver
E. Gallbladder
F. Duodenum
G. Pancreas
H. Ascending colon
I. Cecum
J. Appendix
K. Parotid gland
L. Oropharynx
M. Esophagus
N. Stomach
O. Transverse colon
P. Jejunum
Q. Ileum
R. Sigmoid colon
S. Rectum
T. Anus

Figure 21-2

A. Visceral peritoneum
B. Longitudinal muscle
C. Circular muscle
D. Submucosa
E. Mucosa
F. Submucosa
G. Villi
H. Plica
I. Lymphatic nodule
J. Blood vessels
K. Villus
L. Plica
M. Liver
N. Lesser omentum
O. Stomach
P. Transverse colon
Q. Greater omentum
R. Vesicouterine pouch
S. Urinary bladder
T. Vagina
U. Diaphragm
V. Abdominal cavity
W. Visceral peritoneum
X. Pancreas
Y. Duodenum
Z. Thoracic duct
AA. Mesentery
BB. Small intestine
CC. Uterus
DD. Rectouterine pouch
EE. Rectum

Figure 21-3

A. Superior labial frenulum
B. Hard palate (maxilla)
C. Hard palate (palatine)
D. Uvula
E. Labium
F. Molars
G. Premolars
H. Cuspid
I. Incisors
J. Superior labium
K. Gingiva (gum) or oral vestibule (cavity)
L. Tongue
M. Openings to Wharton's ducts
N. Gingiva (gum) or oral vestibule (cavity)
O. Inferior labial frenulum
P. Inferior labium
Q. Enamel
R. Gingiva
S. Dentin
T. Apical foramen
U. Crown
V. Neck
W. Root

Figure 21-4

A. Mucosa
B. Submucosa
C. Muscularis
D. Connective tissue
E. Artery in submucosa
F. Vein in submucosa
G. Gastric pit
H. Opening to gastric pit
I. Simple columnar epithelium
J. Mucous neck cells
K. Gastric gland

Figure 21-5

A. Transverse colon
B. Ascending colon
C. Cecum
D. Appendix
E. Descending colon
F. Sigmoid colon
G. Rectum

Figure 21-6

A. Right lobe
B. Gallbladder
C. Minor duodenal papilla
D. Major duodenal papilla
E. Duodenum
F. Head of pancreas
G. Body of pancreas
H. Tail of pancreas
I. Pancreatic duct
J. Accessory pancreatic duct
K. Common bile duct
L. Common hepatic duct
M. Cystic duct
N. Left hepatic duct
O. Right hepatic duct
P. Kupffer cell
Q. Hepatocyte
R. Hemopoietic stem cell
S. Endothelial cell
T. Liver lobule
U. Central canal

Figure 22-1

A. Inferior vena cava
B. Renal vein
C. Common iliac artery
D. Common iliac vein
E. Urinary bladder
F. Urethra
G. Abdominal aorta
H. Adrenal gland
I. Left kidney
J. Renal pelvis
K. Ureter
L. Ureteral orifice

Figure 22-2

A. Renal cortex
B. Renal pyramid
C. Renal artery
D. Renal vein
E. Renal pelvis
F. Ureter
G. Minor calyx
H. Renal capsule
I. Renal capillaries
J. Glomerulus
K. Glomerular (Bowman's capsule)
L. Proximal convoluted tubule
M. Distal convoluted tubule
N. Arcuate artery
O. Arcuate vein
P. Collecting duct
Q. Nephron loop (of Henle)
R. Descending limb
S. Ascending limb
T. Renal papilla

Figure 22-3

A. Glomerular capsule
B. Glomerulus
C. Distal convoluted tubule
D. Collecting duct
E. Cortical radiate (interlobular) artery
F. Cortical radiate (interlobular) vein
G. Proximal convoluted tubule
H. Descending limb
I. Ascending limb
J. Arcuate vein
K. Arcuate artery
L. Distal convoluted tubule
M. Juxtaglomerular apparatus
N. Afferent arteriole
O. Smooth muscle
P. Endothelial cell
Q. Fenestrations
R. Podocyte
S. Proximal convoluted tubule
T. Glomerular capsule
U. Parietal cells of glomerular capsule

Figure 23-1

A. Suspensory ligament
B. Fimbria
C. Ovary
D. Ovarian ligament
E. Round ligament
F. Broad ligament
G. Vagina
H. External os
I. Body of uterus
J. Epithelium of ovary
K. Ovarian follicles
L. Secondary oocyte
M. Oocyte in infundibulum
N. Corpus luteum
O. Oviduct
P. Fundus of uterus
Q. Oviduct
R. Urinary bladder
S. Pubic bone
T. Urethra

U. Clitoris
V. Minor labium
W. Vaginal orifice
X. Ovary
Y. Pelvic cavity
Z. Uterus
AA. Fornix
BB. Vagina
CC. Rectum
DD. Anus

Figure 23-2

A. Ovarian ligament
B. Primary follicle
C. Secondary follicles
D. Graafian follicle
E. Corpus luteum
F. Corpus albicans

Figure 23-3

A. Clitoris
B. Minor labium
C. Major labium
D. Anus
E. External urethral orifice
F. Hymen
G. Vaginal orifice

Figure 23-4

A. Vas deferens
B. Ampulla
C. Seminal vesicle
D. Prostate gland
E. Epididymis
F. Seminiferous tubule
G. Spongy (penile) urethra
H. Prepuce (foreskin)
I. Glans
J. Corpus cavernosum
K. Testis
L. Epididymis
M. Bulbourethral (Cowper's) gland
N. Ureter
O. Ureter
P. Urinary bladder
Q. Seminal vesicle
R. Rectum
S. Prostate gland
T. Bulbourethral gland
U. Bulb of penis
V. Epididymis
W. Testis
X. Spongy urethra
Y. Penis
Z. Corpus cavernosum
AA. Pubic bone
BB. Vas deferens
CC. Pubic cavity
DD. Urinary bladder
EE. Ureter

Figure 23-5

A. Head of epididymis
B. Efferent ductule
C. Vas deferens
D. Rete testis
E. Straight tubule
F. Body of epididymis
G. Tail of epididymis
H. Cavity of tunica vaginalis
I. Tunica vaginalis (parietal and visceral layers)
J. Tunica albuginea
K. Septum
L. Lobule
M. Seminiferous tubule
N. Blood vessels in spermatic cord
O. Spermatic cord

Figure 24-2

A. Oogonium
B. Mitosis/cytokinesis
C. Oogonium
D. DNA replication
E. Primary oocyte in prophase I (at time of birth)
F. Primary oocyte in meiosis I
G. Cytokinesis I
H. 1st polar body
I. Meiosis II in 1st polar (may not occur)
J. 2nd polar body
K. 2nd polar body from secondary oocyte's cytokinesis II
L. Ootid (becomes ovum); zygote when pronuclei fuse
M. Completion of meiosis II and cytokinesis II; contains sperm nucleus
N. Fertilized secondary oocyte in meiosis II
O. Fertilization of secondary oocyte; introduction of sperm nucleus triggers meiosis II

P. Ovulation of secondary oocyte
Q. Primary follicle with primary oocyte becoming secondary oocyte within a secondary follicle becoming a Graafian follicle
R. primordial follicle with primary oocyte in prophase I
S. Primordial follicle containing oogonium
T. Oogonium
U. Ovarian follicle in ovary before birth
V. Seminiferous tubule
W. Spermatogonium
X. Mitosis and cytokinesis
Y. DNA replication
Z. Beginning of meiosis I
AA. Completion of meiosis I and cytokinesis I
BB. Completion of meiosis II and cytokinesis II
CC. Spermatogenesis (up to formation of spermatids)
DD. Spermiogenesis (spermatids become flagellated)
EE. Head of spermatozoan with acrosome
FF. Acrosome within developing head
GG. Spermatid
HH. Secondary spermatocyte
II. Primary spermatocyte in meiosis I
JJ. Primary spermatocyte in interphase
KK. New spermatogonium

Figure 24-3

A. Secondary oocyte and 1st polar body surrounded by oolemma and corona radiata
B. Fertilization
C. Fertilized oocyte with pronuclei
D. Zygote
E. Results of mitosis and cytokinesis
F. 2-cell stage
G. 4-cell stage
H. 8-cell stage
I. Morula
J. Early blastocyst
K. Late blastocyst implanting in endometrium

Figure 24-4

A. Endometrial epithelium
B. Inner cell mass
C. Trophoblast
D. Blastocoel
E. Notochord
F. Amniotic cavity
G. Ectoderm
H. Extraembryonic mesoderm
I. Chorion (outer white layer)
J. Amnion (pink layer)
K. Embryo
L. Allantois
M. Umbilical cord
N. Yolk sac
O. Chorionic villus

Figure 24-5

A. Suspensory ligament
B. Areola
C. Nipple (receives milk from lactiferous ducts)
D. Mammary gland tissue